Peter Wooding • Graham Burton

Comparative Placentation

Structures, Functions and Evolution

 Springer

F.B.P. Wooding
G.J. Burton

Centre for Trophoblast Research
Department of Physiology, Development and Neuroscience
Cambridge University
Downing Street
Cambridge
CB2 3EG
United Kingdom
fbpw2@cam.ac.uk
gjb2@cam.ac.uk

ISBN 978-3-540-78796-9   e-ISBN 978-3-540-78797-6

Library of Congress Control Number: 2008923543

Cover design: WMXDesign GmbH, Heidelberg, Germany

Printed on acid-free paper

9 8 7 6 5 4 3 2 1

springer.com

Comparative Placentation

Structures, Functions and Evolution

Acknowledgements

We are indebted to the following colleagues who provided the recent essential expertise and advice at various stages of preparation of this book: Dr. S.L. Adamson, University of Toronto, Prof. J.D. Aplin, Manchester University, Prof. A.M. Carter, University of Southern Denmark, Prof. J.C. Cross University of Calgary, Prof. A.C. Enders, University of California, Davis, Prof. A.L. Fowden, University of Cambridge, Dr. C.J.P. Jones, Manchester University, Prof. R. Leiser, University of Giessen, Prof. C. Loke and Dr. A. Moffett, University of Cambridge, Prof. R. Pijnenborg, University of Leuven, Prof. M.B. Thompson, University of Sydney, Prof. J.R. Stewart, Tennessee State University, Prof. U. Zeller, Institute for Systematic Biology, Berlin. We also thank those who provided micrographs, all acknowledged in figure legends, my son Dr. Steven Wooding for his endless patience and advice on computer detailing.

Contents

Chapter 1
Placentation Fundamentals

1.1 Introduction

Viviparity and the development of a placenta are two of the major reasons for the success of the vertebrates in colonising all habitats, both terrestrial and aquatic.

The placenta is an apposition of fetal to maternal tissue which has two main functions: to maximise oxygen and nutrient acquisition from the mother, but to minimise immunological rejection by the maternal immune system. These competing drives have resulted in an organ of uniquely varied cellular structure across species whereas all other organs such as the eye or the kidney for example, are remarkably uniform in this respect.

The reasons for the range of placental structures and their evolution presents a fascinating scientific puzzle. They are also of immediate medical relevance since in man and animals impaired placental growth produces a weakened neonate and has also recently been shown to be associated with an increased incidence of cardiovascular, metabolic and other diseases in later life (Gluckman and Hanson 2005, 2006; Chap. 10.3) Comparative placental studies will facilitate an understanding of what factors control this impaired growth in humans and also allow informed selection of valid animal models.

This book is mainly concerned with providing pictorial and descriptive data to facilitate broad comparisons of structures and secretions both during development and in their final form with a view of relating them to the functional activity of the placenta. In the viviparous vertebrate amniotes, both reptilian and mammalian, the remarkably wide variety of structures in the mature placenta develops from a similar repertoire of extraembryonic membranes as typified in the amniote egg. The diversity can be further simplified by use of straightforward structural criteria. The major problem is to relate those criteria to the wide range of functions served by the placenta. An appreciable number of viviparous vertebrate non amniotes, fish and amphibia, have developed analogous structures which serve similar functions to the extraembryonic membranes, These will be discussed briefly as they offer further insights into the requirements for successful fetal development. Even a few invertebrate species utilise placental analogues, a tribute to the advantages of viviparity (see Chap. 9.2) in particular environments (Hagan 1951; Woollacott and Zimmer 1975; Amoroso et al. 1979; Bone et al. 1985).

P. Wooding, G. Burton, *Comparative Placentation*,
© Springer-Verlag Berlin Heidelberg 2008

   The placenta developed as an organ of physiological exchange to take advantage of the protection and economy of egg production provided by the adoption of viviparity. The initially small area of fetomaternal apposition is usually enormously increased by folding and refolding as it proliferates in parallel with the growth of the fetus. The main functions of the placenta are to act as a surrogate fetal lung, gut and kidney and to camouflage the usually allogeneic fetus from the mother's immune system (Lala et al. 1983; Gill and Wegmann 1987). It also has an important endocrine role, influencing maternal metabolism and maintaining the uterus in a quiescent state.

   The surrogacy requires efficient exchange of a necessarily wide range of metabolites and hormones using many different types of transport processes (passive and facilitated diffusion, active transport, pinocytosis and phagocytosis), all favoured by a reduction in the separation of fetal and maternal blood flows (Moll 1985; Munro 1986).

   The immunological camouflage is vital to avoid recognition by the multifactorial array of cellular and hormonal mechanisms mediating sensitization and immune rejection (see Chap. 9) and is best served by the provision of a barrier between the two circulations (Amoroso and Perry 1975).

   Since the surrogacy and camouflage requirements are so complex and mutually antagonistic (it being very difficult to adequately conceal and feed an invading cellular army) it is perhaps not surprising that such a wide variety of structures has evolved as solutions to these two multifaceted major problems. In addition, a system for rapid disposal of the placenta with minimum trauma is essential and there are constraints of gross size, habitat (arboreal gliding versus aquatic diving) and lifestyle (secure den versus run-as-soon-as-born) further modifying the requirements defining a successful placenta.

   Perhaps because of these various and often conflicting pressures, there seems to be no straightforward progression to a unique 'best', most efficient placenta in parallel with the generally accepted course(s) of evolution of placental animals (see Chap. 9). Different groups of mammals have evolved different structural solutions, all originally from a common pattern of membranes modified by a general trend to a reduction in separation between fetal and maternal circulations. The greater the understanding of these structural solutions the easier it should be to design experiments to elucidate the manifold functions of the placenta, to establish how best to judge the 'efficiency' of a placenta and to plan strategies to reduce the considerable prenatal embryo losses (Wilmut et al. 1986; Goff 2002).

1.2   General Characteristics and Definitions

1.2.1   Placenta

The most comprehensive definition is that of Mossman (1937): 'an apposition of parental, (usually maternal) and fetal tissue for the purposes of physiological exchange'. However, Steven's suggestion of 'an arrangement of two or more transporting epithelia between fetal and maternal circulations' more vividly encapsulates the functional aspects.

Mossman's definition is wide enough to cover the situations found in a variety of non-amniotes: tunicates, seahorses, some insects, fishes and amphibians as well as the amniotes (reptiles and mammals). The degree of apposition in vivo is sometimes difficult to establish because of technical problems of fixation and tissue preparation. There is good reason to believe that it is very close, often including microvillar interdigitation between fetal and maternal epithelia in the amniotes, but a shell or shell membrane remnant may persist in most reptiles and some fish (Chap. 3).

The tissue on the maternal side of the apposition is usually a derivate of either the epithelium or connective tissue of the ovary, oviduct or uterus.

The fetal component is invariably a derivative of the cellular ectodermal epithelium (trophoblast or trophectoderm) which formed the blastocyst surface. This epithelium together with internal membranes and blood vessels from the conceptus eventually form the choriovitelline and chorioallantoic placental structures.

Physiological exchange covers a range of transport processes from simple diffusion of O_2 and CO_2 to the uptake of uterine secretions and maternal red blood cells. There is increasing evidence that mature placentas have different regions for the different forms of uptake (Chap. 2) and also that different types of placenta have evolved different methods for the uptake of particular nutrients (e.g. iron, see Chap. 2).

1.2.2 Extraembryonic Membranes

With the exception of the trophectoderm component of the chorion, the extraembryonic membranes of the vertebrate conceptus form from, and are in continuity with, the three germ layers (ectoderm, mesoderm and endoderm) of the embryo. 'Extra' here is used in the sense 'outside of' The extraembryonic membranes provide the basis for the formation of placental structures essential for the physiological maintenance and protection of embryos within the egg or uterus (Fig. 1.1; Mossman 1937, 1987; Steven 1975a).

1.2.2.1 Yolk Sac

This is formed by the enclosure of the yolk initially by cells from the embryonic disc which form a single cellular layer of ectoderm (unilaminar yolk sac, vitelline sac or omphalopleur) (Fig. 1.1). Subsequent migration of a layer of endoderm internal to the ectoderm forms a bilaminar yolk sac, and finally growth of mesoderm between the first two produces a trilaminar yolk sac. If no yolk is present the last two layers develop in the same way after a hollow unilaminar blastocyst is produced from the solid ball of cells forming the morula.

In the fishes and amphibians the mesoderm of the trilaminar yolk sac forms the blood vessels which vascularize the yolk sac and transport its contents to the growing

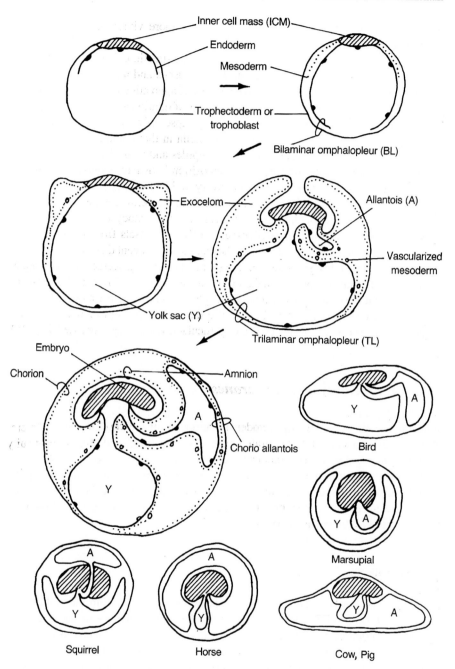

Fig. 1.1 Development of the basic extraembryonic membranes, with diagrams of the definitive disposition of yolk sac and allantois in various animals

embryo. This vascularized yolk sac persists in some fishes after the release of the embryo from the investments of the egg. Retention of the developing embryo within the mother allows the possibility of modifying the vascularized yolk sac into an organ for direct fetomaternal exchange. Several of the cartilaginous fishes (elasmobranchs – skates and rays) have developed such a placental system whereas bony fishes (Osteichthyes) and amphibia have modified other parts of their body surfaces for the same purpose (Chap. 3).

In reptiles and mammals the initial spread of the mesoderm converts only the top part of the bilaminar yolk sac into the vascular trilaminar form. This is the basis for the development of the other extraembryonic membranes typical of the amniotes: chorion, amnion and allantois (Fig. 1.1).

1.2.2.2   Chorion, Amnion and Allantois

Growth of these membranes starts when the mesoderm of the partly trilaminar yolk sac splits to form the exocoelom in continuity with the embryonic coelom (Fig. 1.1). The subsequent development of the mesoderm depends on whether it is associated with ectoderm or endoderm. Mesoderm plus endoderm (gut wall, splanchnopleur = definitive yolk sac wall) vascularizes very readily, whereas mesoderm plus trophectoderm (body wall, somatopleur = chorion) only rarely produces blood vessels.

At this stage the trophectoderm and the inner cell mass (ICM, which will form the embryo) are already separate and non interconvertible cell lineages. For example, in the mouse Cdx2 is expressed predominantly in the trophoblast and Oct3/4 only in the ICM (Selwood and Johnson 2006). Many animals – squirrels, ruminants, carnivores – lose the polar trophectoderm cells over the ICM (squirrel in Fig. 1.2a) in which case the chorion (trophectoderm plus nonvascular mesoderm) produces folds that enclose the embryo to form the fluid-filled amniotic sac (Fig. 1.2a). Inside this the embryo can develop free of asymmetric constraints, in a shock-absorbing and compression-resistant environment. The amniotic membrane is thus derived directly from the chorion. In animals where the trophectoderm persists over the ICM, the amniotic cavity forms within the ICM tissue by what is generally described as cavitation but the exact cellular mechanism is not yet established (mouse, guinea pig and man Fig. 1.2a). In this case the amniotic membrane is derived from the ICM. In both cases the structure of the amniotic membrane is identical. Once the amnion is formed an endodermal vesicle grows from the hind end of the embryonic gut to form the allantoic vesicle, covered from its inception by vascular splanchnopleuric mesoderm. The allantois is in continuity with the developing urogenital system of the embryo and acts as a site for waste deposition. Its rich blood supply also allows it to develop as an organ of respiratory exchange when it grows through the exocoelom to fuse with the chorion which surrounds the embryo (Figs. 1.1, 1.2 and 8.3). In species with a solid allantoic endodermal outgrowth (mouse, guinea pig and man, Fig. 1.2b) waste materials are removed continuously via the placenta.

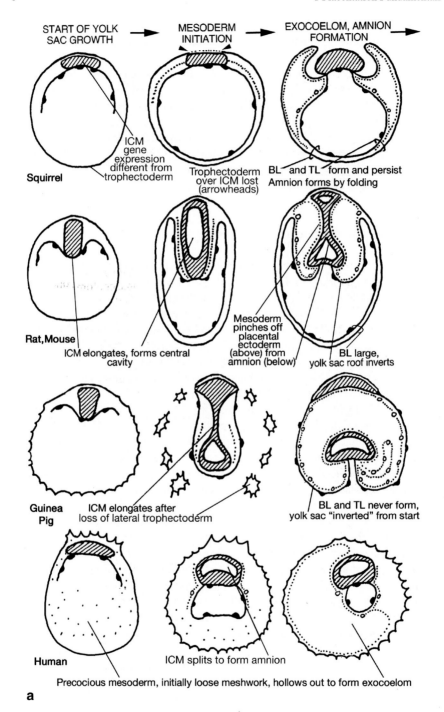

Fig. 1.2 **(a,b)** Development of placental membranes in squirrels, rat and mouse, guinea pig and human. For identification of individual membranes see Fig. 1.1

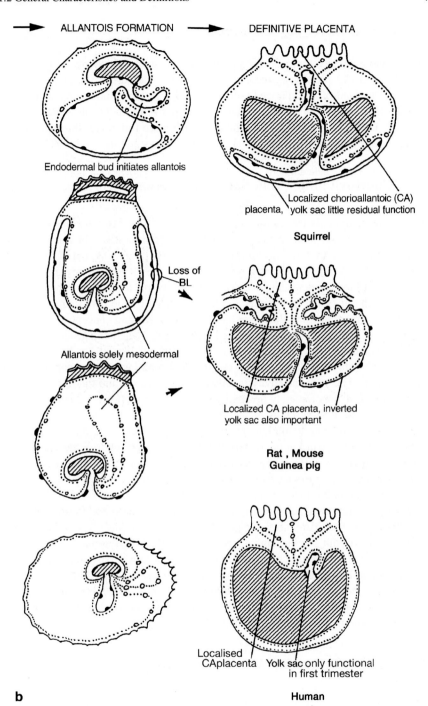

b

Fig. 1.2 b (continued)

1.2.2.3 Variations on the Basic System

These membrane developments provide the basic systems which are common to all amniotes and can usually be recognized at least at the earlier stages of placental growth. The band of vascularized yolk sac fused with the ectoderm forms the fetal basis for the choriovitelline placenta and the vascularized allantois plus chorion forms the fetal part of the chorioallantoic placenta. Both are functional through most of prenatal development in reptiles (Chap. 3), monotremes and marsupials (Chap. 4).

1.3 Choriovitelline (Yolk Sac) Placentation

In eutherian mammals the choriovitelline (yolk sac) placenta functions only for a short time and varies considerably in extent in different orders. In carnivores and some ungulates it may form an extensive area in early pregnancy, but eventually the vascularized yolk sac is completely separated from the chorion by expansion of the exocoelom. In most rodents, lagomorphs and insectivores formation of the trilaminar choriovitelline placenta, if it occurs at all, is restricted by the persistence of a large bilaminar segment forming the lower half of the yolk sac and conceptus. A very different sort of yolk sac placenta is formed subsequently when expansion of the exocoelom invaginates the vascularized embryonic (top) half of the yolk sac into the bilaminar abembryonic (bottom) half to provide a multilayered cup apposed to the endometrium. The yolk sac cavity is reduced to a mere slit and the resultant layers referred to as an "incompletely inverted yolk sac placenta" (Figs. 1.2 and 1.3).

Loss of the non-vascularized bilaminar yolk sac layers adjacent to the endometrium produces "complete inversion" (Figs. 1.2 and 1.3). What is 'inverted' is the sequence of tissues at the maternofetal interface. The choriovitelline placenta has maternal tissue: fetal ectoderm, mesoderm, endoderm; the inverted yolk sac placenta has maternal tissue: fetal endoderm, mesoderm.

The choriovitelline placenta is usually transitory, but the inverted yolk sac type may persist to term and form part of the wide range of mature placental structures developed from the extraembryonic membranes. For example, the ungulates have a mere relic of a yolk sac but an enormous chorioallantois; the marsupials retain a large functional trilaminar choriovitelline placenta with a persistent bilaminar area and an allantois that does not usually reach the chorion; some insectivores and rodents have both inverted yolk sac and chorioallantoic placentas which function up to parturition (Figs. 1.1–1.3). The type of implantation can also greatly affect extraembryonic membrane development.

1.3.1 Implantation and Membrane Development

Implantation of the blastocyst, the initial apposition of fetal and maternal surfaces, usually occurs on the uterine luminal surface. However, some animals have very small blastocysts which develop in a narrow endometrial cleft (mouse, rat) or

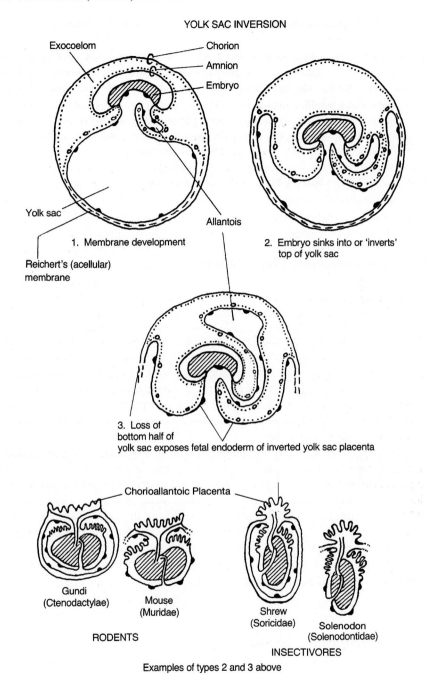

Fig. 1.3 Development of the inverted yolk sac, with examples of definitive yolk sac placentation in rodents and insectivores

actively invade the endometrium (guinea pig, man: referred to as interstitial implantation) (see Chapter 2.1).

Presumably, because of lack of space round the conceptus, the development of the embryonic membranes is considerably modified in species with interstitial implantation although the basic pattern described above can be recognized if sufficient stages are available. The major differences are that the original outer trophectodermal layer over the inner cell mass persists; the amniotic sac forms by 'cavitation' in a solid mass of cells rather than by folding; and the allantoic endodermal vesicle may be very small or absent so that the allantois is replaced by a solid rod of mesoderm. Development of the placental membranes is compressed in both time and space and the homologies of the membranes can be appreciated best from comparative diagrams (Fig. 1.2).

1.4 Chorioallantoic Placentation

1.4.1 Introduction

The great diversity among chorioallantoic placental structures has led to many attempts to devise a classification which would categorize only those common structural features which may be used as a valid basis for evolutionary comparisons and experimental investigations of function. No classification scheme based on a single criterion has proved adequate and the current practice is to use several in parallel, which produces convenient and possibly instructive groupings rather than necessarily exclusive categories.

The simplest criterion is the shape of the term placenta, but potentially more relevant to physiological exchange is the histological structure and organization of the layers separating fetal and maternal blood circulations at the fetomaternal interface (Grosser classification, Amoroso 1952; Steven 1975a). The structure of areas outside or bordering the main placenta (accessory or paraplacental regions) and the amount of maternal tissue shed at parturition are also characteristic. These criteria all refer to the structure at term but the developmental history of the placental membranes may also provide important clues to functional relationships.

1.4.2 Classification by Shape

The fetomaternal exchange area is enormously increased by elaboration of villi or folds. The distribution of the chorionic folds or villi is characteristic among families and has four main categories (Fig. 1.4).

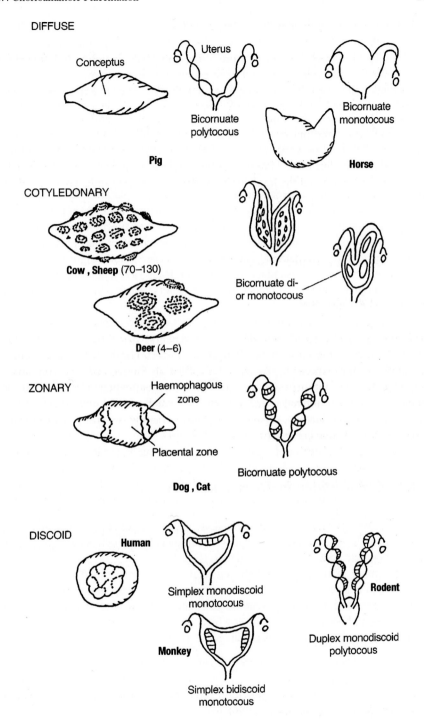

Fig. 1.4 Schematic placental classification based on external form of the placenta and uterus

1. Diffuse: folds/villi over entire surface, e.g. pig, horse.
2. Cotyledonary/Placentomal: chorionic villi grouped into a characteristic number of discrete tufts which can vary from 5 (deer) to 150 (giraffe), e.g. ruminants.
3. Zonary: complex folds restricted to an equatorial band or path, e.g. carnivores.
4. Discoid: villi restricted to a single or double disc, e.g. rodents, insectivores, anthropoids.

This provides a useful simplification but within orders there are invariably exceptions outside the 'usual' category. Most insectivores have a discoid, but the American mole has a diffuse placenta; most of the ruminant placentas are placentomal, but that of the mouse deer is diffuse; carnivore placentas vary from the usual zonary to rare discoid (brown bear) and non-carnivores are found with a zonary placenta (hyrax, dugong); a few placentas start diffuse and are discoid when mature (man, hyrax). These facts indicate that the four categories are convenient rather than functionally distinct.

Placentas with villi concentrated in limited areas (compact placentas) do have a 5–10 times greater fetomaternal exchange surface per gram weight of fetus plus placenta at term than diffuse placentas (Fig. 2.9; Baur 1977), but whether this represents a higher transfer efficiency or lower safety margin in diffuse placentas remains to be established.

The uterine shape and number of conceptuses are also characteristics of a species. During development, the uterus is formed by (partial) fusion of the two Mullerian ducts to form a Y-shaped structure (Fig. 1.4). The degree of fusion varies. Rats and rabbits have completely separate 'cornua' or 'horns' and this is defined as a 'duplex' uterus. Ruminants have partial fusion into common uterine body ('bicornuate'). In man and monkeys the horns are very small and the uterine body forms a 'simplex' uterus. Species producing a single offspring are designated monotocous, two ditocous, many polytocous (see Fig. 1.4).

1.4.3 Classification by Histological Structure

In 1909 Grosser pointed out that mammalian placentas could be grouped according to how many layers of maternal tissue were removed by the persistent chorion during development (cited in Amoroso 1952) (Fig. 1.5).

The number of tissue layers between fetal and maternal blood stabilizes after implantation and early differentiation of organ systems, and would directly affect the physiological exchange between the circulations. Grosser classified the placentas by defining first the maternal layer which was in apposition to the chorion in the definitive placenta, producing four categories (Fig. 1.5):

1. Epitheliochorial: No layers removed; uterine epithelium in contact with the chorion, e.g. pig, horse.
2. Syndesmochorial: uterine epithelium removed, maternal connective tissue in contact with the chorion, e.g. ruminants.
3. Endotheliochorial: maternal uterine epithelium and connective tissue removed, maternal endothelial basement membrane in contact with the chorion, e.g. carnivores.

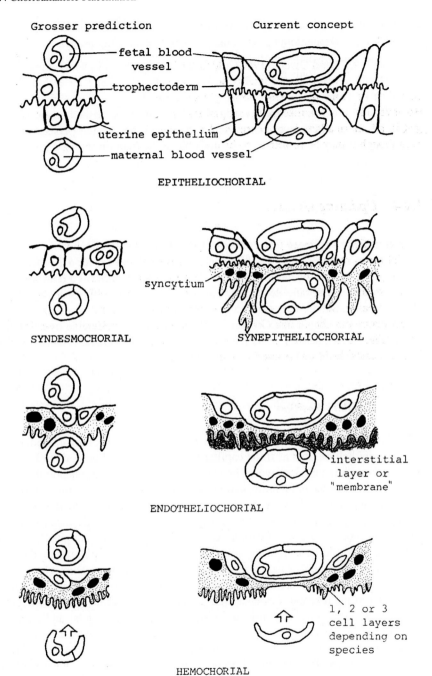

Fig. 1.5 Placental classification suggested by Grosser on the left, revised to take account of recent research on the right

4. Haemochorial: all maternal tissue layers removed, chorion bathed directly in circulating maternal blood, e.g. rodents, insectivores, anthropoids.

Mossman (1937) and Owers (1960) added further categories in which the chorion was lost either from an endotheliochorial placenta (producing an endothelioendothelial) or from the haemochorial placenta (producing a 'hemoendothelial'). However, electron microscope investigations have since established unequivocally that in all chorioallantoic placentas so far examined the chorion persists to term even though it may be so thin as to be invisible at the light microscope level.

1.4.4   Updating Grosser

Grosser initially suggested (see Amoroso 1952) that the epitheliochorial to haemochorial series represented an evolutionary, anatomical and functional progression and proposed that, as the number of layers in what was regarded as a homogeneous semipermeable membrane decreased, the permeability and efficiency of the placenta increased. Subsequent work on capillary position, regional specialization and ultrastructure has shown that Grosser's categories are a considerable oversimplification (Fig. 1.5; see below). More recently molecular phylogenetic evidence (Chap. 9.3) also casts doubt on Grosser's simple scenario.

1.4.4.1   Capillary Position

By differential growth during development fetal capillaries sink into and indent the chorionic epithelium. It is as if the capillary net expanded and pushed hard against a soft and yielding cellular layer. The capillaries always remain outside the basement membrane of the epithelium so that they are not anatomically 'intraepithelial' as sometimes described. They do, however, come much closer to the maternal blood, and in epithelio- and synepitheliochorial (previously syndesmochorial, see below) placentas the maternal capillaries often indent the uterine epithelium [the analogy here would be a tight meshwork of string (the capillaries) around a soft parcel (the fetal villus)]. The extent to which the indentation occurs varies both within and between placental categories but generally increases toward term. There is no electron microscopic evidence that the process ever reduces the number of tissue membrane layers but the diffusion distance between the fetal and maternal circulations can be considerably reduced. This may explain why although the sodium ion permeability follows the Grosser prediction with the haemochorial (few layered) much higher than the epitheliochorial (multilayered) placentas, the gas-diffusing capacities of the two types of placenta are not significantly different (Moll 1985). The sodium ion requires a transport system across each membrane, whereas gas diffusion depends solely on path length and concentration gradient.

1.4.4.2 Regional Specialization

A wide variety of physiological, molecular biological, immunocytochemical and structural studies have all combined to demonstrate that the placental layers can have very different characteristics in different areas of the same placenta or in equivalent layers within the same placental category. The placenta has been shown to be a very active organ metabolically and contains a wide range of selective transport systems and thus cannot be considered a homogeneous semipermeable membrane (Munro 1986; Mossman 1987; Enders et al. 1998; Atkinson et al. 2006; Jones et al. 2007).

1.4.4.3 EM Evidence for Validity of the Categories

Syndesmochorial/Synepitheliochorial

Grosser defined the mature syndesmochorial placenta as having lost the maternal uterine epithelium, thereby bringing the fetal chorion into contact with the maternal endometrial connective tissue. This was based on light microscope studies which indicated that the characteristic chorionic binucleate cell of ruminants was directly involved in erosion of the uterine epithelium. Subsequent electron microscope studies could not confirm this but did demonstrate that a microvillar junction was present in all ruminants investigated, with either a syncytium (sheep, goat) or a cellular epithelium (cow, deer) on its maternal side interdigitating with the fetal chorion. This was taken as evidence that the uterine epithelium persisted (Ludwig 1962) albeit in syncytial form in some species, and that the syndesmochorial category was therefore more accurately described as epitheliochorial with the microvillar junction being also the fetomaternal junction.

However, Wooding (1982a,b, 1984, 1992) and Lee et al. (1985) have demonstrated that the sheep and goat syncytia are formed largely from the fetal chorionic ruminant binucleate cells which migrate across the microvillar junction to fuse into and form the syncytium. The layer in contact with the uterine connective tissue is therefore a fetal chorionic derivative and the placenta is legitimately described as syndesmochorial. At implantation, however, the syncytium originates from the fusion of a fetal chorionic binucleate cell with a uterine epithelial cell and the relative contributions of fetal and maternal tissue to the mature sheep or goat syncytium is still uncertain.

In cow and deer a fetomaternal syncytium is formed in the same way at implantation, but is subsequently displaced by regrowth of the residual cellular uterine epithelium. Continued binucleate cell migration across the microvillar junction produces fetomaternal trinucleate cells throughout pregnancy. These minisyncytia do not spread further but die and are resorbed after releasing their fetally produced granules to the maternal connective tissue (see Chap. 6 and Wooding 1982a,b, 1992).

All ruminants so far studied have binucleate cells and/or mini- or more extensive syncytial formations throughout pregnancy. Production of fetomaternal hybrid tissue in contact with the maternal connective tissue thus appears to be characteristic of ruminants. It is so unlike anything that is found in the other placental categories that it seems useful to maintain the distinction between epitheliochorial and syndesmochorial placentas.

The prefix 'syndesmo' was used by Grosser as a rather obscure synonym for connective tissue (Borland's Medical Dictionary: syndesmo, a continuing form denoting a relationship to connective tissue). It is now clear that the uterine epithelium is not removed by the chorion but more or less extensively modified by fusion with binucleate cells from implantation to term. The chorion is thus in contact with a fetomaternal syncytium and not with connective tissue. A usefully descriptive term for this situation which retains a reference to the older usage would be 'synepitheliochorial' (Wooding 1992) (Chap. 6).

Haemochorial: Enders (1965), in an elegant comparative study, demonstrated that the haemochorial placenta does not always have a single layer of trophoblast. Different species show a constant pattern of one, two or even three separate layers between maternal blood and fetal endothelium. The terms haemomonochorial, -dichorial and -trichorial were introduced to accommodate this (Chap. 8.6). Further complexity was evident in that any layer could be either cellular or syncytial and some layers had macroscopic pores through their entire thickness. Most layers were thin, but this varied considerably. Thus, again the Grosser category covered a much wider range of structure than originally realized. There is an equivalent variation in functional capacity; the gas-exchange values for haemochorial placentas cover such a range that they include the values for epitheliochorial placentas (Moll 1985).

Although many of the initial assumptions and predictions of the functional significance of the Grosser classification have been shown to be incorrect, the basic categories are still easily recognisable and nothing more convenient has yet been found to replace it (Fig. 1.5).

Grosser also separated placentas into 'villous' or 'labyrinthine' types depending on the way in which the continuous increase in the fetomaternal junctional area throughout pregnancy was organized. In villous placentas the chorion initially forms simple finger-like projections which subsequently may become profusely branched but remain separate. Conventionally 'primary' villi are formed by a solid mass of ectoderm. These are said to be 'secondary' villi when mesoderm grows into their core and 'tertiary' villi when this core vascularizes (Steven 1975a; Leiser and Kaufmann 1994; Burton et al. 2007).

In labyrinthine placentas the fetomaternal junction forms a network. This can be produced by fusion between the branches of an initially villous placenta, by a hollowing out of spaces in solid trophoblast or by the engulfment of a meshwork of maternal capillaries by the invasive chorion which replaces the maternal connective tissue. Although the structural development is usually villous to labyrinthine there are instances, such in the human, where the initially labyrinthine placenta becomes villous secondarily.

1.4.5   Classification by Tissue Lost at Parturition

In the 'deciduate' type a variable amount of maternal tissue is lost with the fetal membranes at birth. The non-deciduate (or adeciduate) placenta separates at the fetomaternal junction. There is a reasonable correlation with the Grosser classification in that if there is any erosion of the maternal layers during placental development then the placenta is deciduate and is shed together with some maternal tissue at term.

Since the data are relatively easy to collect, this classification can serve as an initial indication of structure. In contradeciduate placentas some fetal tissue is retained and resorbed after parturition. This has been reported in only a very few epitheliochorial placentas (European mole, lemurs; Amoroso 1952).

It was originally thought that the specialized tissue (decidua) formed early in pregnancy from stromal cells in the endometrium of species with haemochorial placentation was related to the eventual placental separation zone. However, the extent of its production is variable, it frequently disappears before parturition and is completely absent in 'deciduate' endotheliochorial placentas. It therefore seems very unlikely that decidual tissue plays a major role at parturition. For a more detailed discussion of decidua see later in this chapter.

1.4.6   Classification by Accessory Placental Structures

These do not form a comprehensive scheme but are consistent features and provide a basis for further characterizing categories in the other schemes such as Grosser's. These structures are normally outside or at the edge of the main placental region and may be very different in structure.

1.4.6.1   Haemophagous Zones/Organs

These structures are also known as haematomas, but this term is already used in other less specific contexts and is best avoided; they occur in many endotheliochorial and some epitheliochorial placentas, normally around the margin of the main placenta but occasionally centrally (Ulysses et al. 1972; Burton 1982; Enders and Carter 2006; Enders et al. 2006).

The haemophagous zone is a specialized area where the chorion and maternal tissue are widely separated by stagnant, but not necessarily clotted, maternal blood. There is no circulation of blood; the supply is by intermittent leakage and the control not understood. The chorionic trophoblast cells are typically very tall and columnar and full of large vacuoles containing phagocytosed red blood cells in various stages of digestion. It is assumed that by breakdown of maternal haemoglobin these zones provide a principal source of fetal iron.

Haemophagous zones are characteristically found at the edge or centre of the zonary endotheliochorial labyrinthine placentas of carnivores, but also occur in some endotheliochorial placentas in elephants, insectivores and bats (see Chap. 7). Some synepitheliochorial placentas also show haemophagous zones at the base of the fetal villi in the cotyledons (sheep and goat, Chap. 6), but none has been reported in epitheliochorial and only three in haemochorial placentas (shrew, King et al. 1978; tenrec, Mossman 1987; Enders et al. 2007).

1.4.6.2   Areolae

Areolae consist of small areas of the chorion overlying the mouths of the uterine glands whose secretion separates the two epithelia. The chorionic surface is normally elaborated in small folds carrying tall columnar cells showing evidence of active phagocytotic uptake of uterine secretions (Figs. 20, 21 and 35; Enders and Carter 2006). There is evidence of both merocrine and holocrine secretory modes from the uterine glands and epithelium. In the pig there is ultrastructural evidence for secretion of the iron transport protein uteroferrin from the glands and uptake by the chorionic areolar epithelium (Fig. 2.6; Raub et al. 1985). Enzymes related to sodium transport processes have also been localized to the areolar chorion (Firth et al. 1986a,b).

Areolae are characteristic of epitheliochorial placentas and occur in large numbers over the entire surface of the placentas of pigs (see Chap. 5) and horses (8,000–9,000 in pig, Amoroso 1952). They are also common in the interplacentomal areas of synepitheliochorial placentas (ruminants). In both groups large amounts of uterine gland secretion (also known as uterine milk or histotrophe) are produced and the areolae represent a localization and concentration of the uptake function common to all chorions. Their prominence in epithelio- and synepitheliochorial placentas may be correlated with the relative impermeability of such multi-layer structures to non-diffusible substances; they are much simpler and less frequent in paraplacental regions in the cat and dog (endotheliochorial) placentas.

1.4.6.3   Chorionic Vesicles

These may be derivatives of the chorionic roof of an areola invaginated into the allantoic sac (Fig. 5.2c). They range from vesicles lined with well-preserved chorionic epithelium containing material identical to that present between the epithelia in an areola, to sacs with no discernible internal structure (Amoroso 1952; Mossman 1987). These possibly represent the initial and final stages of such structures. They occur in placentas with areolae but are so infrequent and sporadic in their distribution that it seems unlikely that they have any specific function other than removal of excess secretion from between the fetomaternal layers. This is corroborated by the observation that in the horse the remnants of the endometrial cups (see below) are removed from the uterine surface and stored in a very similar structure (Fig. 5.7b).

1.4.6.4   Chorionic Girdle and Endometrial Cups

In equids before implantation, a girdle of binucleate cells develops in the trophectoderm surrounding the conceptus (Figs. 5.7–5.11). These fetal cells migrate across the
uterine epithelium to form a ring of solid structures in the uterine subepithelial connective tissue known as endometrial cups (Figs. 5.7, 5.8 and 5.13). These aggregates
of fetal cells produce large amounts of equine chorionic gonadotrophin (eCG), previously known as pregnant mare serum gonadotrophin (PMSG), during the second
quarter of pregnancy. The cells are subsequently killed by maternal leucocytes and
the remnants are encapsulated by the chorion and invaginated into the allantoic cavity.
This remarkable system (Fig. 5.7) is found only in the Equidae, but the binucleate
nature, migration and hormone production of these cells are very reminiscent of the
binucleate cells in the ruminants (For further details and references, see Chap. 5).

1.4.6.5   Yolk Sac Structures

As discussed above, the yolk sac membranes display a wide range of structural
organization throughout gestation in eutherian mammals. In the present context it
is the usually transient presence of the choriovitelline placenta and the structure and
position of the yolk sac at term which are used as diagnostic features.

The choriovitelline placenta (vascularized trilaminar yolk sac apposed to uterine tissue) forms the definitive placental structure in metatheria (marsupials) and is temporar-
ily present in some families of nearly all eutherian orders whatever their eventual
chorioallantoic placental type. It is important at implantation in subsequently endotheliochorial carnivores but much more transient in ungulates. In both, the vascularized
yolk sac is soon separated from the chorion by the expanding exocoelom, and by mid-
gestation it is regressing in size. It soon becomes vestigial in ungulates but may retain
some (unknown) function in carnivores and some bats (Figs. 1.1 and 1.2).

In the largely haemochorial rodents, insectivores and lagomorphs the choriovitelline placenta may be present during development, but these orders are characterized
from an early stage until term by the presence of an inverted yolk sac placenta which
may be complete or incomplete (Fig. 1.3). This is the only route of transfer for the
first third of pregnancy before the chorioallantoic placenta develops. The incompletely inverted type has many layers but is still reported to be active in histotrophic
absorption in some sciurid rodents (squirrels) (Mossman 1987). The completely
inverted type has fewer layers and in the second half of pregnancy may serve general
histotrophic functions. Several studies have demonstrated its active role in macromolecular uptake including specific transport of immunoglobulins, in mouse, rat and
guinea pig. Typically the relationship with the maternal tissues is far less intimate
than in the choriovitelline placenta, the yolk sac tissue absorbing secretions like a
fetal areolar epithelium by endocytosis from the uterine lumen (Bainter 1986).

An exception to this is the "intraplacental yolk sac" area where the yolk sac is
carried into the fetal base of the chorioallantoic placenta as narrow sleeves around
the initial branches of the umbilical vessels. This brings the yolk sac epithelium

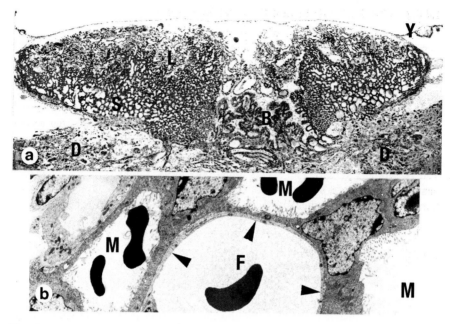

Fig. 1.6 Hemomonochorial guinea pig placenta. (**a**) Light micrograph of a transverse section through the whole placental disc illustrating the different regions of the placenta. *B* suplacenta, *D* decidua, *L* labyrinth, *S* spongiotrophoblast, *Y* (inverted) yolk sac. (**b**) Electron micrograph showing the labyrinth structure. Note the variation in thickness (*arrowheads*) of the diffusion path between maternal and fetal circulations. *F* fetal capillary, *M* maternal blood spaces. (**a**) 23 dpc; ×20. (**b**) 39 dpc; ×5,200

close to the maternal blood inflow channels where it is thought to be involved in fetal calcium uptake (Kovacs et al. 2002 and see Chap. 2).

1.4.6.6 Subplacenta

Unique to hystricomorph rodents, a subplacenta is found on the maternal side in the central area of the placenta (Davies et al. 1961a,b) (Fig. 1.6 and 8.1).

It consists of trophoblast villi which penetrate into the decidualised area of what becomes the placental stalk. The villi are covered by cytotrophoblast underlying a syncytial trophoblast layer of characteristic ultrastructure which proliferates to form closed lacunae not perfused by maternal blood (Figs. 1.6 and 1.7). Recent corrosion cast studies (Miglino et al. 2004) have shown that the villi are well vascularised by fetal capillaries in the second half of pregnancy.

This absence of circulating maternal blood and the ultrastructure make the subplacental syncytium quite unlike that in the rest of the placenta. A secretory rather than absorptive function has been suggested (Davies et al. 1961a,b). The fetal capillary network is said to resemble that found in an endocrine gland (Miglino et al. 2004; Rodrigues et al. 2006) possibly secreting mediators for placental regulation but nothing is known of the synthetic capacity of this region.

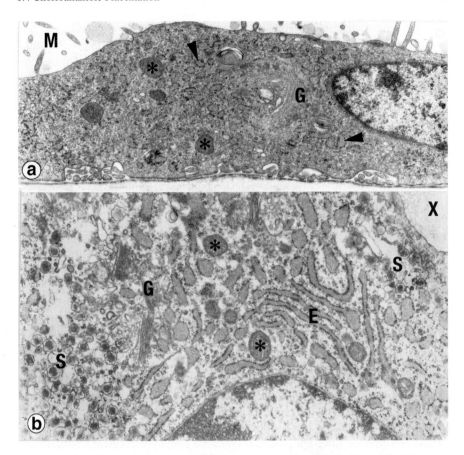

Fig. 1.7 Hemomonochorial guinea pig placenta. (**a**) The spongiotrophoblast fine structure is very similar to, and continuous with the trophectodermal syncytium of the labyrinth. Note the large mitochondria (*asterisks*), vesicular rough endoplasmic reticulum (*arrowheads*) and golgi body; *M* maternal blood space. (**b**) The ultrastructure of the subplacental syncytium is markedly different from other placental syncytium. It has arrays of endoplasmic reticulum (E) and numerous characteristic small secretory granules (S) produced from the golgi body (G). X, intrasyncytial lacuna which usually is edged with microvilli. (**a**) 39 dpc; ×13,400. (**b**) 39 dpc; ×14,000

1.4.7 Definitive Placentation

The 'mature' or 'definitive' placenta may be characterized as the one whose structure shows no significant differences in number or type of layers between fetal and maternal blood when compared with the term placenta. If two types coexist (synepitheliochorial and epitheliochorial in sheep; haemochorial and inverted yolk sac in rodents) then usually one of them has a considerably larger fetomaternal exchange area and is quoted as the definitive form.

Often a functional yolk sac precedes the development of the chorioallantoic placenta. The growth of the fetus also has two phases: an *embryonic* period when the organ systems differentiate and a subsequent *fetal* period characterised by an exponential growth phase. A prerequisite for this rapid growth is a vast increase in the placental capacity for nutritive and gas exchange. It is to meet this demand that the chorioallantoic placenta has attained its characteristic or definitive structure usually by the second half of pregnancy and frequently much earlier, but always by the time that the maximum rate of fetal growth occurs. Baur (1977) has shown for a variety of placental structures that the absolute area of the fetomaternal junction increases continuously throughout pregnancy although the area per unit weight of fetus usually decreases significantly just before term. This indicates that the existing area of the placenta needs to be used with ever-increasing efficiency, and in most placentas the layers between the fetal and maternal circulations do get progressively thinner, which would be one way of achieving this. The fetal and maternal capillaries approach more closely by indentation of the layers and may also increase in width and frequency. The type and number of layers remain, however, unchanged, the modification being merely a refinement of a pre-existing structure (Fig. 5.6).

It has been suggested that, ontogenetically, haemochorial placentation is developed via epitheliochorial and endotheliochorial stages (Amoroso 1952). However, this is only approximately true even at the invading front of the (usually syncytial) trophoblast because of the speed and lack of any synchrony in the development. There is no stage at which there is a significant area of (for instance) endotheliochorial placental structure in human or rodent haemochorial development. Behind the invading 'front' the definitive placental structure is immediately established and apart from 'fine tuning' in respect of diffusion distances no change in basic or definitive structure subsequently occurs, merely an increase in area. This again emphasizes that the different placental types are equally successful solutions to a particular cluster of environmental pressures. If one type of chorioallantoic placenta was more efficient, as the demands of the fetus increased it might be expected that this type would predominate since the placental area is growing continuously after establishment until term and thus is susceptible to continuous modification.

The definitive placental structures may thus be defined as those found in the main placenta close to term. Accessory placental structures should also be considered although they do not normally constitute a significant proportion of the total area. In the bovine placentomal placenta, for example, the interplacentomal area represents less than 2% of the total area available for exchange at term (Baur 1972) but it may have an essential functional role.

1.5 Types of Yolk Sac Placentation

Four basic yolk sac types can be characterized by reference to the origin of the fetal cellular layers they present to the maternal tissues. The association with the maternal tissues is rarely as close as in the chorioallantoic placenta, absorption being from the fluid (usually uterine) at the fetal surface. The fetal layers are different in each of the four categories and serve as important if usually transient sites of physiological exchange between mother and fetus.

1.5.1 Bilaminar Yolk Sac or Bilaminar Omphalopleure

This is present in all species initially as the layers surrounding the yolk or forming the blastocyst (Fig. 1.1). The trophoblast forms the outermost layer and there are ultrastructural indications of absorption by endocytosis. Delayed implantation blastocysts can survive up to 6 months at this stage in the uterine lumen. The trophoblast is usually cellular but may also form a syncytium and initiate implantation as in lagomorphs. The pole opposite the embryonic disc may persist as a functional yolk sac zone to term in marsupials and some insectivores (Fig. 1.3).

1.5.2 Trilaminar Yolk Sac (Trilaminar Omphalopleure or Choriovitelline Placenta)

The central mesoderm is richly vascularized and the external trophoblast usually forms an intimate apposition with uterine tissues so a true placenta is formed. This arrangement is found in certain fish (Figs. 3.1 and 3.2), reptiles (Fig. 3.3) and marsupials (Fig. 4.1) as the major placental structure. Eutherians frequently develop an area of this structure which may be important at early stages of development following implantation but never persists as a functional structure to term (Fig. 1.2). The external trophoblast is usually cellular, but occasionally this forms a syncytium also, for example at implantation in carnivores. Monotremes and marsupials (Metatheria) typically have exclusively cellular bilaminar and trilaminar areas which persist to the time of release from the egg or term (Fig. 4.1). The bilaminar areas show intracellular structures indicating endocytotic uptake, while the trilaminar areas are structured to minimize the diffusion distance between fetal and maternal circulations, thereby facilitating respiratory exchange (see Chap. 2).

1.5.3 Incomplete/Partially Inverted Yolk Sac

Invagination of the yolk sac sphere to form a cup brings the vascularized endodermal top (or embryonic) half of the sac against the outer non-vascularized bilaminar half, obliterating the yolk sac lumen (Fig. 1.3). The bilaminar segment may persist up to term in some insectivores, bats and rodents but usually thins and degenerates before this. A characteristic of the bilaminar segment is the presence of a robust acellular membrane – Reichert's membrane - between trophoblast and endoderm

(Fig. 1.3). In many cases this is the only continuous layer of the three. In the rat it is probably composed of basement membrane material produced by the endodermal layer (Fatemi 1987; Mazariegos et al. 1987). The trophoblast may transform into a discontinuous giant cell layer and invade the endometrium as in the rat and mouse or persist as a columnar epithelium as in the squirrels (Fig. 1.2). There is no firm evidence that the bilaminar layer acts as anything other than a coarse filter between the mother and fetus. Large protein molecules pass across freely. The top segment of the yolk sac does function as a barrier with selective transport capability as described below.

1.5.4 Completely Inverted Yolk Sac

In many rodents, insectivores and bats the bilaminar segment of the yolk sac is either never formed or disappears at an early stage. This exposes the vascularized columnar endodermal cell layer ('inverted yolk sac', see Fig. 1.3) to the uterine lumen or endometrium. A number of studies have demonstrated selective absorption and transmission of proteins across these cells (Bainter 1986; Brent and Fawcett 1998), which are sealed at their apices by tight junctions. This is the main route for transfer of intact maternal immunoglobulins for fetal immunoprotection in rodents, rabbits and guinea pig. There is no evidence that their coexistent chorioallantoic placentas have any similar function. Conversely in primates the chorioallantoic trophectoderm is the immunoglobulin transporting epithelium and the yolk sac although important for general nutrient uptake in the first trimester (see Sect. 1.5.5 below) regresses and has no function after 12 weeks of pregnancy.

There is a considerable range of elaboration of the 'inverted' endodermal layer in the various orders from a simple and loose association with a flat uterine epithelium (mouse, rat) to a profusely branched villous array very closely apposed to corresponding uterine projections – a true placental structure only found so far in one rodent, Perognathus (Mossman 1987). Such placentas play an essential role in early to mid gestation but probably only a subsidiary role during later gestation.

Another region with a specialised function is the "intraplacental yolk sac or "sinus of Duval" area in rats and mice where the yolk sac is carried into the fetal base of the chorioallantoic placenta as narrow sleeves around the initial branches of the umbilical vessels. This brings the yolk sac epithelium close to the maternal blood inflow channels in the chorioallantoic placenta. This yolk sac region has recently been shown (Kovacs et al. 2002) to be a major site for calcium ion transport between mother and fetus, 90% of which occurs in the final five days of pregnancy. Over the same period the amount of the 9 kD calcium binding protein, a marker of calcium transport, increases by one hundred and thirty times in the intraplacental yolk sac cells (Mathieu et al. 1989; Glazier et al. 1992) and the area also increases (Ogura et al. 1998). Hormones and receptors involved in calcium transport are also concentrated in this region (Kovacs et al. 2002).

1.5.5   The Situation in Man and Higher Primates

In man the secondary yolk sac never makes contact with the chorion, being separated by the extensive exocoelomic cavity (Fig. 1.2). Consequently, a choriovitelline placenta is never formed in the classical sense, and it has generally been assumed that the yolk sac plays no role in nutrition of the early embryo. However, the outer mesodermal layer of the yolk sac possesses all the morphological features of an absorptive epithelium (Enders and King 1993), and the recent demonstration of the presence of specific transport proteins indicates that it may serve an important function (Jauniaux et al. 2004). Furthermore, the exocoelomic fluid is rich in amino acids and other nutrients (Jauniaux and Gulbis 2000), Thus in higher primates there may be a physiological choriovitelline placenta, although not an anatomical one.

1.6   Types of Chorioallantoic Placentation

In the chorioallantoic placenta throughout gestation the outermost fetal layer is invariably the trophoblast (synonyms; 'trophectoderm' or 'chorion' if in association with somatopleuric mesoderm on its inner side) or a trophoblast derivative. It is a remarkably versatile epithelium showing great capacity for invasion, cell fusion, hormone production, specific nutrient absorption, selective transport, active metabolism and ability to resist maternal immunological attack.

1.6.1   Cyto- and Syncytiotrophoblast

All trophoblast is initially cellular. In endotheliochorial and haemochorial placentas this cytotrophoblast produces a continuous mantle of syncytiotrophoblast by division and fusion of some of its daughter cells into an overlying syncytium. Subsequent division of the residual cytotrophoblast provides one cell for fusion into the syncytium and the other persists as a stem cell for the syncytium. In haemochorial placentas the daughter cells may develop into individual giant cells which can migrate into the maternal tissues; in synepitheliochorial placentas they may differentiate into binucleate cells which form fetomaternal syncytia with uterine epithelial cells throughout pregnancy.

Evidence from in vivo and in vitro growth studies shows that syncytiotrophoblast can form only by fusion of cytotrophoblast cells. No nuclear division has ever been clearly demonstrated in placental syncytium in any species. The fusion process requires expression of fusogenic proteins (probably derived from endogenous retroviral genes) as the initiators of a sequence of fusion into and differentiation of a fully functional syncytium. In long pregnancies the sequence may be completed by compaction and aggregation of aged nuclei and their removal by localised syncytial fragmentation – producing "syncytial knots" in man (Boyd and Hamilton 1970; Potgens et al. 2002; Black et al. 2004).

In the human placenta all the syncytial nuclei investigated were shown to be diploid, whereas the cytotrophoblast nuclei ranged from diploid to tetraploid as would be expected of a dividing cell population (Galton 1962). However Klisch et al. (2000) have shown in the ruminant synepitheliochorial placenta that the binucleate cells which form from cytotrophoblast become polyploid before they form the maternofetal syncytium characteristic of this type.

The formation of syncytiotrophoblast exclusively from cytotrophoblast means the latter needs to be available throughout pregnancy to provide for the enormous increase in area found in all placentas examined. In most haemomonochorial placentas the cytotrophoblast (which lies between the syncytiotrophoblast and its basement membrane) is discontinuous from an early stage but is always available to support expansion. Cytotrophoblast may also be found in layers or cell aggregates usually close to the junction of the placenta with the endometrium for example the cytotrophoblast shell in the human or the spongiotrophoblast in the mouse. Where the cytotrophoblast itself forms the (cellular) haemomonochorial placenta, growth in area is by simple division [in Zapus the jumping mouse (Fig. 8.20b), Tadarida, a bat (Fig. 1.8d), and hyrax (Fig. 8.20a)].

Haemodichorial placentas (rabbit; Chap. 3.6.4) can theoretically accommodate expansion once formed as long as one layer has some cellular constituents which can divide and fuse into either layer to support growth (Figs. 76, 81 and 98). However, there is no experimental evidence for the production of one layer from the other and the relationship between the two is very different from that between cyto- and syncytiotrophoblast in the haemomonochorial placenta.

In the haemotrichorial placenta, of which all so far investigated have two inner syncytial layers and one outer cellular layer in contact with maternal blood (Figs. 76 and 100) (Enders 1965; King and Hastings 1977), there is a similar growth problem. No evidence for any contribution from the cellular layer to the adjacent syncytial (middle) layer has been reported. There is good evidence that the three layers are initiated at the very start of labyrinth development (Carpenter 1972) and that there is an enormous increase in area of all three layers as this chorioallantoic placental labyrinth grows. To support this growth it seems likely that in both haemodi- and trichorial placentas each syncytial layer grows independently from occasional persistent cellular insertions or interruptions. In the absence of evidence for any nuclear division in the syncytium this is the simplest hypothesis but requires more definite confirmation than the occasional observations currently available (Davies and Glasser 1968; Metz 1980).

Once formed, the haemochorial placentas show increasing attenuation of their constituent layers as pregnancy progresses, as do all other types of placenta. Near term all show areas of trophoblast, syncytial or cellular or both, which are so thin that organelles are excluded (Figs. 8.7 and 8.15). These areas usually overlie deeply indenting fetal capillaries and, presumably, greatly facilitate rapid fetomaternal exchange (Mayhew et al. 1986). The trophoblast also has a considerable capacity for endocytotic uptake of non-diffusible molecules, and the normal ultrastructural marker for this is coated vesicle formation from caveolae at the relevant plasmalemma. In di- and trichorial placentas caveolae are only found on layer II, and not on the loosely apposed fenestrated layer I, which is bathed in maternal blood. Layer I forms no barrier to serum or tracer molecules but does provide a baffle plate slowing down blood flow and producing relatively static regions immediately adjacent to layer II, which may, according to Enders (1965), facilitate endocytosis. Other regions of trophoblast past which maternal blood flow would be sluggish (judging by blood space architecture) also have a high incidence of coated vesicles. In many haemomonochorial placentas a lattice of narrow subsurface intrasyncytial channels is found, and this is in continuity with the maternal blood space (Figs. 1.8a–d and 8.21; Enders 1982).Coated vesicles form from these channels, which may represent an alternative method of providing a static blood serum reservoir for efficient absorption. The channels may be derived initially from the basement membrane of the maternal endothelium, which is displaced by syncytiotrophoblast invasion starting soon after implantation. Subsequently, however the syncytium must synthesize the fibrogranular content since the channels persist and grow in extent as the syncytium increases in area throughout gestation.

In the bat, *Myotis*, autoradiographic studies indicate that the structured glycoprotein/carbohydrate material in the channels (Fig. 8.21) is synthesized by the syncytium (Cukierski 1987). So far in all cases save one, the intrasyncytial channels form a closed system accessible only from the maternal blood space. The exception is the bat *Tadarida*, in which the channel content is clearly continuous with the basement membrane of the fetal syncytiotrophoblast (Fig. 1.9).

Whether this forms a continuous conduit from maternal blood to fetal connective tissue remains to be determined.

The fibrillogranular material in the channels is analogous in origin and appearance to the interstitial layer ('membrane') found at the syncytial surface of the endotheliochorial (carnivore) placenta (Figs. 7.8a, 7.9). Although normally superficial, this interstitial layer is frequently penetrated by syncytiotrophoblast processes. A function exclusively as a fetomaternal barrier therefore seems unlikely and further functions relevant to transport seem feasible. This penetration of the interstitial layer is also seen in the cellular endotheliochorial placenta found in the elephant (Wooding et al. 2005c)

The continuity of a syncytial layer is considered to provide a better barrier to potentially harmful maternal cells and molecules than a cellular layer linked by tight junctions. The rarity of cellular haemomonochorial placentas supports this conclusion, but their existence again emphasizes the lack of any unique advantage of a syncytium.

In haemodi- and trichorial placentas the layers are linked by occasional tight junction remnants, frequent desmosomes and extensive gap junctions (Fig. 8.7) (Metz et al. 1976, 1978). The tight junctions may be remnants of those which linked the cells from which the syncytium formed and the desmosomes are needed to maintain the relationship between the layers. Gap junctions are normally considered to function by allowing the unimpeded passage of molecules below a certain size and thus to link cells electrically for synchronous activity (uterine smooth muscle, for example). It

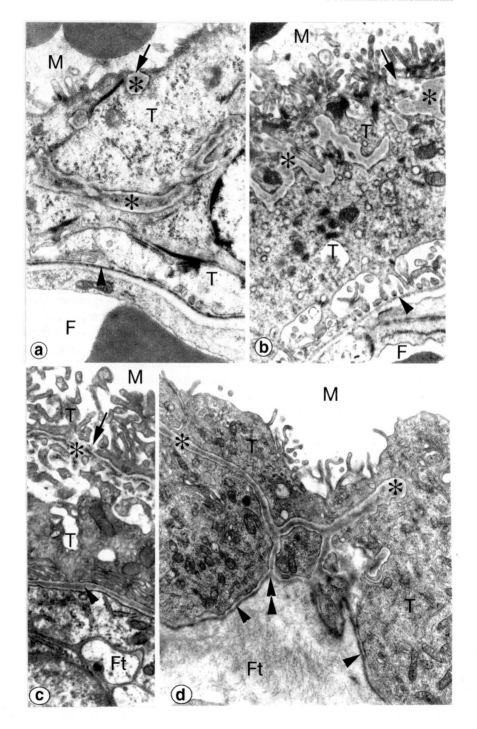

has been suggested that in this placental context the gap junctions facilitate metabolite transfer across the layers. Tracer studies with lanthanum chloride and horseradish peroxidase clearly show that the healthy syncytial layers completely prevent transport in either direction of molecules of this large size (Aoki et al. 1978; Metz et al. 1978), However Shin et al. (1997) have suggested an important role for gap junctions in transport of the essential glucose molecule from mother to fetus and transport of such small molecules by this route seems plausible.

Although the syncytiotrophoblast shows no nuclear division it is very active in metabolism, synthesis and secretion. In man it has been shown to produce a wide variety of steroids and proteins including interferons (Bulmer et al. 1990) and several hormones (Fig. 8.16), in vivo and in vitro (Pattillo et al. 1983). Syncytiotrophoblast in other groups contains an equivalent highly developed range of organelles and so may be supposed to play an equally important metabolic and endocrinological role. However, great care is needed when extrapolating between species. For example, in the mouse the syncytial layers do not synthesise steroid hormones or a chorionic gonadotropin (Malassiné et al. 2003).

Immunologically the syncytiotrophoblast surface facing the maternal blood in man and in mouse (the only two species investigated in detail so far) seems inert with no clear demonstration of any major histocompatibility complex (MHC) determinants. It has been suggested (Billington and Bell 1983) that MHC antigen expression is obligately linked to cell division so that formation of a non-dividing layer would be one way of eliminating this expression with its attendant hazards.

Several advantages of a fetal syncytium in a placental context have been suggested: its plastic invasiveness; its synthetic versatility and capacity to secrete directly into the maternal compartment; its ability to form a continuous selective barrier which can include a considerable degree of regional differentiation or attenuation; and its immunological unresponsiveness and its inability to proliferate even if nucleus- containing fragments are released into maternal blood as they are in the human (Covone et al. 1984). However, there are examples of cellular trophoblasts which apparently perform just as well, so that syncytial change does not seem to be a prerequisite for successful placentation. Its widespread occurrence does argue that it may be better suited for carrying out the multifaceted requirements of a placenta than is a cellular system.

Fig. 1.8 (**a–d**) Hemomonochorial placentas with intrasyncytial lamellae (ISL, *asterisks*). In (**a**) the carnivore, Hyena and in (**b**) the sciuromorph rodent Marmot, the lamella (*asterisks*) is frequently open (*arrows*) to the maternal blood space (M) but no connection to the basement membrane (*arrowheads*) has been reported. In the sciuromorph Chipmunk (**c**) the ISL (*asterisk*) is only open (*arrow*) to the maternal blood space (M) but not to the basement membrane (**c**, *arrowhead*) In (**d**) Tadarida, a Molossid bat, the ISL (*asterisk*) and basement membrane (*arrowhead*, **b**) are continuous (*double arrowhead*) but here there is no evidence for any connection of the ISL directly to the maternal blood space (M). F, fetal capillary. (**a**) Hyena, midpregnancy; ×16,000, (**b**) Marmota, late pregnancy; ×15,200. (**b**) from Enders (1965), (**c**) Chipmunk, late pregnancy; ×21,500, from Enders (1982), (**d**) Tadarida, late pregnancy, ×12,000, from Stephens (1969)

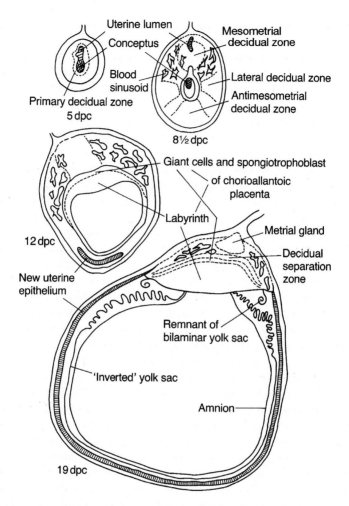

Fig. 1.9 Drawings of the development of decidual and placental tissue in the mouse all at the same magnification, based on Bell (1983) and Rugh (1975)

1.6.2 Giant Cells

Mononuclear giant cells are characteristic of rodent placentation and are formed from the surface cytotrophoblast of the conceptus during implantation and pregnancy. They are phagocytic and invasive and initially play an important role in enlarging the implantation chamber by eroding the superficial endometrial and decidual tissue including the capillary endothelium. Subsequent to implantation they form a loose sheath two to five cells thick around the conceptus through the interstices of which, on the antimesometrial side where they form the outer layer of the bilaminar yolk sac, maternal blood circulates. They are characteristically, though not exclusively, large (50–100 μm diameter) with large polyploid nuclei (16–32 N, but 850 N has been reported) (Barlow and Sherman 1972; Ilgren 1983). Visualization of individual chromosomes indicates that the giant cells usually only have the diploid number but each chromosome consists of multiple unseparated copies of its own DNA. This is polyteny rather than polyploidy and the distinction may have important functional consequences (Snow and Ansell 1974; Bower 1987; Varmuza et al. 1988).

No evidence has been observed for cell division in giant cells with or without prior colchicine treatment. Incorporation of radioactive thymidine into giant cell nuclei is considered to be evidence of endomitosis, DNA replication without subsequent cell division.

Primary giant cells are defined as trophoblast cells which transform during and just after implantation on the antimesometrial and lateral (mural) aspects of the rodent blastocyst external to Reichert's membrane. Secondary giant cells are formed a little later from the edge of the trophoblastic ectoplacental cone (M.H. Kaufmann 1983). This is situated mesometrially and is a thickened cytotrophoblastic cellular plate which develops into the chorioallantoic placenta, producing from its growing edge secondary giant cells which form a layer between spongiotrophoblast and the decidua (Figs 1.8; 8.8).

There is great variation between rodent and lagomorph families in giant cell production, with murine rodents showing the largest and most characteristic forms. Their ultrastructure is unspecialized apart from the massive nucleus and quite different from the adjacent spongiotrophoblast (Fig. 8.8, rabbit). They frequently show evidence of phagocytosis, have little glycogen, contain both rough and smooth endoplasmic reticulum and a small Golgi apparatus. No cell junctions to other cells have been reported (Jollie 1969; Kaufmann 1983). The secondary giant cells form the boundary of the developing chorioallantoic placenta and are central to the selective invasion and phagocytotic erosion of the maternal tissues and remodelling of the maternal vasculature essential for implantation and subsequent chorioallantoic placental growth. They also form the only layer in one haemomonochorial placenta, that of Zapus (jumping mouse) (Figs. 8.4 and 8.20b).

Another important giant cell role throughout pregnancy is to secrete a remarkable range of paracrine and endocrine effectors which control placental growth and development and maternal reactions to pregnancy. They produce a wide and changing variety of Prolactin family hormones (Hall and Talamantes 1984; Soares et al. 1991; Yamaguchi et al. 1992; Soares et al. 1996, 2006; Simmons et al. 2007), steroids (Deane et al. 1962), angiogenic and vasodilatory factors and decidualisation stimuli(see refs in Simmons et al. 2007).

Rodent giant cells are immunologically specialized in expressing none of the MHC determinants on their plasma membranes (Lala et al. 1983) and could therefore represent an immunologically neutral buffer zone when they form a continuous shell around the conceptus in early pregnancy.

The primary and secondary giants described above are found at the edge of the placenta adjacent to the endometrium. Simmons et al. (2007) have recently identified three other populations of cells within the mouse placenta in the second half of pregnancy which are mononucleate, polyploid and produce hormones of the Prolactin family indicating the versatility of the giant cell lineage (Sect. 1.7.5.4).

In humans all so-called 'giant cells' found in the maternofetal interface junction zone are in fact minisycytia. They produce placental lactogen and chorionic gonadotropin and are probably formed by fusion of the diploid extravillous invasive 'interstitial' trophoblast cells found throughout the decidua (Hoffmann and Wooding 1993; Al-Lamki et al. 1999). These extravillous trophoblast cells do express the fusogenic endogenous retroviral protein, syncytin (Frendo et al. 2003).

Other genera, notably the (syn)epitheliochorial type, have specialised populations of large hormone producing trophoblast cells such as the binucleate/giant cells in ruminants (producing ruminant placental lactogen and steroids (Wooding 1992)); the binucleate chorionic girdle/endometrial cup cells in equids (equine chorionic gonadotropin (Allen and Stewart 2001)); and the giant cells in camelids (steroids. Wooding et al. 2003)). All "giant" cells produce hormones to facilitate maternofetal accommodation and many share other characteristics such as migratory ability poyploidy, and MHC nonexpression with the rodent Giants but there are considerable differences and these will be detailed in later sections.

1.6.3 *Spongiotrophoblast*

Rodent and Lagomorph placentas usually have a definite zone between the giant cell layer and the base of the labyrinth referred to as the 'spongiotrophoblast', 'trophospongium' or 'junctional zone'. The tissue, cellular in mouse and rat, syncytial in the rabbit and guinea pig (Figs. 1.6 and 1.7), forms channels draining the maternal blood from the placenta but fetal capillaries do not penetrate into it (Figs. 8.1, 8.5 and 8.8).

The spongiotrophoblast is derived from the outer layers of the original ectoplacental cone and contains stem/pluripotent cells which produce the giants and the basic trophospongioblast layer including occasional small aggregates of glycogen cells. Like the giant cell layer the spongiotrophoblast is perfused by maternal blood from a very early stage of chorioallantoic placenta formation. This is an excellent location from which to secrete fetal hormones or other effector molecules solely into the maternal circulation and the spongiotrophoblast has been shown to produce a wide range of Prolactin like proteins (Ain et al. 2003; Soares et al. 2006). Mossman (1987) suggests that it is homologous with the cytotrophoblastic shell at the base of the placental disc in primates. An immunological role has been postulated because in rat, mouse and man trophoblast in this position is said to bear MHC antigens absent from the vast bulk of the haemochorial placental syncytiotrophoblast (Faulk and Hsi 1983; Kanbour et al. 1987), although there are dissenting voices (Chatterjee et al. 1982).

1.6.4 Invasive and Endovascular Trophoblast

The main effectors of the endometrial arterial remodelling essential for successful pregnancy in the hemochorial placentas, including mouse man and rat, are the invasive cytotrophoblast cells from the margin of the placenta which infiltrate the decidua and migrate into the mouths of the decidual arteries. Unlike the rest of the trophoblast, these cells express MHC subsets G, C and E and a unique range of cytokines and receptors (see Chap. 9.1; Hanna et al. 2006; Hannan et al. 2006). The remodelling consists of replacing the elastic layers in, and the muscle sheath around the decidual and spiral arteries producing widened channels lined with fibrinoid. Such channels can provide the ever increasing volume of maternal blood at a lower pulse pressure required by the placenta in the second half of pregnancy. Decidualisation initiates the changes: the swelling and disorganisation of the muscle and elastic layers. This is then reinforced by cytotrophoblast cells migrating into the wall from the arterial lumen (endovascular) and/or from the surrounding decidual tissue (interstitial).

The cytotrophoblasts are involved in the elimination of the endothelium and the elastic and muscle layers of the arteries, replacing them eventually with cytotrophoblast cells and their syncytial derivatives dispersed between tubular lamellae of fibrinoid and lining the lumen. Regulation of this process depends on complex interactions between invading cytotrophoblasts, uterine Natural Killer (uNK) and other immune cells in the decidua and their cytokine and endocrine networks (Saito 2000; Moffett-King and Loke 2006). The regulation is as yet little understood and there are considerable differences of detail between remodelling in the mouse, rat and man (Ain et al. 2003; Caluwaerts et al. 2005; Pijnenborg et al. 2006). For example, in mice the remodelling is mostly by interstitial cytotrophoblast and limited to the shallow decidual zone, but in rat and human both interstitial and endovascular remodelling is extended well into the myometrial spiral arteries in the second half of pregnancy. In Knock Out mice lacking uNK cells there is no remodelling and pregnancy is deleteriously affected (Croy et al. 2006), whereas in the potentially lethal human condition preeclampsia, characterised by little remodelling, an increased number of uNK cells has been reported. In humans there is blocking of the mouths of the partially eroded arteries for as long as the tenth to twelfth week of pregnancy by plugs of extravillous cytotrophoblasts. This prevents circulation of blood through the placenta in the first trimester favouring the low oxygen environment necessary for early development of the conceptus. Removal of the plugs allows blood to flow and completion of the remodelling at the start of the second trimester. There is a similar arterial blocking by cytotrophoblast in the Rhesus and Baboon (Enders and King 1991) but this block is removed by 28–30 dpc (Blankenship and Enders 2003). This is at ~18% of pregnancy length compared with ~33% in human.

The endovascular cytotrophoblast (EVT) is remarkably invasive in several bat species (Badwaik et al. 1998; Badwaik and Rasweiler 2000). In Molossus ater interstitial invasion produces individual cells and aggregates deep in the decidua and between the maternal vessel endothelium and basement membrane right through the myometrium layer. The EVT cells have even been identified immunocytochemically in the corpus luteum stroma.

In the bat endometrium the arteries "dilate tremendously" when the discoid placenta they supply is functional in the second half of pregnancy (Badwaik and Rasweiler 2000).

1.6.5   *Fetomaternal Hybrid Syncytiotrophoblast*

There is good evidence for fetomaternal cell fusion in three cases so far. Such hybrid tissue persists only briefly at rabbit implantation and in the chorioallantoic placenta of the marsupial *Perameles* (bandicoot) but is formed from implantation to term in all ruminants so far investigated.

In rabbit the blastocyst trophoblast forms a fetal syncytial 'peg' which fuses with a uterine epithelial cell or syncytium formed from the uterine epithelium (Fig. 2.2) (Larsen 1961; Enders and Schlafke 1971a). By this means the blastocyst gains access to the endometrium to initiate formation of the chorioallantoic placenta. The maternal nuclei in his hybrid syncytium are said to disappear rapidly (although this aspect has not been closely studied) and the fetomaternal fusion is seen simply as a means of penetrating the uterine epithelial barrier.

Perameles is one of the few marsupials to develop a chorioallantoic placenta, and this forms only in the last 3 days of a 12 day gestation. At this stage the uterine epithelium consists of syncytia of limited extent (plaques), each containing several identical small nuclei. It is apposed to trophoblast cells each with a morphologically very different, large nucleus. Subsequently, the cellular trophoblast disappears and the syncytial plaques are found to contain both sizes and sorts of nuclei (Figs. 4.1, 4.2 and 4.3) (Padykula and Taylor 1976). No direct ultrastructural evidence for fusion of the apposed cellular and syncytial layers have been observed but this is the simplest explanation. The fusion allows the fetal and maternal blood capillaries to approach much more closely (see Figs. 4.1, 4.2 and 4.3). The postulated fetomaternal tissue persists for the last 2 days of gestation when the fetus grows at a very rapid rate.

Fetomaternal fusion in ruminants is a much more persistent phenomenon (Wooding 1992; Chap. 6). Binucleate cells develop in the predominantly uninucleate trophoblast just prior to implantation. They form 15–20% of this layer throughout gestation. When mature (i.e. with a full complement of their characteristic granules) they migrate out of the trophectoderm and fuse with the apposed uterine epithelium to produce a transient trinucleate cell(cow) or persistent fetomaternal syncytium(ewe) at the interface with the maternal endometrium (Figs. 6.1, 6.4 and 6.5). This process continues throughout pregnancy (Fig. 6.2), contributing to an integral layer of the placenta and/or delivering to the maternal tissue the granules which contain placental lactogen among other constituents (Fig. 6.2). Further details and a discussion of the relevance of this system can be found in Chap. 6 on synepitheliochorial placentation below and Wooding (1992). Other examples of fetomaternal fusion may well occur in other placentas, but without unequivocal markers it is difficult to recognize and verify such unexpected events.

1.7 Decidual Tissue, Structure and Functions

1.7.1 Origins of Decidua

Decidual tissue is characteristically found in those species with invasive, haemo-chorial placentation. It was initially defined as that part of the maternal endometrium which was shed at parturition. However, at partus little is left of this complex tissue which is now known to play a vital role from the start of success-ful conceptus development, and the definition has been widened considerably to take account of this (Fig. 1.9; Bell 1983; Keams and Lala 1983; Lala et al. 1983; Glasser 1990).

By definition decidual cells are produced by steroid hormone-dependent divi-sion, enlargement and differentiation of maternal uterine stromal fibroblasts (Glasser and McCormack 1980; Tarachand 1986). They form decidual cell zones separating the invasive trophoblast from the myometrium in the hemochorial pla-centas of insectivores, bats, rodents, lagomorphs and anthropoids (Wynn 1967). Less developed zones are also reported in the minimally invasive synepitheliochorial ovine placenta (Johnson et al. 2003). In hemochorial placentas the areas are most prominent in early stages of pregnancy but vary considerably in extent in the different genera. In man there is a very extensive development, but in the closely related baboon the initial decidualization is morphologically much less obvious (Ramsey et al. 1976), but all primates have recogniseable decidua by mid-pregnancy (Enders 1991).

Decidualisation involves increased gland secretion and the stromal decidual cells can produce a wide range of hormones, cytokines, growth factors and immu-nomodulatory molecules involved in the recruitment of the limited but specific immune cell populations and development and growth of the placenta (Jones et al. 2006a,b).

In most primates together with Phyllostomid and Molossid bats decidualisation starts before fertilisation and the tissue is shed in non pregnant cycles.

In most other mammals decidual development is initiated by the arrival of a blastocyst or some artificial stimulus (oil drop, suture thread, electrical stimulation) in the hormonally primed (oestrogen plus progesterone) uterine lumen and follows a specific sequence (Fig. 1.9).

In rodents the antimesometrial decidual cells develop first and form a very compact avascular zone which excludes most blood proteins from the uterine lumen containing the implanting conceptus (6–9 dpc) (Parr and Parr 1986; Parr et al. 1986; Christofferson and Nillson 1988). Subsequently, the more loosely packed mesometrial decidual cells are closely involved in the fetomaternal interactions as the trophoblast invades to form the hemochorial placenta (8–12 dpc). Proliferation of the blood sinuses in the decidualiz-ing mesometrial uterine stroma also facilitates placental growth (7–11 dpc). The areas of decidual cells rapidly increase from the uterine lumen outwards but not necessarily as far as the myometrium and the extent can be conveniently visualized by the characteristic presence of alkaline phosphatase and periodic acid-Schiff (PAS) reactivity in the decidual cells (Bell 1983).

The decidual zone also contains a changing and essential population of immune cells significantly different from that in the non pregnant endometrium (Clark 1985a). In humans in early pregnancy in the deciduas basalis, immune cells form 40% of the total cells. 70% of these immune cells are uterine Natural Killer (uNK) cells, with 18% macrophages, 10% T lymphocytes, 2% dendritic (antigen presenting) cells and an occasional B lymphocyte. Rodents have similar concentrations of equivalent cells with uNKs (often referred to as large granular lymphocytes, LGL) the most frequent (Sargent et al. in press).

As the chorioallantoic placenta develops, invasive trophoblast cells become an increasingly important part of the decidual cell spectrum.

Maximum development of the decidual cell populations occurs early in pregnancy, eventually occupying most of the considerably expanded uterine stroma and in mice and rats, completely surrounding the conceptus (Welsh and Enders 1985). In some mammals the decidua includes newly developed blood vessels which play an essential roles in placental growth and development. In rodents the scattered meshwork of capillaries in the decidual zone is the site for placental labyrinth development (Adamson et al. 2002). In the bat *Molossus ater* a tight meshwork (or tuft) of capillaries forms immediately under the uterine epithelium and this is enveloped by the invading cytotrophoblast to form the framework for development of the discoid chorioallantoic placenta (Badwaik et al. 1998).

Decidual cell division largely ceases in early to midpregnancy and the decidual cell zone is continually reduced by stretching and by programmed cell death. This loss of the decidual cells facilitates the expansion of the conceptus and growth of the placenta. and at term it forms only a thin layer at the base of the placenta through which the placenta separates (Fig. 1.9).

There is increasing evidence for a variety of functions for the decidua at different stages of haemochorial pregnancy: initially to produce new decidual blood vessels which act as templates for placental growth, and then as a physical and/or immunological barrier, a nutrient source, a hormone producer and a facilitator of placental growth and eventual release.

1.7.2 Deciduoma

The decidual tissue formed in response to an artificial stimulus is called a deciduoma, and is grossly similar to that induced by a blastocyst but has fewer regional specializations and with no arterial modifications (Wadsworth et al. 1980; O'Shea et al. 1983; Herington and Bany 2007). Deciduomas regress spontaneously in mice and rat even if pregnancy is mimicked by maintaining the progesterone level artificially. This indicates that cells of the deciduoma have a predetermined life span and undergo apoptosis (programmed cell death) rather than necrosis.

1.7.3   Decidualization and Menstruation

The hormonal changes preceding ovulation produce correlated changes in the uterine stroma and epithelium. Usually such changes are restricted to tissue oedema and structural changes in the uterine epithelial cells (in ruminants and carnivores, for example). However, in humans and Old World monkeys the cyclic changes include considerable stromal to decidual cell development and vascular proliferation to elongate the spiral blood vessels. The subsequent cell necrosis and haemorrhage in this tissue as the progesterone level drops produces the menstrual discharge (Ramsey 1982). The instructive similarities between menstruation, inflammation and decidualization have been pointed out in excellent reviews by Finn (1986) and Jones et al. (2006a, b).

1.7.4   Decidual Cell Structure

1.7.4.1   Stromal Cells

The majority of decidual cells develop from the division in the endometrial stroma of irregularly shaped fibroblast-like cells with little cytoplasm and few organelles. Decidualization starts with the blood vessel pericytes in rabbits and humans and subepithelial stromal cells in rodents and characteristically involves a considerable increase in cell size (Amoroso 1952; Mossman 1987). The cytoplasm may develop meshworks of intermediate filaments, many free ribosomes, much glycogen and/or lipid droplets and numerous gap junctions with adjacent decidual cells. Different areas (see Fig. 1.9) also display their own characteristics. In rats and mice, antimesometrial decidua have very large closely packed cells, frequently binucleate or polyploid, with well-developed rough endoplasmic reticulum and Golgi apparatus and a low level of glycogen and/or lipid stores. Mesometrial decidual cells are smaller, more irregularly shaped and uninucleate, usually with rather more glycogen. Lateral decidual cells, among the rapidly expanding blood sinuses, have massive amounts of glycogen and are only occasionally binucleate (Welsh and Enders 1985; Parr et al. 1986).

In man there is a comprehensive decidualization throughout the uterine stroma but with less regional specialization than in rodents (Wynn 1974; Schiebler and Kaufmann 1981). There are three main types of human decidual cells. There are numerous 'small predecidual' swollen fibroblasts ($50\,\mu m \times 5\,\mu m$) which have an equivalent low level of cytoplasmic organelle development and frequent 'large undifferentiated' decidual cells ($100\,\mu m \times 25\,\mu m$). Scattered among these two predominant types are 'large differentiated' decidual cells ($100\,\mu m \times 25\,\mu m$) with well-developed arrays of rough endoplasmic reticulum, many mitochondria, an extensive Golgi complex and large areas of glycogen and lipid. Human decidual cells are isolated from each other by an investment of fibrinoid and are linked by fewer gap junctions than are found in rodents (Boyd and Hamilton 1970; Schiebler and Kaufmann 1981).

These cells form the basic structural matrix of the decidua. Other cells are invariably present (in pregnancy and deciduomata) and these, the immune cells and invasive cytotrophoblast are dealt with in detail below.

1.7.4.2   Uterine Natural Killer (uNK) Cells

Also known as Large Granular Lymphocytes (LGL) in man and Granulated Metrial Gland (GMG) cells in rodents, these are the most numerous lymphocyte subset found in the decidua. The immature and mature forms circulating in the blood are major constituents in the innate adaptive immune response (Lanier 2006).In man they localise to, and proliferate in, the decidualising endometrium during the estrus cycle but die and are shed at menstruation. In pregnancy they persist and increase in number with full decidualisation of the stroma. In rodents they are recruited after conception and rapidly divide and differentiate but only in the decidua basalis region. The decidualising stromal cells produce the signals (hormones, cytokines and chemokines) which trigger the recruitment and guide the uNKs differentiation to express a different phenotype from the circulating NKs. In man they are $CD^{56\ bright\ 16\ neg}$ (Moffett-King 2002) and rodents show a similar phenotype (Croy et al. 2006). These cells can produce a wide range of cyto/chemokines, and under appropriate stimulation by trophoblast and stromal cell mediators can become either tolerogenic or cytotoxic types (Hanna et al. 2006; Lanier 2006; Moffett-King and Loke 2006; Shigeru et al. 2007). This is further discussed in Chap. 9.

Electron microscope and cell repopulation studies with irradiated mice and rats indicate that uNK cells develop from small round cells which originate from the bone marrow but divide and differentiate in the endometrial stroma (Peel et al. 1983; Peel and Stewart 1986). They can easily be distinguished on structural and staining criteria from blood eosinophils, basophils, neutrophils or mast cells. They are present in pregnant and pseudopregnant uteri (Peel et al. 1979).

Mature uNKs are the most frequent cell type after the fibroblast-derived decidual cells. They are often binucleate, and typically contain glycogen, rough endoplasmic reticulum and a large Golgi body which produces numerous membrane-bound granules. These are usually homogeneously electron dense, but some also have membranous inclusions (monkey, Cardell et al. 1969; rat, Larkin and Flickinger 1969; mouse, I. Stewart and Peel 1977; review, Croy et al. 2006). In the light microscope the granules are eosinophilic and PAS positive and contain lysosomal enzymes, IgG molecules synthesized in the granulocyte, MUC-1, LGL-1 and a protein, perforin, characteristic of killer and cytotoxic lymphocytes. Human uNKs show a weak natural killer activity which is enhanced by interleukin 2 (Ferry et al. 1990) and blocked by interferon-γ but not -α (A. King and Loke 1993).

The uNKs do not show phagocytosis, nor do they form intercellular gap, tight or desmosomal junctions with each other or with disparate cell types. In mice the cells average 50 μm in diameter and have a smaller number of larger (5 μm) granules compared with the rat 25 μm diameter, 2 μm granules) or man (12 μm diameter, 1 μm granules).

In rat and mouse the cells are found in the mesometrial decidua from an early stage (mouse, 8 dpc) and they also collect around the myometrial blood vessels to form an accumulation in the mesometrial triangle between the myometrial layers which is referred to as the metrial gland (Fig. 1.9) (Peel 1989).

The uNK cells reach their maximum numbers at 13–14 dpc in the rat. Between 15 and 21 dpc (term) there is a progressive loss of the cells by swelling and lysis and they release their cell contents including the granules into the connective tissue. This process has been referred to as holocrine secretion (Peel and Stewart 1979) but seems more analogous to the 'programmed cell death' seen in other decidual cells, or cell regression in ovarian luteolysis. This would mean that the functional life of the cells is finished rather than that the release of the cell contents is the ultimate purpose of the process of granule development. The cell debris is removed by macrophages, which are the probable precursors for the cells full of lipid droplets and phagolysosomes which progressively replace the granulocytes in the metrial gland region from 17 dpc. Some granulated cells may persist post term, but most regress long before parturition. There is no evidence for direct transformation of granulocytes to lipid-laden cells.

In man uNK cells are present before implantation and continue to proliferate in the decidua as this tissue expands in the first 12 weeks of pregnancy. From 12 to 20 weeks 40% of the total cells in the decidua are uNKs and then there is a gradual loss as in the rodent with very few persisting to term.

Successful pregnancy depends upon the coordination of the possible interactions between uNKs, invasive cytotrophoblasts and stromal decidual cells to produce a suitable local hormone/cyto/chemokine environment to favour immune tolerance throughout pregnancy and sufficient arterial and decidual remodelling to ensure adequate nutrient transfer. The uNKs may play an important immunological role in their tolerogenic form (see Sect. 9.1) and studies in man and rodents show that they are also necessary for normal arterial remodelling (Croy et al. 2006; Hanna et al. 2006; Pijnenborg et al. 2006; Shigeru et al. 2007).

1.7.4.3 Decidual Macrophages

The evidence for this population comes from histological work and from tissue culture studies of mouse decidua (Hunt et al. 1984). The cells have immunoglobulin (Fc) receptors, an ectoesterase enzyme on their plasma membranes and are phagocytic, suggesting an infiltrating macrophage type. The incidence of this cell type increases to a peak at 15 dpc in mouse, is strain dependent and low numbers seem to be correlated with poor fertility and a high conceptus resorption rate (Bell 1983; Tawfik et al. 1986).

Their main function is the removal of the cell debris from the stromal and arterial remodelling necessary for placental establishment (Mor and Abrahams 2003).

1.7.4.4   Dendritic (Antigen Presenting) Cells

Cells with a high level of Ia expression (MHC class II antigen) which lack lymphocyte or macrophage markers are typically involved in antigen presentation in the immune response sequence. Such cells have been identified in tissue cultures of mouse decidua from 14 dpc onwards (Elcock and Searle 1985) and on sections of human decidua in early pregnancy (Oksenberg et al. 1986). Interaction with uNK and cytotrophoblast cells contributes to the production of local fetal immunotolerance (Blois et al. 2007; Zarnani et al. 2007).

1.7.4.5   T Cells

Cell populations have been isolated from mouse (mesometrial) (Brierley and Clark 1987) and human (basalis) (Daya et al. 1985) decidua which display suppressor activity against cytotoxic lymphocytes (CTLs) in vitro. Decidua lacking such Treg cells cannot usually support a successful pregnancy. Recent work has established in mouse and human that 50% of the T cells in the decidua are $\alpha\beta$ and 50% $\gamma\delta$. The latter have been identified in man and mouse to be important contributors of cytokines which promote the establishment of a decidual immunological environment tolerant to the fetus (Mincheva-Nilsson 2003; Clark et al. 2004).

1.7.4.6   'Decidual' Structures in Non-Haemochorial Placentation

Extensive decidual cell development characterizes haemochorial placentation, but in some synepitheliochorial and endotheliochorial placentas specialized cells originating from uterine stromal cells are consistently incorporated throughout the definitive placental structure at the feto-maternal interface. The best example is the frequently binucleate giant cells in the cat placenta (Fig. 7.8a). Much smaller less obvious cells are also present in an equivalent position in the dog (Amoroso 1952; Mossman 1987). Cells of similar structure are found as giant pericytes around the maternal vasculature in the sheep and goat (Fig. 7.8b). These cells never form a coherent layer as in haemochorial placentas but their consistent presence suggests that they may play a role (producing a specific secretion?) in maintaining the delicate balance between fetal and maternal tissues. Decidual changes have also been reported in the pseudopregnant ferret, but the alterations are in the subepithelial uterine capillary endothelium rather than the stromal cells (Beck and Lowe 1972).

Molecular biological and microarray studies of the gene expression during decidualisation in hemochorial placentas have recently allowed a more accurate identification of the genes specifically involved (Hess et al. 2007; Kashiwagi et al. 2007). Similar studies in the sheep synepitheliochorial placenta on implantation show that three such genes are upregulated on implantation in the endometrial fibroblasts which also divide and enlarge. The extent of this previously unrecognised "decidualisation" in the minimally invasive synepitheliochorial placenta is considerably less than in a hemochorial placenta but greater than in the noninvasive epitheliochorial pig (Johnson et al. 2003).

1.7.4.7 Epithelial Plaque

Decidualization has been defined here as endometrial stromal cell differentiation. Epithelial plaque, which is always cellular, is formed by proliferation and development of uterine luminal and glandular epithelium dependent on the same hormonal regime and enhanced in a similar way by contact with the blastocyst.

Deciduomas, produced by luminal injection of gonadotropin (hCG) at the correct time after ovulation("window of receptivity"), decidualise and form plaque as if pregnant but the reaction is not as localised, The effect is specific, FSH injection produces no reaction (Jones and Fazleabas 2001).

Plaque formation is characteristic of some monkeys whose implantation is superficial and whose early stromal decidualization is minimal (Rossman 1940; Enders et al. 1985; Owiti et al. 1986). The epithelial proliferation is transient, starting at 6 dpc, and the plaque is engulfed and removed by the developing trophoblast by 20–22 dpc in the rhesus monkey for example (Figs. 8.17 and 8.18) (Enders et al. 1985; Enders 2008). Prior to its obliteration it undergoes symplasmic degeneration. This life history is similar to that undergone by the surface epithelium in the uterus of carnivores and lagomorphs (Steer 1971; Schlafke and Enders 1975; Leiser 1979). In these cases there is less initial proliferation but the symplasmic masses have to be absorbed in the same way by the trophoblast during the cellular remodelling involved in the trophoblast invasion of the maternal tissues. The extreme proliferation involved in plaque production can be seen as one of the ways of avoiding too rapid a penetration of the endometrium by the trophoblast (compare Figs. 8.9 and 8.17, 8.18). Plaque allows controlled access to the maternal circulation. Once this has been achieved superficial development of a purely fetal vascularized syncytium to form villi or a labyrinthine meshwork is all that is required to produce the haemochorial primate placenta. There is no need for deeper erosion of the endometrium for in most Old World primates the maternal blood supply is restricted to orifices at the maternal face of the fetal tissue. Epithelial plaque formation is one way of achieving this controlled access to the maternal blood after superficial implantation: by contrast, interstitial implantation induces stromal decidual cell development (Ramsey 1982). Both are systems for establishing a viable fetomaternal dialogue with a minimum of delay and stress to either party.

The variation in the balance between epithelial plaque and stromal decidual transformation (Ramsey et al. 1976) is in sharp contrast to the similarity of the final placental structure. This may be taken as further evidence for the difficulty of rapidly establishing the fetomaternal dialogue: of balancing the nutritional needs of the embryo against the immunological sensitivities of the mother.

The same problems may be solved by what appear superficially to be very different morphological methods. In practice, small changes in mitotic rates of particular cell groups or phagocytotic capacity of the trophoblast might be sufficient to account for the differences.

1.7.5   Functions of the Decidual Cells

Many functions have been proposed for decidua and it seems likely that there may
be several roles it can play; which one is emphasized probably depends upon the
specific requirements of each particular species.

1.7.5.1   Isolation of the Implanting Blastocyst

In rodents with a small blastocyst there is a rapid initial decidual development
of stromal cells immediately around the blastocyst (Fig. 1.9). These cells
become so closely packed, accentuated by both tight (Wang et al. 2004) and gap
junction formation, that maternal blood cells and proteins are prevented from
reaching the blastocyst (Parr et al. 1986). The effect is transient, lasting only
from 6 to 9 dpc in the rat, and subsequently these early decidual cells die, to be
displaced by the expanding conceptus. Reichert's membrane forms between the
yolk sac ectoderm and endoderm coincidentally with the loss of this decidual
barrier, and it has been suggested that this new acellular layer may take over
the function of excluding harmful constituents of the maternal blood from the
embryo (Parr and Parr 1986).

1.7.5.2   Barriers to Trophoblast Invasion

Direct evidence that decidual cells act as barriers to trophoblast invasion comes
from the observation that mouse blastocysts transferred to subepithelial ectopic
sites (kidney, testis, anterior chamber of the eye) penetrate into the non-decidualized connective tissue much more deeply than into uterine decidua (Kirby 1963,
1965; Porter 1967). In human ectopic (usually tubal) pregnancies the endometrial
fibroblasts do not express Prolactin or IGFBP1 which are the normal decidual
markers and there are more invasive trophoblasts and uNK cells (Kemp et al. 2002;
Goffin et al. 2003). This produces a more extensive and much deeper invasion than
normal which can lead to maternal vascular damage, haemorrhage and death showing the importance of a decidual 'buffer' zone.

Corroborating this buffer concept is the fact that skin grafted onto decidualized
uterine stroma is rejected much more slowly than skin in the non-decidualized
uterus. The process of decidualization starts at the luminal surface, and the uterine
epithelium, though rapidly lost, probably plays an essential role in mediating and
amplifying the stimulus to decidualization (Kennedy 1983). Uterine epithelium that
is not receptive (i.e. that is not hormonally primed) is a much more complete barrier
to blastocyst penetration.

A mechanical barrier results from the formation of a coherent decidualized cell
layer by rapid division and expansion of the initially widely separated fibroblasts in
the oedematous endometrial stroma (Glasser and McCormack 1980; Parr et al. 1986).
The resulting decidual cells are often linked by gap junctions and are thus capable of

concerted changes. It has been suggested that such a linked cellular barrier is so effective that expansion of the trophoblast is dependent on prior programmed cell death of the decidual tissue (Beaulaton and Lockshin 1982). Certainly there seems to be a very close relationship between the two tissues in rodents (Welsh and Enders 1985).

In man the decidualized cells are not so closely packed as in rodents but they frequently secrete a dense extracellular matrix consisting of collagen and glycosaminoglycans, which would also hinder trophoblast invasion (Boyd and Hamilton 1970; Schiebler and Kaufmann 1981). Such material is usually referred to as 'fibrinoid' and may form semi-continuous sheets between fetal and maternal domains like Nitabuch's and Rohr's 'striae' in the basal plate of the human placenta (Boyd and Hamilton 1970).

However, such decidual cellular and extra-cellular barriers are rarely complete (Martinek 1970) and elements of the trophoblast frequently penetrate well past the decidualized 'layer'. The interstitial and endovascular trophoblast in the myometrium of humans and rats are examples of this breaching of the 'barrier' (Ain et al. 2003). The decidual layer may form an effective if partial barrier in some species (Bradbury et al. 1965), but even then it is probably only part of an array of accommodations between mother and fetus to promote fetal survival.

1.7.5.3 Nutrient Provision

One of the characteristics of decidualizing cells is the accumulation of glycogen and/or lipid stores, although this is not always found. It is generally assumed that this is produced for the eventual nutrition of the blastocyst. Amoroso (1952) noted that animals without decidual cells usually have many more uterine glands producing histotroph for blastocyst maintenance and growth.

1.7.5.4 Hormone Production

Prolactin

The main role of pituitary prolactin in pregnancy is to maintain the progesterone production by the corpus luteum, but PRL or PRLReceptor null mice cannot sustain pregnancy even with added progesterone. This indicated a placental source, and progesterone-dependent prolactin production by cultured decidual cells has been shown in man (Maslar et al. 1986) and rodents, and histological studies in man showed localization of prolactin to basal decidual cells by immunocytochemistry (Braverman et al. 1984; Al-Timimi and Fox 1986; Bryant Greenwood et al. 1987) and hybridization histochemistry (Wu et al. 1991).

Control of decidual prolactin secretion is significantly different from pituitary prolactin because it is unaffected by bromocriptine (which blocks) or thyrotropin-releasing hormone (which stimulates) prolactin secretion from the pituitary (see Gibori et al. 1984).

In man only one PRL gene is expressed in decidua and syncytiotrophoblast, but in rodents up to 35 members of the PRL superfamily have been identified and localised (Soares 2004, 2006) a few (~3) in the decidual stromal cells but many more in total (~30) in various groupings in four specific trophoblast locations (Simmons et al. 2007). PRL family members control genes which regulate placental angiogenesis and immune function (Bao et al. 2007) and play an important role the complex and changing cellular interactions in the decidua required to initiate and maintain a functional placenta (see Chap. 8).

Prostaglandins and Steroids

Cultures of rodent and human decidual cells have been shown to produce prostaglandins (Lala 1989) and steroids, which have immunological potential as demonstrated by their effects on the mixed lymphocyte reaction (MLR) and cytotoxic lymphocyte (CTL) function. They are also implicated in the control of predecidual cell division, in the initiation and maintenance of decidual differentiation (Glasser and McCormack 1980; Kennedy and Lukash 1982; Clark 1985b), and in the initiation and process of parturition.

Relaxin

Relaxin mRNA has been localised to the human decidua and trophoblast from the first trimester to term (Bogic et al. 1997). The same group also find relaxin receptor protein in those locations (Lowndes et al. 2006) and relaxin is suggested to play an important role weakening fetal membranes immediately prior to birth. This seems to be specific to man, there are no reports of relaxin in rodent decidua or placenta.

Insulin-like Growth Factor Binding Protein

One of the principal secretions of the human decidua is a IGFBP1 (Bell et al. 1988). This secretion will modulate the bioavailability of IGF2, one of the main promoters of placental growth, at the materno-fetal interface. Elevated levels will restrict growth, and its presence has contributed to development of the 'conflict theory' by which maternal and fetal demands during pregnancy are balanced (Haig 1993; see Chap. 9.3).

1.7.5.5   Immunological (Functions) Implications

The decidual tissue forms a matrix in which immunologically active cells can modify the maternal response to the fetus (see Chap. 9.2; Colbern and Main 1991).

In primates and rodents the three main cell types involved are the decidual stromal cells, uterine Natural Killer cells and the invasive cytotrophoblast. Interactions between these three populations produce a local environment tolerant of expressed fetal antigens and capable of considerable local maternal arterial remodelling. The lack of any MHC I or II expression on the fetal syncytial surface facing the maternal blood and the constant presence of the tolerogenic progesterone simplifies the tolerance requirement. However the wide and ever expanding list (from molecular biological and microarray studies (e.g. Shigeru et al. 2007)) of hormones, growth factors, cytokines and surface receptors which each of the three cell types can produce considerably complicates attempts to work out exactly which cell does what. A possible scenario follows.

Progesterone initiates decidualisation. Hormones and cytokines (e.g. PRL, IL11) from the decidua then modify the immune cell content, specifically favouring recruitment and proliferation of uNK cells. Activation of mature uNKs then produces different cytokines (IFNγ, Ang2) which initiate swelling and disruption of the tissue in the walls of the spiral arteries. MHC subsets (C,G) and receptors (KIR) unique to the invasive trophoblast then interact with the uNKs to switch their cytokine production from those (eg IFNγ) favouring the cytotoxic Th1 subtype to those favouring the tolerogenic Th2 phenotype. This can produce a wide variety of cytokines some of which serve to attract and direct the invasive cytotrophoblasts to complete the remodelling of the arteries aided by phagocytosis of the cell debris by the macrophages. The uNKs disappear once the cytotrophoblast has fully invaded the arteries and the Th2 phenotype including T suppressor cells is established in the decidua. This phenotype will be reinforced by more cytokines directing the dendritic and $\gamma\delta$T cells to favour tolerance through the rest of pregnancy (see Chap. 9.1).

This has the advantage of brevity and does fit many of the known details but it ignores many others. This perhaps suggests that there may be several routes through the web of cellular interactions, each leading to a more or less successful placentation.

1.7.6   Release of the Placenta

By definition the decidua is a hormone-dependent tissue layer at the fetomaternal interface autolysing on progesterone withdrawal to facilitate release of the placenta soon after birth of the young. Release of prostanoids by the autolysing cells may also play a part in the initiation and process of parturition (Bell 1983). However, the wide variety of animals without decidua manage to separate their placentas successfully and in the European mole Talpa there is a well-developed decidua but the placenta remains in situ and is resorbed (Amoroso 1952).

Chapter 2
Implantation, Maternofetal Exchange and Vascular Relationships

2.1 Implantation

Implantation is the mechanism by which the blastocyst establishes itself in the uterus, and the trophectoderm develops an intimate relationship with the uterine epithelium (Parr and Parr 1989; Denker 1993). This process is easily disrupted and most early pregnancy loss occurs during this time (Wilmut et al. 1986). The process can also be suspended ('delay of implantation, embryonic diapause') at this stage, which serves to reduce the metabolic load on a lactating mother and to ensure that birth occurs at the most advantageous season (Enders and Given 1988; Given and Enders 1989). During diapause, progesterone secretion must be maintained and implantation can be reactivated by a pulse of estrogen in rodents or LIF secretion controlled by day length in mustelids (Lopes et al. 2004).

The conceptus reaches the uterus in the morula stage surrounded by the acellular zona pellucida. It must then position itself correctly in the uterus, remove the zona often with the aid of endometrial secretions, adhere to the uterine epithelium and possibly invade that epithelium in initiating the development of the characteristic placental structure. Recent studies have indicated that these structural changes are controlled by complex interactions between hormones, growth factors, prostaglandins, cytokines and their receptors (Bowen and Burghardt 2000; Paria et al. 2002; Dey et al. 2004; Lee and Demayo 2004; Spencer et al. 2004a,b; Kennedy et al. 2007).

2.1.1 Position (Spacing) in the Uterus

In duplex uteri (rat, mouse, guinea pig, rabbit) the horns open separately into the cervix and blastocysts cannot pass from one horn to the other. Spacing in each horn is independent. In bicornuate uteri (pig, dog, cat, ruminants) the two horns are continuous and a conceptus ovulated on one side can pass to the other side prior to implantation (Fig. 1.4).

In species with several conceptuses (polytocous), the blastocysts normally implant at evenly spaced sites along a uterine horn (Boving 1971). How this spacing is achieved in the 2 or 3 days available is not clear. In the rabbit (duplex uterus) there appears to be no relationship to any pre-existing structure such as the scars of the previous pregnancy since conceptus number varies but the spacing is always even. If there is only one conceptus per horn it is always in the middle of the horn (Boving 1971). In the pig and cat with bicornuate uteri, animals with a single ovary provide evenly spaced implantation sites in both horns and with pig conceptuses of different phenotypes (colours), if one colour is introduced into each horn, the colours randomize before implanting at an even spacing (Dziuk et al. 1964).

These experimental observations indicate a considerable capacity for blastocyst migration along the uterine lumen and the most likely explanation is an initial mixing followed by a passive movement propelled by the vigorous contractions of the uterine musculature observed at this time (Crane and Martin 1991). Inert beads of a similar size to a rabbit blastocyst are moved along if introduced into the uterine lumen at this time (4–6 dpc; Boving 1971). However, the trauma of surgery makes this an unreliable experimental model. Presumably the conceptus would first need to be roughly spaced by vigorous muscular contractions instituted perhaps by substances present in the oviductal fluid entering with the conceptuses; and subsequently adjusted by local initiation of contractions and/or uterine growth by the individual conceptus.

However, all the above discussion refers to cat, pig or rabbit experiments, in which the blastocyst swells considerably (10–2,000× in the rabbit) during the positioning and adhesive phases. In the mouse, rat and guinea pig the blastocysts do not swell at this stage but are still evenly spaced in the uterus prior to adhesion. This is probably achieved by muscular contraction since both relaxin (Pusey et al. 1980) and changes in uterine prostaglandin synthesis, which affects uterine contractility (Hama et al. 2007) disrupt spacing but there is no direct evidence for such movement of these tiny blastocysts (Boving 1971).

In species with a single conceptus in the uterus (monotocous) or in each uterine horn (ditocous) there is also evidence for preferred sites of implantation. For example, multiple ovulators such as the elephant shrew implant only a single conceptus in each horn, and some African antelopes ovulate on either side but implant predominantly in the right horn. In the bat *Carollia*, only one small area in the uterus is receptive (Oliveira et al. 2000), but no clearly identifiable receptive *structures* have yet been described (Mossman 1987).

Most ruminants normally implant about halfway between the cervix and uterotubal junction but the caruncles at this site have no reported unique characteristics. It has been shown that cow and sheep blastocysts anchor themselves at their preferred site by trophectodermal outgrowths (papillae) into the mouths of the uterine glands (Fig. 6.1) (Drieux and Thiery 1951; Guillomot and Guay 1982; Wooding et al. 1982), but this does not explain how the site is initially selected.

There is now direct evidence for considerable conceptus movement in the uterus prior to implantation: the large horse blastocyst can be located in vivo with ultrasound, and it is reported that it moves freely between the two horns on 11–15 dpc but then implants caudally in one horn at 16–17 days (Ginther 1979, 1983, 1984a,b).

There seems to be clear evidence for free conceptus movement, and also a consensus that positioning is complete before the zona pellucida (or its replacement in horse and rabbit) is lost. Rupture of the zona then allows intimate adhesion between blastocyst and uterine epithelial surfaces. Most rodents implant on the antimesometrial side of the uterus. This is said to be due to circular muscle contractions since the antimesometrial position is more central with respect to the circular muscle in many uteri. Solid pieces of tumour tissue locate antimesometrially in the mouse uterus (Wilson 1963) whereas deformable oil droplets are randomly positioned (Martin 1979).

2.1.2 Adhesion

There are three main requirements for adhesion: loss of the zona pellucida, 'closure' of the uterine lumen, and development of membrane contact between the trophectoderm (cellular or syncytial) and the uterine epithelium.

The zona pellucida is ruptured and/or dissolved by a combination of blastocyst swelling plus the action of glycosidase and protease enzymes from uterine epithelium and blastocyst (Denker 1978; Denker et al. 1978; Denker and Tyndale-Biscoe 1986). The success of ectopic implantation demonstrates the non-specific nature of any uterine enzymes involved.

After rupture of the zona pellucida, 'closure' of the uterine lumen occurs by the resorption of fluid by the uterine epithelium, edematous swelling of the uterine stroma and development of muscular tone. These processes bring the uterine epithelium into an intimate relationship with the trophectoderm and this is accentuated in many species by swelling of the blastocyst.

Uterine closure brings the glycocalyces on the external surface of the trophectoderm and on uterine epithelial cells close together. Since both bear negative charges they will not readily adhere. Several cytochemical investigations have shown a generalized decrease in uterine epithelial surface charge (measured by cationic probes) around the time of implantation (reviewed by Parr and Parr 1989). Direct evidence for change prior to the apposition between blastocyst and uterine epithelium is shown by erosion of the unusually thick glycocalyx in the ferret (Enders and Schlafke 1971b) and in the pig (Fig. 2.1).

Changes in uterine epithelium luminal proteins have been found in the rabbit (T.L. Anderson et al. 1986; Hoffman et al. 1990a) and alterations in the expression of carbohydrate epitopes demonstrated in the mouse (Kimber and Lindenberg 1990; Kimber 2005). Specific lectins have been used to demonstrate pregnancy-dependent regional differences in binding to the blastocyst (Chavez and Enders 1982) and to the uterine epithelium (Anderson and Hoffman 1984; Whyte and Allen 1985; Anderson et al. 1986). More recent molecular biological and gene knockout experiments have identified a very wide range of hormones, growth factors, prostaglandins, cytokines, transcription factors and lipid mediators which are, or may be involved at implantation with much variation in detail among families (Bowen and

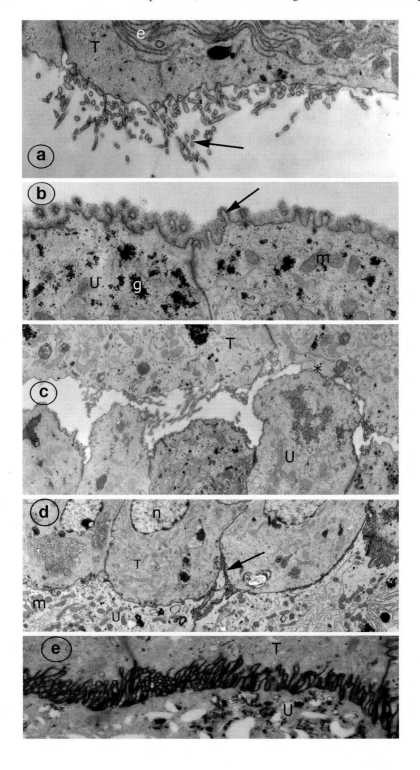

Burghardt 2000; Paria et al. 2002; Dey et al. 2004; Lee and Demayo 2004; Spencer et al. 2004a,b; Kennedy et al. 2007). The problem is that redundancy and compensatory gene expression complicate the recognition of which molecule does what and when in any particular species. However some common themes do emerge. Once a blastocyst is correctly positioned in the uterus it can provide a local stimulus. MUC 1 is a transmembrane glycoprotein expressed apically on the uterine epithelium of many species (DeSouza et al. 1999; Murphy 2004). It has a very large (~350 nm above the plasmalemma Meseguer et al. 1998) extracellular domain to prevent access to pathogens but which also blocks ligands on the trophoblast from binding to uterine plasmamembrane receptors in the glycocalyx (~10 nm above the plasmalemma). To allow attachment, local removal or deglycosylation of MUC 1 is necessary, and the blastocyst is ideally positioned to trigger this. In humans in vitro studies show that the MUC1 removal is restricted to the immediate vicinity of the implanting blastocyst (Meseguer et al. 2001). Coincidentally the cytokine LIF is induced in response to a pulse of maternal oestrogen. This stimulates HB-EGF production in the uterine epithelium triggering full receptivity as well as reacting with receptors on the blastocyst. Both processes promote changes in the integrin proteins expressed on the two surfaces which favour firm adhesion, reinforced by specific linking proteins (galectins in the ewe) in gland secretions. Adhesion triggers further factors (COX2 for prostaglandin production, homeobox gene HOXA 10 and matrix metalloproteases) necessary for decidualisation of the stroma, and some immunoregulatory genes are also changed at this time.

However it is achieved, the fetal and maternal plasmalemmas characteristically finally adhere very closely, but at a constant ~20 μm separation, at the implantation site (Fig. 2.1). There is little clear evidence for fetomaternal tight or desmosomal junction formation between the two apices despite several claims (Reinius 1967; Potts 1968; Schlafke and Enders 1985; Tachi and Tachi 1979). Evidence from freeze-fracture is required to adequately characterize the nature of the adhesion. Examination of the blastocyst attachment to uterine epithelium in vitro may also provide useful information (Lindenberg et al. 1989).

Fig. 2.1 Electron micrographs of early Pig implantation (**a**) 15 days post coitum (dpc), apex of a trophoblast cell (T) shows elongated microvilli (*arrow*) characteristic of the free blastocyst, whereas the uterine epithelium (U) at the same stage (**b**) shows blunt microvilli (*arrow*) with a clear glycocalyx coating all of the plasmalemma. (**c**) 16 dpc, the trophoblast (T) is now close to the uterine epithelium (U). The latter has lost its glycocalyx and most of its microvilli and the trophoblast microvilli are considerably reduced. In places (*asterisk*) the two are in close contact. (**d**) 16 dpc, a different region of the same specimen as in (**c**).Here the trophoblast and the uterine epithelium are in close and flat apposition, with microvillar interdigitation starting in place. (**e**) 20 dpc to Term, the microvillar junction is uniformly developed, maximising the exchange area between trophoblast (T) and uterine epithelium (U). **a,b** ×15,000; **c,d**, ×8,000: Glutaraldehyde/osmium fixation, araldite embedded, uranyl /lead stained. **e**, ×8,000: Glutaraldehyde fixed, araldite embedded, phosphotungstic acid/uranyl stained. *m* mitochondrion, *g* glycogen, *e* endoplasmic reticulum, *n* nucleus

2.1.3   Invasion and/or Placental Development

From the common start of flat apposition between maternal and fetal apical plasma membranes there are several different ways of initiating placental development. Schlafke and Enders (1975) categorized four basic types: simple interdigitation, displacement, fusion and intrusive implantation (Fig. 2.2).

2.1.3.1   Interdigitation

In the pig (Dantzer 1985) the flat apposition develops into an interdigitation of microvilli from trophectoderm and uterine epithelial surfaces (Fig. 2.1). There is no loss of cells or any fetal or maternal syncytial transformation. Development of the placenta is essentially a vast increase in the area of fetomaternal apposed epithelia with their interdigitated microvilli. Similar epitheliochorial placentation is found in camels, lemurs, whales, and dolphins.

2.1.3.2   Displacement

In the mouse, rat and hamster the apposition of the fetal and maternal cellular epithelia results in loss of integrity and subsequent delamination or sloughing of individual cells and sheets of the uterine epithelium which are then phagocytosed by the trophectoderm (Fig. 8.2; Kaufmann 1983).

This sort of concerted delamination of a large area of uterine epithelium has also been reported for the vespertilionid bats Myotis and Glossophaga (Enders and Wimsatt 1968; Rasweiler 1979; Kimura and Uchida 1983, 1984).

2.1.3.3   Fusion

Evidence for fusion between trophectodermal syncytial protrusions and uterine epithelial cells in rabbits has been presented by Larsen (1961) and Enders and Schlafke (1971a). Fusion of cytotrophoblast cells produces a trophectodermal syncytium which forms numerous processes. These penetrate the zona pellucida and indent the apex of the uterine epithelial cells forming initially a flat apposition. Subsequently, the apposed membranes fuse to form a fetomaternal hybrid tissue. Evidence of fusion comes from the observation of chains of vesicles in regions where one would expect flat apposition. Also, cells linked to the adjacent uterine epithelium by the usual apical tight junction complex demonstrate cytoplasmic continuity with the overlying trophectodermal syncytium (Fig. 2.2). Since at the same time the uterine epithelium itself is producing plaques of symplasm by loss of lateral membranes, it is difficult to follow the detailed outcome of the initial fetomaternal fusion. Fusion seems to be necessary to initiate trophectodermal

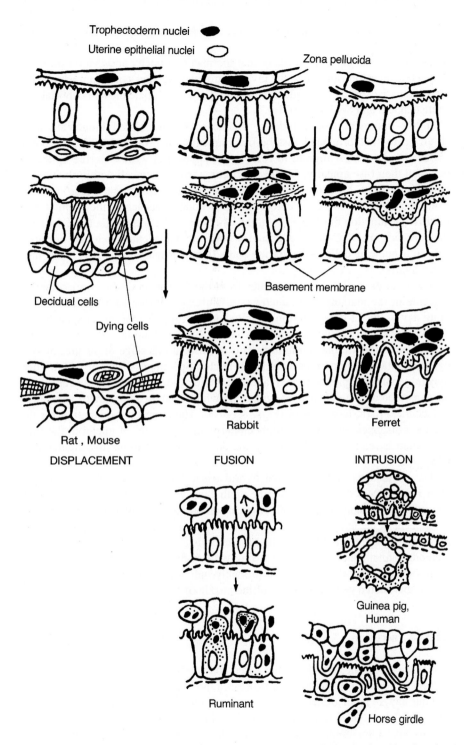

Fig. 2.2 Classification of the different fetomaternal cellular interactions at implantation (based on Schlafke and Enders 1975)

invasion, but it is considered likely that the maternal nuclei and cytoplasm make only a transient contribution to the invasive trophectodermal syncytium which penetrates into the stroma to invade the fetal capillaries to initiate the hemochorial placenta (Enders and Schlafke 1971a).

In the ruminant trophectoderm, characteristic granulated binucleate cells (BNCs) are first produced just before implantation (Wimsatt 1951; Wooding 1982b, 1983). BNCs migrate to form a flat apposition with a uterine epithelial cell, and then fuse apically, finally providing a trinucleate cell within the uterine epithelium (Figs. 2.2 and 6.1) (Wooding 1982b). Evidence for the fusion comes from images of the two cells cytoplasmically continuous across the fetomaternal microvillar junction but linked by tight junctions to their epithelia (Fig. 6.4) and the presence of BNC-derived granules containing placental lactogen hormone only in trinucleate cells in the uterine epithelium (Morgan et al. 1987) (see Chap. 6 for details).

The final example of fusion comes from the formation of the chorioallantoic placenta in the marsupial *Perameles* (Padykula and Taylor 1976). On about the ninth day of the 12-day gestation a limited area of trophectoderm cells, each with a characteristically large nucleus and backed by the allantois, comes into close apposition to the uterine epithelium. This consists at this stage of plaques of syncytium containing numerous very small nuclei. Specimens a day or so later display no separate trophectodermal layer but the fetal and maternal blood capillaries are separated only by plaques of syncytium which contain a mixture of very small and characteristic large nuclei (Figs. 4.1–4.3). The simplest explanation for the apparent disappearance of the trophectoderm and two sizes of nuclei in the final syncytium is fetomaternal fusion but the details have yet to be investigated (see Chap. 4).

The advantages of fetomaternal hybrid tissue formation are obscure. It does offer rapid access through the uterine epithelium while maintaining its structural integrity; presumably some immunological camouflage is provided by the maternal component of the hybrid tissue. However, the problem of dual cytoplasmic control must be formidable, and the fact that the hybrids are produced by such dissimilar methods – fetal syncytium or cell fusing with maternal cell or syncytium – once again indicates that these cases are probably individual answers to specific problems rather than a general solution of the puzzle of fetomaternal accommodation.

2.1.3.4   Intrusion

This type is found in several families and is characterized by firm adhesion followed by penetration of trophectodermal processes through a healthy uterine epithelium without any signs of destruction of, or alteration in, the contiguous uterine cells (Fig. 2.2). In the ferret (Enders and Schlafke 1971b), cat (Leiser 1979), Rhesus(Macaque) monkey (Enders et al. 1983) and marmoset monkey (Enders and Lopata 1999) syncytial processes pass between the cells, clearly forming tight junctions with them. In contrast, in the horse individual binucleated trophectodermal cells are said to indent the cell apices away from the tight junctions and initially pass through the body of the uterine epithelial cell on their way to form the endometrial cups (Allen et al. 1973). In all cases the uterine epithelial cells are subsequently surrounded and phagocytosed to allow a broader trophectodermal access to the endometrium.

The tiny blastocysts of the human and guinea pig form a trophectodermal syncytium at the site of apposition to the uterine epithelium (Enders and Schlafke 1969; Boyd and Hamilton 1970). This seems to facilitate rapid subsequent penetration of the entire conceptus into the endometrium, but the details of the process are unknown. The uterine epithelium never quite covers over the tiny implantation site.

As more EM studies are made it becomes clear that the categories above may not be exclusive. In the mouse, for example, at the light microscope level penetration by fusion with the intact uterine epithelium has been reported after actinomycin D injection (Finn and Bredl 1973). In the marmoset there is some evidence for fusion during the removal of the uterine epithelium (Enders and Lopata 1999).

In some marsupials there is a local penetration of an otherwise unchanged uterine epithelium by cellular trophoblast over the yolk sac which could be by fusion or intrusion (Tyndale-Biscoe and Renfree 1987); a similar process has been reported in *Galago* (Butler 1967). There are now three examples of fusion implantation; only one was known in 1975. This process may be more widespread than has been realized.

2.2   Maternofetal Exchange

All of the substances required for the growth and metabolism of the fetus, including water, oxygen, carbohydrates, most amino acids, non-esterified fatty acids, metal ions and trace elements enter the fetal from the maternal circulation. Products of metabolism such as carbon dioxide and urea, as well as certain other compounds, pass in the opposite direction. Placental intervascular transport (either chorioallantoic or vitelline(yolk sac), or both) is the principal route of fetomaternal exchange (see Atkinson et al. 2006; Jones et al. 2007 for recent comprehensive reviews).

Conventionally there are two sources of fetal nutrition: "haemotroph" - materials carried in and transferred directly from the maternal blood and "histotroph" - extracellular material including cellular debris secreted or released into the space between maternal and fetal surfaces. Phagocytosis of histotroph is considered to be a characteristic of both cellular and syncytial trophoblast. Uptake by this means is usually concentrated in specific regions such as areolae (pig, ruminants) or haemophagous zones (carnivores, sheep).

In an epitheliochorial placenta a molecule of hemotroph passing from maternal to fetal blood requires transport across maternal endothelium, uterine epithelium, trophoblast and fetal endothelium and all four epithelia are sealed by tight junctions. Between the circulations there are eight membrane 'barriers' with a variable amount of cytoplasm between the pairs. The endotheliochorial has six membranes and the hemochorial four (Fig. 1.5). However varied, placental structure has as its principal function the transport of considerable amounts of material. In the fetal lamb late in gestation 7.8 g of carbon per day per kg fetal weight enters the fetal circulation, of which 4.6 g per day per kg is returned as carbon dioxide and urea (Battaglia and Meschia 1978). Estimates of amino acid uptake made in fetal lambs between 118 and 146 days' gestation show that approximately 11 g of amino acids enter the fetal circulation per kg fetal weight per day, of which approximately 4.5 g is accumulated as new tissue (Lemons et al. 1976; Battaglia and Meschia 1978). The mean growth rate of lambs at 130 days' gestation is 36 g per day per kg (Rattray et al. 1974). Bell (1991), Bell et al. 1999 provide an excellent review of the maternal metabolic adaptations to the demands of pregnancy and quantitative aspects of transport to and metabolism in the developing conceptus.

There are three main transport mechanisms: simple diffusion down a concentration gradient (O_2, CO_2, urea); specific membrane transporters, i.e. facilitated diffusion (glucose) or active transport (amino acids); and endo- or phagocytosis (of gland secretions, red blood cells).

To understand how these mechanisms are implemented in vivo, it is necessary to use chronically instrumented animals which are conscious and unanaesthetised, since measurements made during anaesthesia often differ significantly. Most detailed studies of maternofetal transfer have used ewes, cows or horses in late pregnancy, when transport rate across the definitive structure is probably at its peak. Ewe and cow have syepitheliochorial, the horse epitheliochorial placentas and they show instructive differences which help to correlate structure and function. These results are briefly discussed below with comparisons and correlations with haemo- and endotheliochorial placentas where these are available.

The control of transfer is always multifactorial. The rate of simple diffusion depends not only on the concentration difference between placental, maternal and fetal circulations, but also on diffusion distance, utilisation of transported molecules by placental tissue and the rate of and anatomical relationship between the always separate maternal and fetal blood flows. With facilitative diffusion or active transport processes, further factors influencing transfer are the optimal rates, concentrations and affinities of the transporters and their localisations in the several membrane layers separating the two blood flows.

2.2.1   Simple Diffusion

Theoretical considerations indicate that maternal to fetal O_2 transfer is primarily flow-limited due to the high diffusing capacity of O_2 and the high reaction velocity of the exchange reaction with haemoglobin. These predictions are borne out by experimental manipulation of maternal or fetal blood flow in the conscious animal. The flow rate and anatomical relationships between the two circulations (Fig. 2.3) are considered to be more important than the concentration difference, diffusion path length, number of membranes between or surface area of those membranes available for transport between the two flows (Wilkening and Meschia 1983).

2.2.1.1   Blood Flow

Blood flow rates through the placenta differ between maternal and fetal capillary beds and also with the species; for example, in the horse maternal and fetal flow rates are 500 and 100 ml kg^{-1} fetus per min, respectively, compared with 470 and 200 ml^{-1} kg fetus per min, respectively, in the ewe (Fowden and Silver 1995; Fowden et al. 2000). Unfortunately, there are no data available to determine which pair is more 'efficient' for O_2 transfer. Blood flows can vary considerably from minute to minute but mean values from day to day do not vary greatly under normal conditions, although in all species there is a steady increase in absolute flow during pregnancy (Comline and Silver 1970). In the ewe, maternal blood flow has to be decreased to nonphysiological levels before fetal O_2 uptake is significantly reduced (Wilkening and Meschia 1983). Normal transient physiological variation in blood flow would therefore not be expected to compromise the efficiency of O_2 transfer, and the same is probably true of the horse. In ewes, the maternal and fetal haemoglobins are different and have significantly different O_2 saturation curves, with the 2, 3-diphosphoglycerate (DPG) concentration the same in maternal and fetal red blood cells. The difference between haemoglobins makes the O_2 transfer more efficient. In the horse there is no difference between the two haemoglobins, but the lower DPG levels in the fetal compared with maternal red blood cells produce a similar P_{50} to the fetal lamb; therefore, the efficiency of unloading O_2 to fetal tissues is probably equivalent in the two species (Comline and Silver 1974).

The anatomical relationship, or rather the direction of maternal and fetal blood flows with respect to one another (Fig. 2.3 and 2.4), is probably an equally important factor in determining oxygen transfer efficiency.

Horse placental blood vessels are arranged so that maternal blood flows in the opposite direction to fetal blood in the capillary networks that form the exchange vessels, which produces the optimal countercurrent system (Silver and Steven 1975; Chap. 5). In the ewe the capillary architecture is more complex and the consensus, from reconstructions of sectioned material and corrosion cast studies, is that the flows are crosscurrent or even concurrent, both less efficient arrangements than the countercurrent system for transfer in the horse (Silver and Steven 1975; Leiser and Kaufmann 1994).

2.2.1.2   Oxygen Gradients

The PO_2 and PCO_2 differences between maternal uterine and fetal umbilical arteries are smaller in horses than ewes over the normal maternal range found in unstressed pregnancies (Fig. 2.5; Comline and Silver 1970). For any given value of maternal PO_2 between 30 and 100 mmHg fetal PO_2 values rise linearly, but the fetal horse value is much closer to that in the mother than is the ewe.

Solute exchange between tubes
Countercurrent

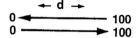

Theoretically 100% transfer is possible if 'd' is long
enough, the tubes are close enough with thin enough
walls, and the flow is slow enough

Concurrent

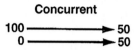

Maximum transfer 50%

Cross-current

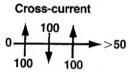

Transfer efficiency depends on numbers of tubes crossing

In vivo in most placentas the gross flow is countercurrent
but at the capillary or microflow level the tubes are
never found to be arranged in parallel

but in apposed, usually random, networks:

resulting in a mixture of counter - and cross-current
flows. It is not known whether there are significant
differences between network patterns in the different
placental types. Modifying the initial convention

Network current

seems best to fit the available evidence as a
general description of placental microflow of blood

Above 100 mmHg, fetal PO_2 plateaus in the lamb, while it rises continuously in the foal such that, at the top of the maternal range, horses can increase the fetal PO_2 significantly (Comline and Silver 1970; Table 1). Under normal maternal arterial PO_2 80–100 mm Hg, the horse has a lower transplacental PO_2 gradient than the ewe and, consequently, a smaller driving force for the movement of O_2 from dam to fetus. However there is a much smaller difference between the maternal and fetal PO_2 values in the venous outflows (horse 3 mmHg, ewe 17 mmHg) indicating a much better extraction rate by the horse fetus. This can probably be attributed to the operation of the more efficient countercurrent system in the horse so that, overall, the two species may be considered equally 'efficient' at transplacental O_2 transfer.

2.2.1.3 Diffusion Pathways

A further constraint on transfer is the diffusion distance between maternal and fetal capillary beds in different animals, but this has yet to be measured accurately and at present relies on purely anecdotal evidence (Silver et al. 1973; Silver and Steven 1975; Allen and Stewart 2001). In all species investigated so far, maternal and fetal capillaries progressively indent the uterine epithelium and trophoblast, respectively, as pregnancy progresses, thereby continuously reducing diffusion distance. Only measurements from placentas successfully perfusion-fixed from both maternal and fetal sides, representing the in vivo situation as closely as possible, are suitable for these measurements. Preliminary results from tissue prepared in this way from sheep, horse and cow, all at equivalent late stages of pregnancy indicate that the circulations are closest in the ewe (mean ± s.e. 1.0 ± 0.3 μm), further apart in the horse (3.0 ± 0.8 μm) and furthest apart in the cow (6.0 ± 2.0 μm) (Wooding and Fowden 2006). Thus the sheep would have the most 'efficient' design in this respect.

A final factor directly affecting measurement of the 'efficiency' of maternal to fetal diffusion is O_2 utilisation by the placental tissue itself. In the ewe, placental tissues use proportionately less O_2 as pregnancy progresses (from 70% of the total transferred from the mother at mid-gestation to 50% at term), but in the horse the same proportion (50%) is used throughout (Comline and Silver 1970; Silver and Comline 1975; Silver and Steven 1975). This is probably related to the demand from the continuous growth and increase in weight of the horse placenta throughout gestation, whereas in the ewe the placenta decreases in weight from a peak at mid-gestation. How these differences are related to O_2 transfer efficiency is unknown at present.

Fig. 2.3 Theoretical considerations of solute exchange between capillaries

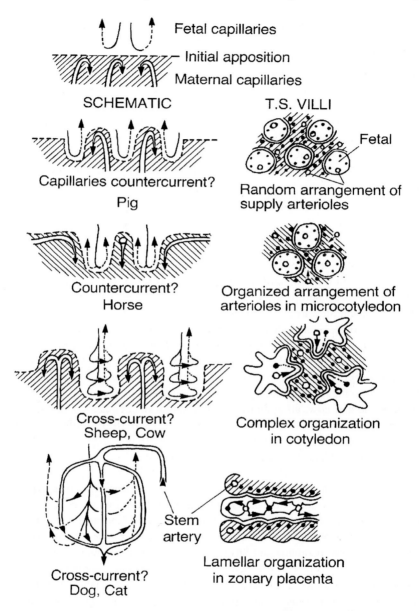

Fig. 2.4 Schematic illustration of the relationship between the direction of capillary blood flows in the fetal and maternal placental circulations in various animals. See also Figs. 2-7, 8-1, 8-12

Oxygen gradient across the placenta

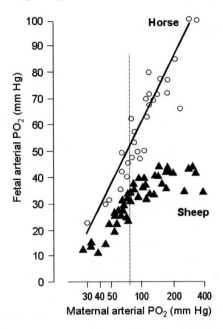

Fig. 2.5 Oxygen gradient across the placenta of the horse and ewe. PO$_2$ values are from the fetal umbilical and maternal uterine arteries over a range of experimental conditions, mostly from acute experiments. Normally quoted values for maternal arterial blood in ewes and horses lie between 80 – 100 mm Hg: mean value is indicated here by a dotted line. The difference between the two circulations is always greater in the sheep than the horse, although the mare can deliver more PO$_2$ at the upper extreme of the maternal range since the fetal blood value in the ewe does not increase after the maternal value reaches 100 mm Hg (from Silver and Comline 1975)

It is clear from the above discussion that that neither the horse nor the ewe placenta shows a consistent advantage in simple diffusion over the other when all the various factors are considered, and both produce viable offspring. There is, as yet, insufficient information to apportion relative values to each factor, but the one feature which seemed initially to correlate with placental efficiency across several species (measured as the birth weight of the fetus produced per gram fetal placental weight) was the relative direction of maternal and fetal blood flows, with countercurrent producing the most fetus per gram placenta (Leiser and Kaufmann 1994). However, the countercurrent rabbit was a clear exception; and recent work with the horse (Wilsher and Allen 2003), also established as countercurrent (Silver and Steven 1975), shows it to have an equivalent 'efficiency' (13–14 for small Thoroughbreds and ponies, calculated from data in Allen et al. 2002; Wilsher and Allen 2003) to the crosscurrent ewe (13, from data in Heasman et al. 1999). It seems unlikely, therefore, that flow directions are of critical importance in determining efficiency.

2.2.2   Membrane Transport

2.2.2.1   Facilitated Diffusion

This depends on transporter molecules (proteins) in the placental membranes, is stereospecific and saturable, requires a higher concentration in maternal than fetal blood, and needs no energy input. Facilitation can speed transport considerably compared with simple diffusion. All factors modifying simple diffusion also affect the rate of facilitated transfer, but the most important factors are type, concentration and kinetic efficiency of the carrier in each of the constant number of membrane layers between the two blood flows in each species. In horse and ewe placentas, there are eight membranes to cross (Fig. 1.5).

Glucose: In the case of glucose, one of the most important fetal nutrients, the increase in rate of transfer with a transporter can be up to 10,000 times that of simple diffusion across a membrane (Elbrink and Bihler 1975). In the horse and ewe, two isoforms of the glucose transporter (GT1 and GT3) are present in the placenta and show different affinities and kinetics. Immunocytochemical electronmicroscopical investigation demonstrates that, in both horse (Fig. 2.6) and ewe, the isoforms are localised to different uterine epithelial and trophoblast membrane layers. A glucose molecule must utilise both isoforms sequentially to cross these placentas (Wooding et al. 2000, 2005a).

   The relevance of this arrangement is at present unclear, but recent work shows that the same sequential system is found in the placenta of other ruminants (cow, deer and goat; Wooding et al. 2005a) and the endotheliochorial elephant (Wooding et al. 2005c). In hemomonochorial human (Illsley 2000) and endotheliochorial carnivore (Wooding et al. 2007b; see Fig. 7.5) placentas only GT1 is present, on both syncytiotrophoblast surfaces, whereas in the haemotrichorial rat GT1 and -3 are located on the same blood-facing membrane layer and glucose molecules can therefore utilise the isoform with the most suitable affinity and kinetics for the prevailing maternal blood glucose concentration (Shin et al. 1997). The transplacental glucose concentration difference between maternal and fetal placental blood flows remains relatively stable on a daily basis in the horse and ewe, but increases significantly in the latter as fetal glucose concentration decreases with increasing gestational age (Hay 1995). This is an overall gradient down which the glucose molecule must travel, but in practice it covers eight individual membrane steps, each of which may or may not have a unique concentration of a particular carrier isoform with individual kinetic properties. Also, a large proportion, 60–70% in ewes and horses (Fowden et al. 2000) but only 30% in rodent, human and rabbit (Hauguel et al. 1988), of the glucose taken up by the placenta is utilised by placental tissues for growth and maintenance of function, thereby complicating estimates of the rate of transport from mother to fetus (Fowden 1997). Little is known about the factors regulating the activity of this complex transport system.

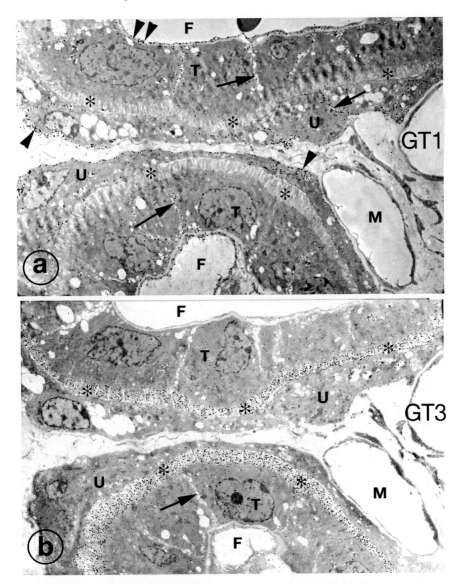

Fig. 2.6 ab Electron micrographs of immunocytochemical preparations of adjacent sections of horse placenta(290 days pregnant). **(a)** The black dots (gold colloid particles) show the localisation of Glucose Transporter 1 (GT1) immunoreactivity (IRT) on the basolateral membranes (arrows and single arrowheads) of trophoblast (T) and uterine epithelium (U) and on both surfaces of the fetal endothelium (F, double arrowheads) but the maternal endothelium (M) and the microvillar junction (asterisks) between maternal and fetal tissues are not labelled. **(b)** The only significant GT3 IRT is on the microvillar junction (asterisks) and although there is a non specific background label scattered on most tissue there is no label on any of the other membranes. Thus maternal blood glucose molecule must utilise both isoforms in sequence to cross to the fetal blood (from Wooding et al. 2000)

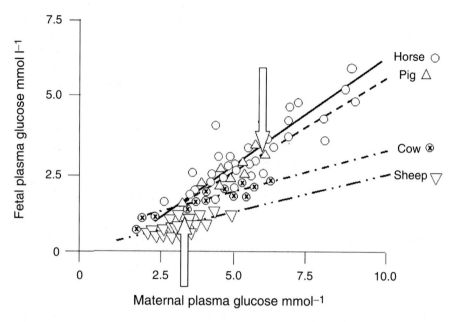

Fig. 2.7 Relationship between fetal and maternal blood glucose concentrations. The physiological range of maternal blood glucose concentrations varies according to species; mean value for horse (0) and pig (o) is indicated by the upper arrow; for cow (o) and sheep (o) by the lower arrow. Equine maternal blood maintains a mean concentration twice as high as that of the ewe because glucose is an important metabolic fuel for the mother and therefore readily available. In sheep metabolism, short chain fatty acids from the rumen replace glucose which has to be specially synthesised in the maternal liver for the fetus. The mare can therefore establish a larger glucose gradient across the placenta than the ewe without too much metabolic adjustment, providing a greater driving force for facilitated diffusion of the molecule (from Fowden 1997)

Transporter distribution and isoforms appear to be very similar in horse and ewe; the horse placenta uses slightly more of the glucose flux for its own metabolism than that of ovids, but the major difference is the force driving the transport, i.e. the higher glucose concentration in maternal compared to fetal blood (Fig. 2.7).

The horse and ewe have similar maternal PO_2 levels, but normal post prandial glucose plasma concentrations are very different, with maternal levels in the horse (pony) at ~6 mmol l^{-1} and fetal levels ~3 mmol l^{-1} being considerably higher than those in the ewe (welsh mountain), with maternal and fetal levels at ~3 and ~1 mmol l^{-1}, respectively. The transplacental gradient is therefore greater (3 mmol l^{-1}) in the horse than in the ewe (2 mmol l^{-1}) and should drive the transfer of glucose more effectively in the horse. The transplacental gradient can be abolished in the ewe by fetal glucose infusion or in the pig by hypoxic stresses and the fetus then supplies the placenta with glucose. The transfer normally operates downhill, but is capable of transferring radioactively labelled glucose in either direction after maternal or fetal injection, illustrating the dynamic nature, flexibility and complexity of the system (Hay 1995; Fowden 1997). Such findings also indicate that there may be a considerable 'pool' of glucose available for placental metabolism but not for transport. In the horse, this weight-specific rate of placental glucose utilisation decreases in late gestation so that proportionately more of the transferred glucose taken up by the uterus is available to the fetus. No similar gestational change in placental glucose consumption occurs in the ewe.

The concentrations of GT1 and -3 proteins in the membrane(s) are of paramount importance in determining the rate of transfer. In the ewe, it is known that levels of GT1 and -3 protein and mRNA per gram of placenta increase several-fold from mid-gestation to term commensurate with the fivefold increase in glucose transport. Furthermore, placental GT1 and -3 mRNA expression is independent of maternal growth and nutritional status. GT1 peaks in concentration at 120 days of pregnancy, but GT3 increases to term (145 days) (Currie et al. 1997; Ehrhardt and Bell 1997). This is a similar pattern to the rat (Zhou and Bondy 1993). In the horse, localisation of GT1 and GT3 is the same as in the ewe, suggesting a similar system, but no quantitative data on gestational trends in the abundance of these isoforms is as yet available. In human, the levels vary little during pregnancy (Illsley 2000).

Lactate: Lactate is an important fetal nutrient (up to 25% of the total energy requirement of the fetus) and requires monocarboxylic acid transporter isoforms from the same 12 transmembrane helix topology family as GT1 and -3 (Halestrap and Price 1999). They have a fairly broad specificity (pyruvate is also carried) and are proton-coupled cotransporters operating down a proton (pH) gradient. In late gestation, the horse placenta produces and transports significantly more per gram of placenta or per gram of fetus than that of the ewe. This is directed far more to the fetal than the maternal circulation. There is no transplacental transfer from the mother (Fowden 1997). In the human placenta there are different isoforms on apical and basal syncytiotrophoblast membranes (Settle et al. 2004), but no localisations have as yet been reported for the placenta of the horse or ewe.

2.2.2.2 Active Transport

Amino acids: The plasma concentrations of most amino acids are higher in the fetus than the mother in most mammals, suggesting that transport across the placenta is active. Recent work has shown that it is mediated by wide variety of specific transporters (some Na-dependent) located in the placental membranes, and requires metabolic energy (Battaglia and Regnault 2001; Jansson 2001; Atkinson et al. 2006).

However, in the horse there is no significant gradient to the fetus, with mean fetomaternal ratios (FMR) for 21 amino acids of 1.0 ± 0.3 (Young et al. 2003). In the ewe, conversely, fetal concentrations are almost twice those in the maternal circulation, with an FMR of 1.8 ± 0.7 (Chung et al. 1998). This suggests that less metabolic energy is needed to transfer amino acids across the placenta in the horse than in the ewe. However, the fact that the human placenta, with only four membranes, has a very similar FMR of 1.7 ± 0.4 (Phillips et al. 1978) and the guinea pig, with four membranes, an FMR of 3.7 ± 1.0 (Hill and Young 1973), indicates that there are other factors involved which may be more important than FMR in controlling the rate of transport. It also illustrates the potential danger in this context of comparing acute observations (unavoidable with the guinea pig) with considerably less stressful chronic results. Acute measurements in the ewe, for example, produce FMR values close to those of the guinea pig due to the stress level modification of the transport system (Schulman et al. 1975).

Another factor possibly more important than the FMR is the concentration of free amino acids in the placenta itself. This is several times higher than concentrations in either maternal or fetal circulations in all cases so far investigated (Battaglia and Regnault 2001), and is produced in haemochorial placentas by the transporters in the apical trophoblast membrane as well as by synthesis of amino acids in the placenta. The high concentration produced in the trophoblast can then lead to simple diffusion of amino acids down the appreciable gradient into the fetal circulation without the need for any active transport across the basal membrane or fetal endothelium. However, different amino acids transfer at very different rates after accumulating in the placenta, such that some transporters are definitely used between placental 'pool' and the fetal circulation (Battaglia and Regnault 2001), but both simple and facilitated diffusion are also possible. In addition, there is a considerable fetoplacental shuttle of amino acids and their metabolites and a few like glutamate are synthesized by the fetoplacental unit and none is required from the maternal circulation. This still requires active transporter molecules on the fetal side. Compared with glucose transport, a further complication is that there are 21 amino acids in different groups, acidic, basic and neutral, and each group has several different transporters with particular characteristics and isoforms (Battaglia and Regnault 2001; Atkinson et al. 2006). In man, identification, localisation and quantification of these transporters using molecular biological and in vitro vesicle transport techniques indicates that there are different populations of transporters at different concentrations on the apical and basal syncytiotrophoblast membranes (maternal and fetal 'sides'), which in the human placenta can be obtained and separated relatively easily for such investigations (Ayuk et al. 2002; Regnault et al. 2005). Such studies are beginning to provide a basis for understanding control and maintenance of the transport systems. Similar techniques have shown that equivalent transporters are present in the rat placenta but no membrane localisations have yet been established (Matthews et al. 1998). In the cat, system beta and system A amino acid transporters have recently been localised immunocytochemically to the basal membrane of the syncytiotrophoblast (Champion et al. 2004). Unfortunately, as yet there has been no identification or localisation of the presumably equivalent numbers and variety of transporters on the eight membranes separating maternal and fetal circulations in the horse and ewe, although many physiological characteristics of the overall in vivo amino acid transport are sufficiently similar to the human situation to indicate that equivalent constraints are operating. For example, in the ewe, the differential rates of transport of amino acids from placenta to the fetal circulation indicate the need for equivalent specific transporters, as in the human placenta (Battaglia and Regnault 2001).

The horse transports sufficient amino acids for optimal fetal protein accretion, whereas in the ewe fetus there is a 50% surplus, which is used for oxidative metabolism. This is reflected in the much lower urea production in the horse fetus (Silver et al. 1994; Fowden 1997). Whether this difference indicates that the horse amino acid transport system is more or less 'efficient' than that of the ewe is impossible to establish until the location, concentration and kinetics of the various transporters have been established in both species.

Ion Transport: Most ions (sodium, potassium, chloride, phosphate, etc.) are transported across placentas at each membrane barrier by a variety of carrier-mediated processes, including active, exchange and diffusion components. A major complication in assessing ion transport is the effect of the essential high water requirement of the fetus, which produces a continuous diffusive flow driven by the osmotic and hydrostatic differences between the two circulations (Stulc 1997). Small non metabolised molecules with no known transporters (Cr EDTA, mannitol) do cross the placenta passively in this flow. The sizes of pores required for this "paracellular" aqueous channel have been calculated as 2.3 nM (man), 3.5 nM (rat) in hemochorial placentas and much smaller (0.15 nM) in the ewe synepitheliochorial placenta. Inhibition of the membrane carriers indicates that this paracellular route for sodium or potassium may be of considerable significance in man but not in rats (Atkinson et al. 2006). These paracellular pores have no convincing ultrastructural correlate in any placenta so far investigated. A further complication is the recent demonstration of placental plasma membrane bound aquaporin molecules in man, mouse and ewe which facilitate water flow into and across the placenta (Atkinson et al. 2006). The relative contributions of these systems to the overall ion and water transport are as yet unknown.

Lipids: The hydrophobic nature of lipids results in a ready diffusion of their component free fatty acids, glycerol and cholesterol across the placenta, since at least in the second half of pregnancy the concentrations in the maternal circulation are considerably higher than the fetal. Appreciable placental metabolism complicates estimates of transfer rates. The essential long chain fatty acids such as docosahexenoic and arachidonic, are necessary components of fetal membrane phospholipids and these probably rely on placental membrane bound fatty acid binding proteins and translocases plus ATP to achieve the higher concentration in the fetal compared with the maternal plasma (Atkinson et al. 2006).

2.2.3 Endocytosis- and Phagocytosis

Epitheliochorial placentas have special trophoblast regions, known as areolae, for histotrophic uptake of endometrial gland secretions by small endocytotic vesicles or much larger phagocytotic vacuoles. In the horse these are located between the bases of the microcotyledons, in the ewe and cow in the intercotyledonary areas and randomly scattered in the pig. They consist of a flat or domed area of tall columnar trophoblast cells separated by a variable distance from the flat uterine epithelium onto which the duct of a uterine gland opens (Figs. 5.1 and 5.2). The gland secretions are essential for the establishment of pregnancy in the ewe (Spencer et al. 2004a,b) and, throughout pregnancy, produce a variety of important micronutrients including calcium ions (horse, ewe) and uteroferrin protein (horse, pig).

2.2.4 Iron Ion Transport

Uteroferrin is the carrier which, in the horse and pig, mediates maternofetal transport of iron, required for haemoglobin production and in other enzyme cofactors in the fetus. Transport can be monitored immunocytochemically (Wooding et al. 2000), secretion in the lumen of the gland is strongly positive, as is the areolar lumen and the trophoblast cells bordering that lumen (Fig. 2.8a) which take up the secretion by endocytosis into the lyso/endosomal system on the fetal side. No uteroferrin is detected anywhere in the horse microcotyledons or pig nonareolar trophoblast, which are the sites of the vast majority of hemotrophic macronutrient transfer by simple and facilitated diffusion and active transport, as detailed above.

A similar transport protein, transferrin, is used for iron transport in hemochorial placentas by endocytosis from the blood over the entire trophoblast surface. In contrast, in endotheliochorial (carnivores) and synepitheliochorial (ewe, cow) placentas fetal iron comes from trophoblast phagocytosis and breakdown of maternal red blood cells from hemophagous zones. These are localized separations of trophoblast and uterine tissue into which blood is intermittently released.

2.2.5 Calcium Ion Transport

The fetal demand for calcium is high and the transport process can be localized because all epithelia across which there is a significant flux of calcium carry the 9 or 28 kD calcium binding protein (CBP) at high concentration in their cytoplasm (Hoenderop et al. 2005). The areolar trophoblast epithelium is the only region of the horse placenta to express the 9CBP protein (Fig. 2.8b; Wooding et al. 2000) and it is therefore assumed that this is an important site of calcium uptake. Similar localised areas for calcium uptake are found in cow (Nikitenko et al. 1998) and ewe (Wooding et al. 1996) placentas, where 9CBP is restricted to the interplacentomal trophoblast epithelium. In the hemochorial human the 9CBP is present throughout the trophoblast syncytium (Belkacemi et al. 2004), but in the rodents it is only found in significant amounts in the intravascular and visceral yolk sac placentas (Bruns et al. 1988). Calcium entry across the apical plasma membrane is probably by facilitated diffusion through a specific calcium channel (Hoenderop et al. 2005). The CBP then binds to and facilitates the diffusion of the ion to the basolateral plasmalemma. This isolates it in transit from the very calcium sensitive cellular metabolic processes and presents it to a plasmamembrane Ca-ATPase to be actively pumped out basally. Ca-ATPases have been localized in this position in human, rat (Borke et al. 1989) and cat (Champion et al. 2003) placentas, and a suitable apical calcium channel is expressed in human syncytiotrophoblast but not localized as yet (Hoenderop et al. 2005).

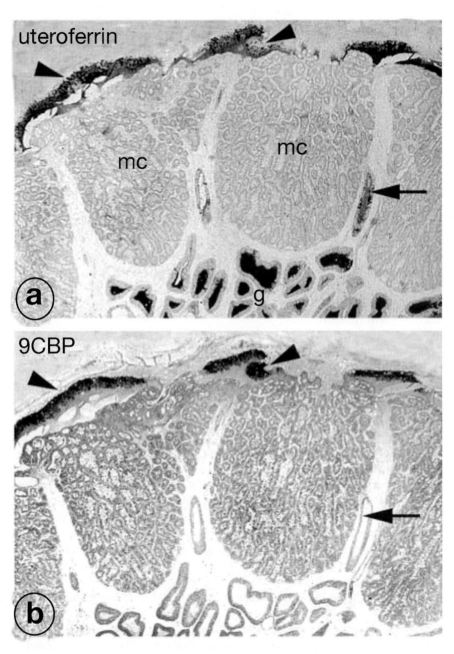

Fig. 2.8 (a, b) Light micrographs of immunocytochemical preparations of adjacent sections of horse placenta (270 days pregnant). (a) Uteroferrin IRT is restricted to the lumen of the glands (g) and vesicles of various sizes in the fetal trophoblast cells of the areolae (arrowheads). It is also present in the ducts between glands and areolae (arrow). No significant localisation is seen in the microcotyledons (mc). (b) Calbindin 9kDa protein (9CBP) IRT is present throughout the areolar trophoblast cell cytoplasm (arrowheads), but no significant localisation is seen in glands or micro-cotyledons. From Wooding et al. 2000

2.2.6   Protein Transport

This is largely limited to specific proteins of particular value to the fetus such as immunogloblins which protect the neonate until its own immune system develops and is only found in hemochorial placentas. Epithelio- and endotheliochorial placentas are impermeable and immunoglobulins are transferred in the colostrum.

In the hemochorial rabbit, IgG from rabbit is transported unchanged across the yolk sac while bovine IgG is degraded in the cells (Hemmings 1957). In human IgG is taken up in late gestation by endocytosis into the syncytiotrophoblast apical endo-tubulovesicular sorting and delivery system. Specific (e.g. Fc) receptors select (IgG) molecules for transcytotic dispatch unchanged across the epithelium, unselected material is broken down in the lysosomal system. The full details of this process have yet to be elucidated (Simister 2003, Atkinson et al. 2006).

2.2.7   Conclusions

The details above emphasize that within the basic constraint of the number of barrier membranes, the varied types of placentas show considerable localized differentiation of transport functions. However the horse placenta uses systems very similar to those of the ewe to acquire the nutrients it needs for transfer to the fetus and for its own maintenance. The equivalent efficiencies in grams of fetus produced per gram of placenta compared with those of the ewe (both 13–14) clearly show that the suggested advantage of horse countercurrent flow must be balanced by other factors in the ewe. Two recent studies (Allen et al. 2002; Wilsher and Allen 2003) have established that foal birthweight is directly proportional to both the weight and area of fetomaternal contact in the term placenta and that this is controlled by interaction between the age, parity and size of the dam and fetal genotype. These studies emphasise the importance of considering the area of fetomaternal contact, i.e. the surface across which all the nutrient transfer occurs, rather than just the weight. This suggests that efficiency per gram of placenta is perhaps not such a relevant measure as the area of maternofetal exchange surface within that gram. The limited information available (Baur 1977) indicates that the horse has a much smaller area per unit volume (of the fetus plus fetal placenta) ($4.4 \, cm^{-2} \, cm^{-3}$) than the cow ($27 \, cm^2 \, cm^{-3}$), but both produce a similar weight of neonate from a similar length of gestation. In the ewe, the area per unit volume is $20–23 \, cm^2 \, cm^{-3}$ (Wooding et al. unpublished data). This is similar to that in the cow and emphasises the fact that the horse is 5–7 times more efficient in transfer per unit area than either cow or ewe. This could be attributed to the countercurrent/crosscurrent difference, but whether this is the only factor explaining the difference in placental efficiency remains to be established.

This difference between horse and ruminant is in fact an example of a wider truth, in that all compact placentas have a larger area per unit volume of placenta than the diffuse placental types (Fig. 2.9; Baur 1977).

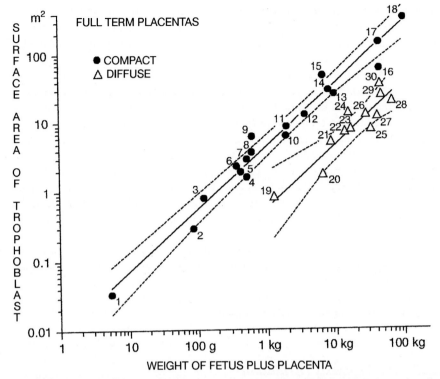

COMPACT PLACENTAS: 1rat 2 guinea pig 3 cat 4 alsatian dog 5 sloth 7 rhesus monkey
8 langur monkey 9 colobus monkey 9 leopard 10 chimpanzee 11gorilla 12 man
13 dwarf cattle 14 seal 15 sea lion 16 giraffe 17 cow 18 elephant
DIFFUSE PLACENTAS: 19 pig 20 dwarf hippo 21 dolphin 22 llama 23 pony 24 dwarf
donkey 25 zebra 26 ass 27 camel 28 rhino 29 horse 30 hippo

Fig. 2.9 Relationship between the surface area of the trophoblast and the weight of the fetus plus
placenta in a wide variety of mammals. The compact and diffuse types of placentas clearly fit to
different lines which are well outside the 95% confidence limits (drawn as *dotted lines*) for each
line. From Baur (1977)

Nor is the smaller relative surface area compensated for by a longer development
time; horse and cow have very similar gestation periods and the (compact) giraffe
($11\,cm^2\,cm^{-3}$) has a longer (450 day) gestation than the (diffuse) camel ($2.6\,cm^2\,cm^{-3}$,
390 days), which has a similar birthweight. Unravelling the reasons for these dis-
parities poses a considerable challenge and emphasises the multifactorial nature of
the demands and constraints in constructing an 'efficient' placenta.

An efficient placenta may be defined as one which produces an optimally
viable neonate capable of growth to a functionally viable adult. Achieving the
optimal maternofetal exchange area, minimising the diffusion distance and
providing sufficient transporter molecules depends on ensuring normal placental differentiation and growth. Studies in rodents and ruminants have shown that

these processes can be compromised by maternal nutrition and hypoxic or behavioural stress which reduce the size of the placenta and directly affect the growth of the basic organ systems of the embryo (Fowden et al. 2006a,b). Modification of fetal growth and development can have deleterious effects, producing a smaller neonate that later in life may suffer organ malfunctions (e.g. heart weaknesses or pancreatic endocrine insufficiency) which have been shown to result from insult to the conceptus. In human subjects and rodents, for example, inadequate remodelling of the maternal placental blood channels in early pregnancy leads to intrauterine growth retardation (IUGR). In the horse if uterine morphology and health is compromised by such factors as age and/or chronic infection (Bracher et al. 1996) or a disparity between sizes of stallion and mare (Allen et al. 2002; Wilsher and Allen 2003), the surface area available for placental growth is considerably reduced and the resulting impaired growth in utero is associated with dysmaturity, maladaptation and poor viability at birth (Rossdale and Ousey 2002).

The incidence of clinical intervention and degree of care necessary during an animals lifespan depends, in part, on the growth, differentiation and functional capacity of the placenta. However, to date, very little is known about the factors regulating these processes or the critical windows in gestation when placental structure and function can be beneficially modified.

2.2.8 Phylogenesis (Evolution)

Intriguingly, it has recently been suggested, mostly on molecular phylogenetic evidence (Vogel 2005), that the horse epitheliochorial and ruminant synepitheliochorial structures are secondary evolutionary developments from a theoretically original primary endothelio- or haemochorial type. In evolutionary terms, this would require the (syn-)epitheliochorial form of placenta to have some clear evolutionary benefit over the endothelio-/haemochorial in the context of ungulate ecological niches. Whether there is an evolutionary or ecological significance to the differences in placental efficiency between ungulates remains speculative at this stage (see Chap. 9.3).

2.3 Vascular Relationships

As we have seen in the previous section, the rate and disposition of the fetal and maternal blood flows are central to any consideration of the efficiency and mechanisms of placental transfer.

Continuous flow rate measurements are possible externally with Doppler ultrasonic probes or internally using electromagnetic or ultrasonic flowmeters placed around vessels, but the uterine arterial/venous figures will always include a variable contribution from the non placental uterine tissue, the umbilical will be solely placental. Experiments with microsphere injections can only give results at a single time.

As the placenta grows the mother in most species has to increase her blood volume and cardiac output between 40 and 50% in order to perfuse the new organ (Silver et al. 1982). In the cow, between 137 and 250 dpc, the uterine flow increases four times and the fetal umbilical flow 21 times (Reynolds and Ferrell 1987). The ewe increases maternal blood flow from 0.4 to 1.21 min⁻¹ between 70 and 131 dpc (term 147 dpc), and the human maternal flow increases two-and-half times in the second half of pregnancy. The umbilical flow velocity increases and resistance decreases in parallel with fetal growth. This indicates that blood flow can be considered an important if not the primary mechanism for controlling transplacental exchange, because the oxygen extraction rate and arteriovenous differences in nutrient concentrations show much less change in the same period (Reynolds et al. 2005, 2006). Supporting this thesis, Meishan pigs show smaller placentas but with increased vascularity and angiogenic VEGF synthesis compared to Yorkshires, and produce a greater fetal weight per uterus (Wilson et al. 1998; Vonnahme and Ford 2004). Work in progress suggests that interventions increasing blood flow in nutritionally compromised ewes can restore fetal weight to normal (Reynolds et al. 2006).

There are large differences unrelated to type of placenta between species in blood flow per gram of placenta. The ewe (synepithelio-) and guinea pig (haemochorial) uterine and umbilical flows are about three times the rates in man and rabbit (both haemochorial). This can partly be explained by the differences in relative sizes of the placentas, since in ewe and guinea pig the ratio at term of the fetal to placental weight is 10:1, in rabbit and man only 4:1 (Silver et al. 1982).

There are at least four theoretical types of relationships between two flows: con-, cross- and countercurrent and mixtures of some or all of these together (Martin 1981; Fig. 2.3). Of these, the countercurrent system is expected to be most efficient because blood leaving the exchange area on one side will tend to equilibrate with that entering it on the other, and a maximum solute transfer of 100% is theoretically possible. For concurrent flow such equilibration can only result in a maximum solute transfer of 50% (Fig. 2.3).

In the placenta it must be remembered that there are two levels to be considered, gross flow and what may be termed microflow. Gross flow is the overall direction which in all placentas so far investigated (with a few exceptions, see below) appears to be countercurrent. Fetal stem arteries run directly to the maternal side and maternal stem arteries to the fetal side before dividing up into capillaries which conduct the blood back to the veins in parallel but opposite directions. The microflow is at capillary level, and this is where the haemotrophic exchange takes place, most efficiently with the least practicable distance between the flows.

In all cases so far examined in sufficient detail, the architecture of both fetal and maternal capillaries or blood spaces is a network, sometimes but not always with a predominant direction). There is no evidence for significant lengths of parallel apposed tubes of the networks which would allow sufficient equilibration of, for example, oxygen concentration as usually illustrated in model systems (Moll 1972; Martin 1981). It may be more realistic to consider that all placental capillary exchange takes place between microflows in networks which vary continuously between cross-, counter- and even concurrent and are best described as 'network current' (Figs. 2.3 and 8.12).

The vascular morphology of the placental exchange region may be investigated using cleared histological sections of ink injected specimens (Fig. 5.4b,d) or by injection and subsequent hardening of plastic or latex followed by removal of the soft tissues and light (Fig. 5.4a,c) or scanning electron microscopic (SEM) (Fig. 5.5) examination of the resulting 'corrosion casts'. SEM studies of corrosion casts give exquisite detail of either maternal or fetal vasculature (Figs. 6.8 and 6.9) but it is very difficult to elucidate the exact relationship between them, since it is not yet possible to distinguish fetal from maternal capillaries unequivocally on the same preparation, nor is there any way of establishing what the flow directions were in the capillary meshworks. However, there are some very convincing remarkably detailed proposals available from Leiser's group (e.g. Figs. 6.8e and 6.10) As more SEM microcorrosion cast studies are published (e.g. Leiser and Kohler 1983, 1984; Burton 1987; Leiser 1987; Christofferson and Nillson 1988; Leiser and Dantzer 1988; Ogura et al. 1991; Leiser et al. 1997; Miglino et al. 2004), it becomes increasingly obvious that a placenta consists essentially of two bloodstreams separated by a minimum of tissue. Most conventional EM and LM sections overemphasize the tissue component. Better perfusion fixation shows that probably more than 50% of the in vivo placental volume is occupied by blood channels (Fig. 6.7), a figure supported by measurements of the total blood volume using tracers in placentas clamped in vivo.

The progression (in morphological but not necessarily evolutionary terms) from epitheliochorial to haemochorial placentation occurs with increasing loss of uterine tissue, with important consequences for vascular organization (Fig. 2.4). Epitheliochorial placentas are formed by a mutual growth of short fetal and maternal villi to form a structure consisting principally of two complex and extensive capillary beds. Synepitheliochorial and endotheliochorial placentas develop similarly but have much longer villi or lamellae with great modification or elimination of the uterine epithelium.

The maternal connective tissue remains an important constituent of maternal placental structure. Haemochorial placentas have no residual uterine epithelium, blood capillaries or connective tissue, but need to deliver maternal blood up to the fetal side of the placenta from where it drains back through entirely fetal tissue. To accomplish this the main maternal arteries persist, usually passing through the full depth of the fetal villi or labyrinth (Fig. 2.4). The endothelium of such arteries is usually replaced by invading fetal trophectoderm early in gestation, but the initial maternal blood conduits are retained, lengthened and enlarged throughout pregnancy (Fig. 8.1). The only exceptions to this are the Old World monkeys and anthropoids (Chap. 8 and Fig. 8.1). Thus, in late pregnancy there is virtually no maternal tissue left in a haemochorial placenta above the decidual layer or myometrium.

2.3.1 Placental Vascular Anatomy

The following descriptions of placental vascular anatomy will be limited to the mature placentas of late pregnancy. It should be noted that placental vasculature develops throughout gestation as the fetal nutrient and blood gas requirements increase, and that total placental blood flows and chorionic villous areas increase with increasing fetal mass (Baur 1977). In the pig's definitive placenta short blunt fetal villi (length 150 μm, width 30 μm, Tsutsumi 1962) are inserted into maternal crypts. Each fetal villus has a peripheral meshwork of non-fenestrated capillaries supplied by an arteriole running up centrally to the top of the villus before branching significantly into capillaries draining down to the base. The maternal crypts have a very similar arrangement, but of fenestrated capillaries, with arterial supply to the top and draining around the base (Figs. 2.8, 5.4 and 5.5). The predominant capillary flow is therefore down both fetal villous and maternal crypt, in opposite directions. As can be seen from the diagram (Fig. 2.4), this produces a gross countercurrent flow between the capillaries, which indent their epithelia deeply (Goldstein 1926; Steven 1983; Figs. 5.3b and 5.6b) and are presumably the main site of haemotrophic fetomaternal exchange. The corrosion cast evidence (Tsutsumi 1962; MacDonald 1975, 1981; Leiser and Dantzer 1988) shows (Figs. 5.4 and 5.5) that the microflow is between two capillary networks not parallel tubes. Microflow is therefore the mixture of cross and countercurrent, here called network flow (Fig. 2.3). Leiser and Dantzer also suggest there may be significant capillary flow around many fetal villi which would make even the gross flow partly crosscurrent. The vascularity of the pig placenta can differ significantly between breeds, with the Meishan producing a smaller but more vascularised placenta than the Yorkshire, and thus able to produce a larger litter from a similarly sized uterus(Wilson et al. 1998; Vonnahme and Ford 2004).

   In contrast to the pig, the vascular structure of the diffuse epitheliochorial placenta of the horse is more complex (Tsutsumi 1962; Silver and Steven 1975; Steven and Samuel 1975; Abd-Elnaeim et al. 2006) because of the formation of longer fetal villi and their organization into groups called microcotyledons (Chap. 5). In these fetal villi, fetal arterioles run centrally to the tips before branching into a peripheral capillary network 75–225 μm in length (Tsutsumi 1962), which carries blood back to veins at the base. The capillary blood therefore flows from the tips to the bases of the villi. Maternal blood, entering each microcotyledon over the rim of the cup-like structure, divides into a capillary network vascularizing maternal septa which have their tips between the bases of the chorionic villi. The flow is from the tips to the base of the septa; venous blood is collected in veins running round the base of the microcotyledon. In this species, therefore, there are clear morphological grounds for considering the gross blood flow to be solely countercurrent (Tsutsumi 1962), but the microflow would still be the cross and countercurrent system characteristic of apposed networks (Tsutsumi 1962).

   In the ruminants the sheep and goat have 10- to 100-fold longer villi forming much larger cotyledons (placentomes) than in the horse. Both maternal septa and fetal stem villi have central supply arteries which give off some branches as they run in opposite directions to their tips. Fetal and maternal blood passes back via capillary meshworks about 250 µm long in grossly countercurrent flow (Tsutsumi 1962; Steven 1966; Makowski 1968).

The microflow arrangement has been controversial. It was initially suggested to be countercurrent (Barcroft and Barron 1946) but reports of axial venules in the fetal terminal villi and maternal septa indicated that a crosscurrent system was equally likely (Fig. 2.4; Tsutsumi 1962; Steven 1966; Makowski 1968). Corrosion cast studies of the goat and cow placental vasculature have clearly demonstrated that the maternal and fetal capillaries are arranged in mixed cross and countercurrent microflow systems, network flow again (Figs. 6.8–6.10; Leiser 1987; Leiser et al. 1997). Measurements of transport using the non-metabolizable, freely diffusing substance, antipyrine, in chronically catheterized preparations have confirmed that the relationship is not exclusively countercurrent (see Battaglia and Meschia 1986; Wilkening and Meschia 1992).

The placenta of the cow is, in many respects, similar to that of the sheep (Tsutsumi 1962) except that the placentomes are convex instead of concave, because of the greater complexity of the branching of terminal fetal villi (Chap. 6; Figs. 6.9 and 6.10). As in the ovine placenta, fetal villi are vascularized by central arterioles running to their tips, branching to form many small lateral capillary beds, length 150–300 µm, which are drained by central venules. In cross-section these branches, which are longer and more prominent than those in sheep, resemble the arms of a starfish (Figs. 2.4, 6.9 and 6.10; Tsutsumi 1962; Leiser et al. 1997). The peripheral capillary bed in the maternal villi is supplied with blood from central arterioles and drained into venules at either the base of the villus or its centre. The relationship between maternal and fetal blood flows in the cow is therefore similar to that in the sheep and goat, with the exception that some maternal blood drains via central venules, whereas in the sheep the maternal capillary cascade between individual terminal villi (Silver and Steven 1975; Fig. 6.8e) is said to be the only continuous channel back to the basal veins.

A detailed study of the endotheliochorial labyrinthine zonary placenta of the cat (Fig. 7.6; Leiser and Kohler 1983, 1984) and mink (Fig. 7.7; Krebs et al. 1997) using scanning electron microscopy of corrosion casts shows that the maternal stem arterial system delivers blood to opposite side of the placenta before it drains back through a capillary lamellar mesh. However although the fetal capillary lamellae extend to the maternal side, the arterial supply to each lamella is lateral not apical and the resulting capillary flow is across the maternal flow direction (Fig. 2.4). These labyrinthine placentas would have gross flow cross-current (Leiser and Kohler 1984; Leiser and Kaufmann 1994; Krebs et al. 1997) but still with network microflow.

The mature rodent hemotrichorial placenta has three zones: labyrinth, spongiotrophoblast, and decidua (see Figs. 1.9 and 8.1). The labyrinth consists only of fetal trophoblast villi each containing a branch of the umbilical artery which passes directly to the spongiotrophoblast boundary. Corrosion cast studies show that there it turns back and divides into a columnar open meshwork of trophoblast covered capillaries which finally recombine to form an umbilical venule at the base of the villus (Adamson et al. 2002; see Fig. 8.5). The capillary meshwork of each unit is separate but the units are laterally held together to form the labyrinth, by layer I of their trophoblast covering which is continuous with the spongiotrophoblast zone. Blood flow in the fetal capillaries is thus from the maternal to the fetal side. The maternal blood crosses the labyrinth in four or five wide vascular channels which empty into a sinus/ subchorial lake at the fetal side of the labyrinth. It then flows back to the maternal side through the interstices of the fetal tissue meshwork to venous channels in the spongiotrophoblast therefore forming a truly countercurrent system The maternal vascular channels are formed from widened and trophoblast lined extensions of the endometrial spiral arteries (Adamson et al. 2002; Hemberger et al. 2003). The guinea pig and the rabbit have very similar labyrinthine countercurrent blood flow systems.

In the haemomonochorial villous placentas of man and Old World monkeys, as in the synepithelio- or endotheliochorial systems, the fetal arteries run to the tips of the fetal villi, here anchored in the maternal decidua, before draining back through a capillary network. However, unlike all the other placental arrangements, there are in man no channels to carry the maternal blood to the fetal surface, the maternal spiral arteries end abruptly at the maternal surface (Fig. 8.12). A plausible explanation of how such a system works is as follows. The fetal villi are organized into groups called lobules or cotyledons, each of which has a hollow centre where villi are less densely packed. The lobules are situated over the openings of the maternal spiral arteries, which deliver blood into the hollow centre; in order to leave the intralobular spaces the blood flows radially over the villi into spaces between lobules (Fig 8.13; Wigglesworth 1969). From these spaces venous blood leaves the placenta through venous openings in the decidual plate and interlobular septa (Fig. 8.12). By angiography, puffs of maternal blood can be seen entering the intervillous space from the spiral arteries (Ramsey 1967); pressure of blood may lead to a hollow-centred lobule by pushing villi outwards. Growth of the villi is thought to occur from the centre of the lobule towards the periphery, in the direction of blood flow; the most peripheral villi are reported to be more highly differentiated (Schuhmann 1982). Oxygen and solutes are removed from the maternal blood as it passes from the spiral arteries to the endometrial veins, thereby establishing a concentration gradient from the point of entry into the lobule to the exit. This is supported by the finding that antioxidant enzyme expression and activity are greatest in the centre than the periphery of a lobule, the magnitude of the differences being proportional to the radius (Hempstock et al. 2003). The diffusion barrier thickness of the human vasculosyncytial membrane does appear to be significantly reduced in regions where oxygen and solute concentrations would be predicted to be lowest (Critchley and Burton 1987; Burton and Tham 1992). Arteriovenous shunting and blood stasis in the intervillous space are prevented by the velocity and momentum with which maternal blood enters this space and the villus architecture, which ensures mixing of highly oxygenated blood with that already present (Ramsey and Donner 1980).

It has been suggested that the human placental villi represent a mop passively suspended in a bucket, but this does not explain the movement of maternal blood around the fetal villi in each lobule. It would be more realistic to imagine a brief separation to capillary dimension of the villi caused by each spurt of maternal blood, followed by closure of the intervillus 'capillary' spaces as the pressure drops – rather like water flow through fish gills. This system would prevent significant backflow towards the arterial orifices.

The structure of the fetal vessels in each villus is such that the terminal villi which protrude from the intermediate villi at random are vascularized by a network of peripherally located anastomosing capillaries arising from the central primary arteriole and draining into the central primary vein (Habashi et al. 1983). There is no evidence that the capillaries in the terminal villi have any consistent spatial relationship to the mean direction of blood flow in the intervillous space, so the maternal and fetal blood microflows presumably have a mixed or random directional relationship to one another (Fig. 8.12).

2.3.2 Placental Vascular Architecture and Placental Function

Correlation of structure and function in the different placental morphological types should be simplest with diffusible gases and solutes where, given similar concentration gradients, the rate of transfer depends directly on the interhaemal diffusion distance and the area of the exchange surface.

The permeability of a placenta to blood gases can be determined directly by measuring the transfer of non-metabolizable carbon monoxide or estimated from morphological measurements (Mayhew et al. 1984). The results show that the rate of fetal oxygen utilization is never limited by the oxygen diffusion capacity of the normal placenta. The range of values for oxygen diffusion found in different species (7–45 ml min^{-1} mmHg^{-1} per kg of placenta) bears no simple relationship to the range of placental interhaemal structure. One problem here is that it is very difficult to determine the average interhaemal distance since the degree of indentation of fetal (or maternal if present) layers by the capillaries varies considerably both within the same placenta (Figs. 5.6 and 6.7) and between placentas of different species. This indentation means that the *minimum* interhaemal distance in the epitheliochorial six-layered pig placenta may be very similar to that in the human haemomonochorial three-layered placenta, but the *average* interhaemal distance is much greater in pig than in man. This difficulty can be overcome by measurement of the Harmonic Mean Thickness which produces comparable figures for different placentas. Harmonic Mean Thickness is defined as the reciprocal of the mean of the reciprocals of all measurements of the distance between the blood flows (see Coan et al. 2004 for an example of its use in mouse and Critchley and Burton 1987; Burton and Tham 1992 in human).

There is some evidence for a local specialization of structure and function in the human placenta (Chap. 8). The areas of minimum interhaemal separation, the vasculosyncytial membrane, are more extensive in regions where the oxygen concentration is plausibly suggested to be lower (Critchley and Burton 1987). Whether such an arrangement is also found in other types of placenta remains to be demonstrated.

Differences in the transport capacity of various placental types clearly do not determine fetal growth rates, because these are as high in the pig and the mare (epitheliochorial) as in primates or the guinea pig (haemomonochorial). Other factors that potentially influence placental efficiency are villous surface area, the size of the placenta relative to the fetus and placental blood flow rates. However, none of these parameters alone appears to correlate with fetal growth rate. Calculation of villous surface area at term per kg (fetus plus placenta) shows that this parameter varies widely, independently of fetal growth rate, from $300\,cm^2\,kg^{-1}$ in the mare to $10{,}400\,cm^2\,kg^{-1}$ in the cat (Fig.2.9; Baur 1977).

Fetuses with compact placentas do not consistently grow faster than those with diffuse placentas, although compact placentas always have a higher villous surface area per gram of tissue.

Among the species in which fetal blood gases can be measured accurately, the mare has a markedly higher rate of transfer of oxygen into fetal blood, as judged by PO_2 values (maternal vein minus fetal vein; Fig. 2.5) than ruminants or the pig. This cannot be accounted for by a particularly high oxygen affinity of fetal haemoglobin compared with maternal, or to a high fetal haemoglobin concentration, or a high fetal arterial oxygen saturation, since these three parameters are similar in all species investigated. Unlike other species, neither the horse (Kitchen and Bunn 1975) nor the pig has a special 'fetal' form of haemoglobin. It has been suggested (Silver and Steven 1975; Steven and Samuel 1975) that the efficiency of gas exchange in the equine placenta reflects the exclusively countercurrent nature of the vascular architecture.

The similar high gas-exchange efficiency of the guinea pig placenta may also be due to the suggested countercurrent flow (Kaufmann 1981a) plus the high fetal-maternal haemoglobin-oxygen affinity ratio. These are probably both adaptations to the lower oxygen availability at the high altitude at which these animals live in the wild in South America.

As the placenta grows the mother has to increase her blood volume and cardiac output in order to perfuse the new organ. The higher the blood flow through the maternal placenta in millilitres per minute per kg of fetus the greater the demand on the maternal vasculature. Animals with haemochorial placentas have the advantage here since they consistently show lower maternal placental flow rates (100–$160\,ml\ min^{-1}\ kg^{-1}$ fetus) than those with epithelio- or synepitheliochorial placentas (250–$330\,ml\ min^{-1}\ kg^{-1}$). The last two types must also provide extra maternal energy to fuel active transport across the two maternal layers, endothelium and uterine epithelium, which are absent in haemochorial placentas. In all placental types on the fetal side the blood flows in endothelium-lined channels. Since the range of umbilical blood flow rates in haemochorial placentas from guinea pig ($64\,ml\ min^{-1}\ kg^{-1}$ fetus) to rhesus monkey ($208\,ml\ min^{-1}\ kg^{-1}$) spans nearly all of the synepithelio- and epitheliochorial values from pig ($115\,ml\ min^{-1}\ kg^{-1}$) to cow ($232\,ml\ min^{-1}\ kg^{-1}$) (Wooding and Flint 1994) there would seem to be no simple relationship between structure and function in this respect.

As well as total blood flow, the rate of fetomaternal exchange will depend, at least to some extent, on the speed of the flow through the placenta. It has been suggested that the slower the rate the more efficient the transfer, and structures have been identified in haemochorial placentas which are said to be designed to produce local maternal blood stasis (Enders 1965). There is no doubt that in many haemochorial placentas the total maternal blood in flow is slowed by the development of arterial sinuses (rabbit) or grossly widened arterial vessels (guinea pig, man). There is also some evidence for intermittent rather than continuous blood flow in primate placentas. How relevant these flow rate changes are to efficiency of maternofetal transfer and whether there is an optimal flow rate remains to be determined.

Transport of hydrophilic, non-diffusible compounds requires carrier systems to allow them to cross each membrane barrier. Rates of transfer have been shown to be inversely proportional to the number of interhaemal membranes, which would suggest a similar frequency or equivalent capacity of carriers in each membrane. Sodium, for example, is transported 300 times faster in human (four membranes) than in pig (eight membranes) with the cat (six membranes) in an intermediate position (Moll 1985).

However, what direct evidence there is indicates local specialization rather than uniform distribution. In the pig, Firth et al. (1986a,b) have shown that the Na^+, K^+-ATPase-based sodium transport system appears to be concentrated in the apical trophectodermal membranes of the fetal areolae. In human there are differences in carrier systems between microvillar and basal membranes of the synctiotrophoblast (Shennan and Boyd 1987; Vanderpuy and Smith 1987) and the latter shows considerable local elaboration (Fig. 8.15b). Transport of calcium and glucose is in separate areas of the ruminant placenta (Figs. 2.6 and 2.8). Thus within the basic constraint of the number of barrier membranes, there may be considerable localized differentiation of transport functions which are only now becoming apparent.

2.3.3   Control of Placental Blood Flow

Most organs are capable of regulating their blood flow (the process known as autoregulation), so as to maintain a constant rate (ml/min) of perfusion during periods of altered blood pressure. This process which involves local control over dilatation or constriction of arterial precapillary sphincters, would not be expected to occur in haemochorial placentas, which lack post-myometrial vessels with such sphincters. However, the premyometrial uterine arteries are responsive to systemic agents acting on vascular beds (Greiss and Anderson 1970; Ford et al. 1977) and, since maternal placental flow accounts for a high proportion of uterine blood flow, maternal placental flow can be modulated through effects on these vessels. Thus, Venuto et al. (1976) showed that maternal placental blood flow was maintained at approximately 25 ml min^{-1} in haemochorial rabbits when mean arterial pressure was reduced from 135 to 65 mmHg, and Bell (1972, 1974) has described uterine vasodilatation following uterine nerve stimulation in the pregnant haemochorial guinea pig. However, in the synepitheliochorial sheep placenta, which has both precapillary sphincters and systemic agents affecting premyometrial vessels, there is no evidence for maintenance of a constant rate of perfusion; Greiss (1966) showed that, during the last 30 days of gestation in the synepitheliochorial sheep, uterine blood flow is linearly related to central arterial pressure over the range 10–110 mmHg (arterial minus venous pressures), and Anderson and Faber (1982) showed that this is also true for fetal placental flow over the range 25–45 mmHg, in chronically catheterized fetal lambs.

Although uterine arterial blood flow may not change with changing maternal blood pressure in all species, there is evidence that local factors may control the ratio of maternal to fetal blood flows in the placenta. Rankin and his colleagues (see Rankin and McLaughlin 1979) have drawn an analogy between the placenta and the lung in this respect; in the lung compensatory mechanisms exist to maintain a constant ventilation-perfusion ratio following local vascular embolism or hypoxic alveolus. It is clear that mechanisms serving a similar function exist to control blood flow in the placenta; when part of the umbilical blood supply is occluded in sheep and rabbits, maternal blood flow to that region (as measured with microspheres) eventually falls (Rankin and Phernetton 1976); furthermore in the normal sheep placenta maternal-fetal perfusion ratios are similar throughout the organ although the gross flow to various parts of the placenta varies (Rankin et al. 1970; Rankin and Schneider 1975). Rankin (1978) and Rankin and McLaughlin (1979) have suggested that PGE_2, which is produced in large quantities by the fetal placenta in the sheep (Challis et al. 1976; Mitchell and Flint 1978a), may be involved in regulating the ratio of maternal to fetal flow, by exerting opposite effects on different sides of the placenta. Such a mechanism would not work in a haemochorial placenta with no arterial precapillary sphincters capable of responding to vasoactive compounds; in synepitheliochorial placentas, such as that of the sheep, it would depend on the PGE_2 diffusing very considerable distances from the fetal vasculature. An alternative function for fetal placental PGE_2 may be in maintaining the patency of the fetal ductus arteriosus (Coceani et al. 1978). Other mechanisms by which placental blood flow may be altered over long periods of time must include angiogenesis and local necrosis; the latter could explain the effects of partial umbilical occlusion and the former the relatively constant maternal-fetal blood flow ratios referred to above.

Because of the proximity of the maternal and fetal placental vessels it is possible that a transient rise in pressure on one side might lead to occlusion of the other. This has been investigated experimentally in the near-term sheep placenta, in which it has been shown that, while an effect can be demonstrated in the isolated perfused placenta by applying large changes in pressure (Bissonnette and Farrell 1973; Power and Longo 1973; Power and Gilbert 1977), this cannot be confirmed in sheep with chronically implanted catheters (Thornburg et al. 1976). Raising the pressure in the intervillous space of the human placenta in vitro can also compress the fetal capillaries within terminal villi (Karimu and Burton 1994). A second form of mechanical control of placental blood flow, which has been investigated principally in man and rhesus monkeys, is that exerted by myometrial contractility; individual contractions of the kind associated with the early stages of labour lead to compression of the veins running through the myometrium so that flow of blood into the intervillous space is intermittently interrupted, with potentially dramatic adverse effects on fetal well-being. The most usually monitored parameter which reflects this form of stress during labour is fetal heart rate, the transient fetal bradycardia associated with increased uterine pressure known as type II dip (or late deceleration) being an indication of fetal distress (see Ramsey 1967; Carter 1975).

Chapter 3
Fish, Amphibian, Bird and Reptile Placentation

3.1 Fish

3.1.1 Introduction

The fishes are non-amniotes, which have evolved a wide variety of modifications for successful viviparity (Amoroso 1960; Hogarth 1976; Wourms 1981; Mossman 1987). The most widely used system is an increased secretion of nutrients into the ovarian or uterine cavity and absorption via the fetal gut or skin surface, but a number of genera have produced specialized structures, both embryonic and extraembryonic, some of which closely parallel amniote placentas.

3.1.2 Classification

Fishes:

Class Chondrichthyes (cartilaginous fish)
Subclass Elasmobranchii, selachians

About half the 100 families have viviparous members; only two of these families (Carcharinidae and Syphonidae, including sharks, skates and rays) have either a yolk sac placenta (Fig. 3.1) or trophonemata. The latter are long extensions of the uterine epithelium producing abundant secretion (Wourms 1981).

Class Osteichthyes (bony fish)
Subclass Actinopterygii

This includes more than 99% of all living bony fish. Fourteen out of the 425 families are viviparous, and five of these have structures ranging from fin elaborations, gut extensions (trophotaeniae) and expansions to extra-embryonic pericardial sac elaborations analogous to the vertebrate amnion and chorion (Fig. 3.2).

P. Wooding, G. Burton, *Comparative Placentation*,
© Springer-Verlag Berlin Heidelberg 2008

Table 3.1 Dry weight increases during gestation in fishes

Species	% Change in dry weight of conceptus	Reproductive pattern	Uterine secretion
Chondrichthyes; Cartilaginous fish Oviparous			
Scyliorhinus caniculus-20 Viviparous		Oviparous	None
Squalus acanthias	−40	Yolky egg	Serous
Mustelus mustelus	370	Yolky egg	Mucous
Dasyatis violacea	1,600	Trophonemata	Lipid
Mustelus canis	1,050	Yolk sac placenta	Mucous
Gymnura micrura	4,900	Trophonemata	Lipid
Odontaspis Taurus	1,200,000	Intrauterine cannibalism	
Scoliodon laticaudus	5,800,000	Yolk sac placenta and appendiculae	Wourms (1993)
Osteichthyes; Bony fish Oviparous			
Salmo vivideus Viviparous	−40	Oviparous	
Poeciliopsis monacha	−40	Yolky egg	
Poecilia (lebistes) sp.	0	Yolk, small pericardial sac	
Heterandria formosa	4,000	Large smooth pericardial sac, simple	
Anableps dowei	800,000	Follicular placenta, large specialized pericardial sac	
Ameca splendens	15,000	Villous follicular placenta and	
Embiotoca lateralis	20,400	Trophotaeniae from anus	
Jennynsia sp.	24,000	Fin extensions, gut expansion and trophonemata from ovarian wall	

Data from Wourms (1981).

It is clear that fish viviparity has evolved many times in many different genera (Amoroso 1960; Wourms 1981) and even three times in a single genus (Reznick et al. 2002). A decrease in the amount of yolk after retention in the body of the mother necessitated an efficient system of nutrient transfer. In fishes, the placenta is only one, and not necessarily the most efficient, solution to the problem. In fish physiology a rough measure of the efficiency of viviparity is taken as the percentage increase or decrease in dry weight (mostly organic material) between egg and neonate (Table 3.1).

3.1.3   Chondrichthyes

In Table 3.1, for the cartilaginous fish it can be seen that retention in the uterus does not necessarily offer any nutrient transfer advantage *(Scyliorhinus* versus *Squalus)* unless the organic content of the uterine secretion is increased *(Squalus* versus *Mustelus mustelus).* Development of trophonemata, uterine epithelial extensions that are in close relationship with the fetal mouth or gills, further augments the percentage increase (*M. mustelus* versus *Dasyatis*) and a true trilaminar yolk sac closely apposed to an expanded semivillous uterine epithelium is almost as effective (Cate-Hoedemaker 1933; Schlernitzauer and Gilbert 1966) (*M. canis* versus *Dasyatis*). However, a richer uterine secretion plus trophonemata can be three or four times more efficient than a yolk sac placenta(*Gymnura* versus *M. canis*), and a species in which the surviving uterine tenant has eaten all its originally numerous siblings (*Odontaspis*) has an even greater nutrient transfer advantage. However the yolk sac placenta and appendiculae(see below) of *Scoliodon* performs even better than *Odontaspis* but whether the placenta is epitheliochorial or hemochorial has yet to be established (Wourms 1993).

3.1.3.1   Yolk Sac Placenta

In several of the skates and rays after the yolk is exhausted the trilaminar vascularized yolk sac is then modified to serve a placental function. Such species have very lengthy gestation times. In *Mustelus canis (laevis),* which has been most closely investigated structurally (Cate-Hoedemaker 1933), the placenta forms after a 3-month dependence on yolk and is functional for another 9–10 months (Fig. 3.1).

The uterus develops shallow folds around each embryo and within the fold the yolk sac trophoblast forms with the uterine epithelium a localized intimate association, which subsequently increases in area by folding. A shell membrane, or derivative thereof, usually persists between the maternal and fetal epithelia until term, but there is a considerable decrease during pregnancy in the thickness of the epithelia and the shell membrane. Both maternal and fetal blood capillaries eventually deeply indent their epithelia, reducing still further the separation of fetal and maternal blood (Fig. 3.1). In a good review Wourms et al. (1988), have reported species which lose one or both epithelia in forming the definitive placenta but still retain a shell, and in *Scoliodon* suggest the possible formation of a hemochorial placenta where maternal blood irrigates the trophoblast sheathed fetal capillaries (Wourms 1993). However there are no published micrographs supporting any of these claims.

A tracer and ultrastructural study of the yolk sac placenta of *Carcharhinus* (Hamlett and Wourms 1984; Hamlett et al. 1985) demonstrated the ability of the trophoblast to absorb proteins endocytotically and of the uterine epithelium to mediate transfer, but capillary indentation of the epithelia was not so marked as in *M. canis*. There is a complete range from the structurally modified yolk sac placenta of *Mustelus canis* with a very thin shell to the transient non-specialized association of

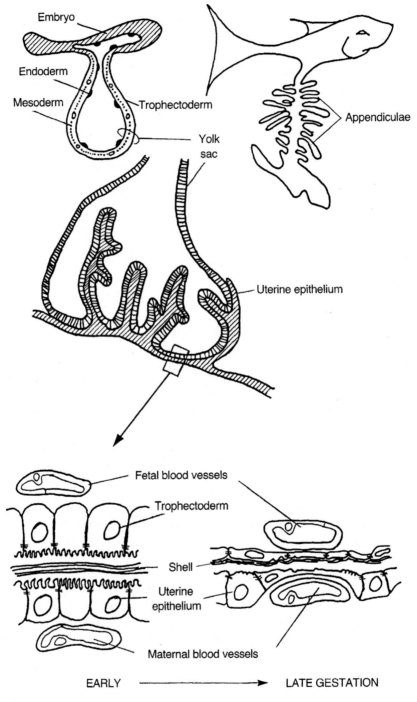

Fig. 3.1 Placental development in the cartilaginous fishes; Skates, Rays and sharks (Chondricthyes)

yolk sac trophectoderm and uterine epithelium (across a fairly thick shell membrane) of *Squalus acanthis* (Jollie and Jollie 1967). In some cases (*Scoliodon,Rhizoprionod on*) very complex elaborations of the yolk sac stalk surface (appendiculae) are present as well as a placenta (Wourms 1993) (Fig. 3.1). These fetal appendiculae, like the glandular extensions of the maternal uterine epithelium (trophonemata), may also play an important role in fetomaternal nutritive exchange in other families but, since the tissue association is very loose, as far as it has been investigated, a placental relationship cannot be assumed (for a review see Hamlett et al. 1993).

Thus, in the Chondrichthyes a true yolk placenta with considerable specialization for haemotrophic exchange has been developed, but the vast majority of viviparous species rely on histotrophic nutrition with only a few structural adaptations to facilitate this.

Recent investigations show some possible parallels with mammalian placentation in sharks. *Rhizoprionodon*, with a persistent shell, has uterine natural killer cells and lymphoid aggregates in the uterine stroma which increase in number as pregnancy progresses (Haines et al. 2006). *Mustelus canis* has the constituents of the interleukin 1 system present in the epithelium of both uterus and yolk sac (Cateni et al. 2003) and also a wide range of glycosylated molecules on the membranes and the shell (Jones and Hamlett 2004).

3.1.4   Osteichthyes

The bony fish do not have true uteri and viviparous development takes place either in the ovarian follicle or, after follicular rupture, in the ovarian cavity (Turner 1947; Amoroso 1960; Wourms 1981). Table 3.1 demonstrates that again the level of maternal organic provision is greatest when facilitated by specialized fetal and/or maternal structures. Unlike the Chondrichthyes the yolk sac is not developed further than its function as a yolk store; but in parallel with the Chondricthyes some species do have gland-rich trophonemata which reach from the uterine surface into the mouth or gill cleft of the embryo. The other structures, pericardial sac, trophotaeniae or fin extensions, are unique to a few families of the Osteichthyes (Fig. 3.2) (Schindler and Hamlett 1993).

3.1.4.1   Pericardial Sac

In three families, (Anablepidae, Poeciliidae and Goodeidae), gestation starts in the follicle. In early development the coelomic cavity around the heart extends to form an extraembryonic (exocoelomic) sac whose wall consists of somatopleuric mesoderm bounded by trophectoderm. This sac enlarges to a variable degree in different genera enclosing the entire head in some cases (Fig. 3.2). It is richly vascularized and blood from the body flows through this portal system before

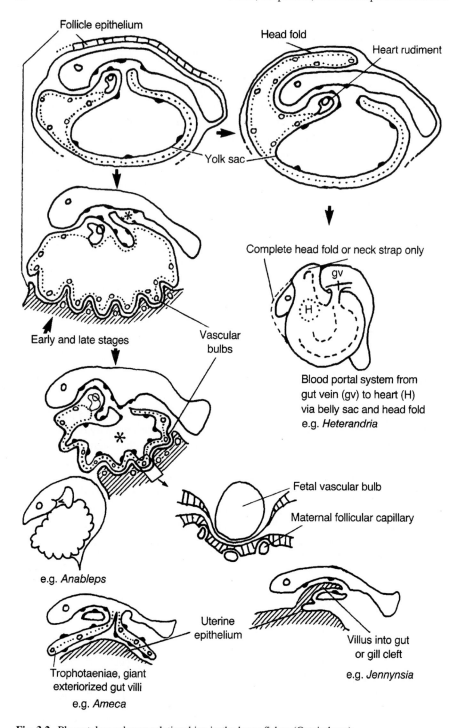

Fig. 3.2 Placental membrane relationships in the bony fishes (Osteicthyes)

returning to the heart. The trophectoderm is closely apposed to a well-vascularized follicular epithelium, and capillary indentation of fetal and maternal epithelial layers is observed in some species in later gestation. The degree of development of the pericardial sac varies considerably in different genera. *Lebistes* (Kunz 1971) and *Poecilia* (Tavolga and Rugh 1947) still retain a large yolk sac but the development of a small pericardial sac increases the efficiency of the organic transfer (Table 3.1). *Heterandria* has a much larger pericardial sac covering the entire head and most of the body at one stage with only a small yolk sac (Fraser and Renton 1940; Grove and Wourms 1991) and this combination produces a considerable increase (4,000%) in organic material during gestation. In *Anableps* the pericardial sac does not cover the head but expands ventrally into the area normally occupied by the yolk sac (vestigial in *Anableps*). The pericardial sac develops vascular swellings ('bulbs') under the trophoectoderm surface which is closely apposed to a proliferation of slender, richly vascularized villi developed from the follicular epithelium (Fig. 3.2). Fetal and maternal epithelia are tightly adherent and the efficiency of this true placental structure is exemplified by the vast increase in organic transfer (800,000%, Table 3.1) during intrafollicular development (Knight et al. 1985).

The Anablepidae also show considerable embryonic gut expansion in the last third of pregnancy so that the gut wall comes to lie close to the pericardial sac with its placental function (Fig. 3.2). The relevance of this juxtaposition to increased fetomaternal exchange is unknown. This association of a vascularized endodermal sac (the hind gut) with the trophectoderm of the pericardial sac is directly analogous to the formation of the allantochorion in amniotes. Many other viviparous genera show similar gut expansion and direct absorption of follicular or ovarian fluid via mouth or anus has been suggested (Wourms 1981).

3.1.4.2 Trophotaeniae

These are extensions from the hind gut epithelium (endoderm) which are richly vascularized by splanchnic mesoderm (Fig. 3.2). They range from small rosettes to groups of ribbon-like appendages as long as the embryo.

Trophotaeniae are present in several unrelated families, but are best developed in the Goodeidae, of which *Ameca splendens* has been best investigated (Lombardi and Wourms 1985a–c; Wichtrup and Greven 1985). Embryos are released early from the follicle into the ovarian lumen; there may be a brief pericardial sac and yolk sac stage before the development of the trophotaeniae. There is a close relationship between the trophotaeniae and the ovarian epithelium but never a permanent adherence. Ultrastructural and tracer studies indicate that the trophotaenial epithelial cells are greatly specialized for endocytotic absorption (Schindler and DeVries 1987) and the ovarian epithelial cells for secretion. The efficiency of the system is indicated by the 15,000% increase in fetal organic material during gestation (Table 3.1).

3.1.4.3   Other Systems

Two other families show large dry weight increases during gestation (Wourms 1981). Embiotocidae (20,000% increase) have spatulate fin extensions, an enlarged hind gut and lie close to an actively secreting ovarian epithelium; and *Jennynsia* (24,000% increase) has ovarian trophonemata extending into the mouth and buccal cavity (Fig. 3.2). Nothing is known of the functional cytology of these ostensibly efficient associations.

3.1.5   *Homologies Between Fish and Amniote Placentas*

Based on embryological derivation and intimacy of apposition the chondricthyan yolk sac placenta is directly homologous to the trilaminar choriovitelline placenta of mammals. Functionally there may be differences since the most specialized fish placenta appears to have a considerable respiratory function, judging by the extensive capillary indentation and attenuation of the fetal and maternal epithelia (Cate-Hoedemaker 1933; Schlernitzauer and Gilbert 1966). Similar modifications are reported for the persistent trilaminar yolk sac of most marsupial families (Luckett 1976; Tyndale-Biscoe and Renfree 1987). This contrasts sharply with eutherians, which have tall columnar trophectodermal cells over the transient trilaminar yolk sac designed more for macromolecular uptake rather than for gas exchange.

Trophotaeniae are essentially exteriorized giant gut villi, with an endodermal epithelium vascularized by splanchnic (gut) mesoderm (Fig. 3.2) (Lombardi and Wourms 1985a–c). They are thus homologous to the inverted fetal yolk sacs of rodents and insectivores (Fig. 1.3) and like them specialized for endocytotic uptake rather than gas exchange and with a similar loose association with the maternal epithelium.

The pericardial sac membrane of Poeciliidae and Anablepidae consists of somatopleuric mesoderm underlying trophectoderm. The extension and folding of the sac over the embryonic head (Grove and Wourms 1991) indicates that it is functionally homologous with the chorionic headfold which produces the amniotic cavity in some amniotes. In amniotes the somatopleuric chorion would not normally be vascularized until reached by the allantoic mesoderm, but in the Poeciliidae vascularization spreads from the adjacent yolk sac portal circulation as this yolk sac is superseded by the pericardial sac (Turner 1947; Knight et al. 1985). In Anablepidae the modified pericardial sac and villous elaborations of the maternal follicular epithelium are firmly adherent and Wourms (1981) seems entirely justified in defining this as a true follicular epitheliochorial placenta, not a 'pseudo' placenta (Turner 1947). It is much more efficient than a yolk sac placenta in facilitating embryonic dry weight increase and possibly represents a parallel development to the replacement of the yolk sac by the chorioallantoic placenta in mammals. How specialized the apposed epithelia of the follicular placenta are with respect to gas exchange and macromolecular uptake remains to be determined.

Clearly the fish have evolved placental structures remarkably analogous to those in the amniotes. Apart from intrauterine cannibalism, proliferation of a yolk sac or a non-yolk sac extra-embryonic membrane in intimate association with maternal tissue seems to be the most efficient solution to the problem of nutrient transfer.

3.2   Amphibia

3.2.1   Introduction

Like the fish, all of the Anuran amphibians have external fertilisation but the Urodeles, although having fewer species, have developed internal fertilisation and have evolved many viviparous forms (Wake 1993). Most provide internal nutrition by secretions of the oviduct or ovary. There is little evidence of fetal modifications for fetomaternal transfer. Some species do show expansions of a highly vascularized tail or gill extensions, but how closely these are related to maternal tissues remains to be demonstrated.

3.2.2   Classification

Class Amphibia
Order Anura, Frogs and Toads

This order has some remarkable adaptations to viviparity (Hogarth 1976). *Rheobatrachus* broods the young in its stomach, digestion being suspended for the 7 month duration. In *Rhinoderma* the male incubates the young in its vocal chord pouch for a similar period. The dorsal skin modifies into individual pouches in which the young develop in *Pipa* and *Gastrotheca*; the former has an expanded vascularized tail, the latter leaf-like gill extensions wrapped around the embryo (Del Pino et al. 1975). Similar modifications are seen in many free-living juvenile forms and further investigation is necessary to establish if the association with the maternal tissue is close enough to be considered placental. *Nectophrynoides* is, more conventionally, retained in the oviduct and depends on the epithelial secretions for maintenance and growth (Xavier 1976; Wake 1993).

Order Urodela, Salamanders

Nearly all the salamanders have internal fertilization, but only one of the 400 species, *S. atra,* is obligately viviparous. It relies on ovarian/oviductal secretions for growth (Wake 1993).

Order Gymnophiona (Apoda, legless burrowers).

Only 25 of the 170 species have been investigated; half of these are viviparous, supported by lipid-rich ovarian/oviductal secretions with no obvious fetal modification to facilitate absorption. Some have a long gestation period, depend on yolk for the first 3 months but then develop on ovarian secretions for a further 9 months. Viviparity is normally associated with extreme conditions of altitude or desiccation (Wake 1993).

3.3   Aves

The only class with no viviparous forms. All birds have the full complement of embryonic membranes in the egg with very little variation despite enormous differences of habitat and size. The massive amount of yolk in the yolk sac is the initial and continuing source of nutrition; the allantois with deeply indented capillaries is the organ of respiration within the egg (see Mossman 1987).

3.4   Reptilia

3.4.1   Introduction

The reptilia are mostly egg-laying amniotes with a full complement of extraembryonic membranes. Two orders, Crocodilia and Chelonia have extraembryonic membranes like the birds and all are oviparous. The Squamata, lizards and snakes, although with basically the same membranes as birds have a unique yolk sac modification. Most are oviparous but have evolved viviparity over 100 times within the order and 20% of the extant species are viviparous. The adoption of viviparity may be associated with extremes of habitat (Panigel 1956; Shine and Bull 1979; Thompson and Speake 2006). A few genera today have oviparous and viviparous species in different climatic areas and there are also three species of lizards that are said to be reproductively bimodal having both oviparous and viviparous populations (see references in Blackburn 2000).

Oviparous species depend upon large amounts of yolk and a protective, calcified shell. Many viviparous species also rely on a significant amount of yolk, but the egg is retained in the oviduct or uterus for up to 9 months; the shell may be reduced to an acellular fibrous layer (shell 'membrane') which fragments in utero in some, and a few ovulate eggs with very little yolk. The extraembryonic membranes have developed into structures facilitating fetomaternal exchange (Stewart and Blackburn 1988; Blackburn 1993; Stewart 1993; Stewart and Thompson 2000; Thompson and Speake 2006; (Fig. 3.3)).

3.4.2   Classification

Class Reptilia
Orders Chelonia (turtles), Crocodilia
All species oviparous.
Order Squamata
Suborder Lacertiliformes (lizards)
Most families oviparous with an occasional viviparous lineage. The Family Scincidae
    have the most viviparous forms such as Chalcides, Eulamprus, Mabuya, and
    Xantusia, but far more skinks are oviparous like Bassiana or Morethia.
Suborder Serpentiformes (snakes)

Most species are oviparous, only about 20% are viviparous. Some examples of viviparous snakes are to be found in the Families Colubridae (Thamnophis, garter snake); Hydrophiidae (Hydrophis, sea snake) and Viperidae (Viper).

3.4.3 Suborder Lacertiliformes (Lizards)

All lizards have yolk sac and chorioallantoic extraembryonic membranes which persist to hatching/term but show a wide variety of sizes and structures.

3.4.3.1 Yolk Sac

If the yolk sac contains a significant amount of yolk, a bilaminar (ectoderm plus endoderm, bilaminar omphalopleur) non vascular area persists on the abembryonic side of the conceptus. This remains because the mesodermal derivative, which would normally insert between the bilayer to provide the potential for forming a vascularised trilaminar yolk sac, instead grows into and isolates a bottom (abembryonic) crescent of yolk against the bilayer called the isolated yolk mass (Stewart 1993) (Fig. 3.3). This assemblage (uterine epithelium – eggshell – bilayer – isolated yolk mass – mesoderm) is referred to as the omphaloplacenta, is unique to, but present in, virtually all the Squamata and does not develop a blood supply. The upper part of this mesodermal ingrowth or yolk sac cleft (see Fig. 3.3) is vascularised later in pregnancy to facilitate mobilisation of the upper mass of yolk for embryonic growth. In oviparous species the omphaloplacenta may play a role in histotrophic or water uptake prior to oviposition, but the yolk has practically disappeared before hatching, crowded out by growth of the allantois (Fig. 3.3). Viviparous species with shell 'membrane' loss early in pregnancy (*Pseudemoia spp*) develop the omphaloplacenta at a later stage in utero and some retain it in late pregnancy as forty percent or more of the placental surface with uterine and yolk sac epithelial proliferation indicating evidence of histological transfer (Stewart and Thompson 2000; Thompson and Speake 2006). However in the viviparous *Chalcides* it regresses well before birth (Blackburn and Callard 1997) and in *Mabuya spp.* there is no histological evidence for its formation (Blackburn and Vitt 2002; Jerez and Ramirez-Pinilla 2003).

3.4.3.2 Chorioallantois

The chorioallantoic placenta is extensive in lizards, with highly vascularized fetal and maternal epithelia always separated by a shell in oviparous forms, but in close apposition in viviparous forms in later gestation when the shell disappears. The early work of Weekes (1935) defined three structural types of chorioallantoic placentas, recent work has added a fourth (Blackburn and Vitt 2002) and provided measurements of the dry weight increase from egg to neonate (Stewart and Thompson 2000) to facilitate comparisons of placental transfer (or "efficiency") as in the fish.

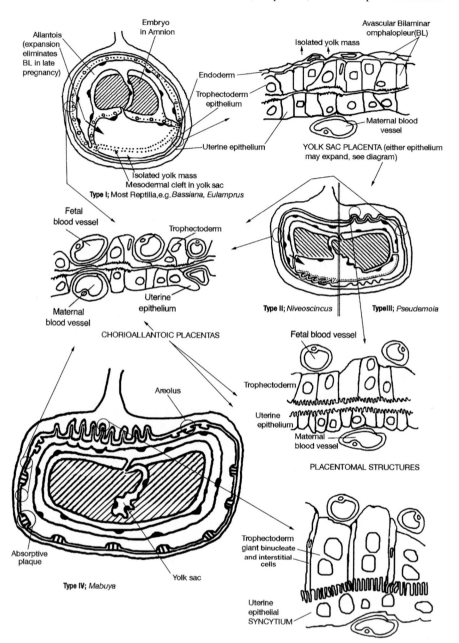

Fig. 3.3 Placental membranes in reptiles. Not shown is the very thin acellular shell membrane which persists between the trophectoderm and uterine epithelium in many reptiles

Type I includes the vast majority of oviparous and viviparous species. The trophectoderm is a smooth cellular layer deeply indented by the allantoic capillaries and closely apposed to the shell. In viviparous forms the shell disappears during gestation and the trophectoderm is then apposed to a similarly thin cellular uterine epithelium with indenting capillaries forming a structure optimising respiratory exchange (Fig. 3.3) and as shown in *Xantusia* also a site capable of considerable amino acid uptake (Yaron 1977).

Electron microscope and high resolution light microscope studies (Stewart and Thompson 2004) have established that all Type I trophectoderm and uterine epithelia so far investigated are cellular not syncytial as suggested by Weekes.

The chorioallantoic placenta covers the embryonic half of the conceptus at mid pregnancy. In some species this area increases as expansion of the allantois and yolk utilisation reduces the yolk sac placenta to an insignificant area by late pregnancy (Fig. 3.3). However in many other species the yolk sac placenta hypertrophies and persists throughout gestation (Stewart 1993). Neonatal dry weight is 10–25% lower than the egg in both oviparous (e.g. *Bassiana*, 15% loss) and viviparous (*Eulamprus*, 16% loss) species, so there is no appreciable increase in placental organic transfer in viviparous type I species.

Type II, all viviparous with early shell loss, show a more structured, ridged chorioallantoic placenta, still with respiratory exchange areas but also with more variation in the trophectodermal cell types but no significant increase in maternofetal transfer (*Niveoscincus metallicus*, 20% loss). However *Niveoscincus ocellatus* manages a 170% increase in dry weight, the increased transfer presumably resulting from the proliferation of the omphalopleur in mid pregnancy in this species, since both have similar chorioallantoic placental structure.

Type III may also show abembryonic omphalopleuric proliferation (*Pseudemoia*) but in addition develop an oval expansion at the embryonic side, of chorioallantoic villous folds of loosely apposed columnar trophectoderm and uterine epithelial cells called a placentome (Fig. 3.3). There is structural evidence to indicate that there is considerable placentomal histotrophic transfer to the trophectoderm from the uterine epithelium. The non placentomal chorioallantois shows the same respiratory specialisation as seen in Types I and II. The placentome in *Pseudemoia spp.* does not produce any great increase in transfer over the Type II of N. ocellatus, with values ranging from 130 to 216% (Thompson and Speake 2006).

Type IV placentas in *Mabuya spp.* however, do support a massive 39,000–47,000% increase in dry weight. Starting with a tiny yolk poor egg at least 50 times smaller than any found in Types I, II or III, *Mabuya* relies on four diversified chorioallantoic structures which together cover all of the conceptus surface (Jerez and Ramirez-Pinilla 2001). The yolk sac placenta regresses before any significant transfer occurs (Blackburn et al. 1984). The first structural modification is areas for respiratory exchange, as in Types I, II and III. The second is areolas, local dome shaped expansions of columnar absorptive trophectoderm over actively secreting uterine gland mouths. Gland secretions may be available in the other Types but no correlated trophectodermal differentiations have been reported. The third feature is numerous (~50) small (~0.5 mm diameter at stage 40) "absorptive plaques" scattered over the abembryonic half of the chorioallantoic placenta and the fourth a much larger and more complex placentome (~7 mm diameter) than Type III, occupying most of the top of the embryonic half of the conceptus (Ramirez-Pinilla 2006). Excellent recent light and electron microscopical studies (Blackburn and Vitt 2002; Ramirez-Pinilla et al. 2006) show that while the plaque and placentomal trophectoderm shows similar cell types—large columnar binucleate cells interspersed with very thin columnar uninucleate interstitial cells (Fig. 3.3)—the placentomal cells are 5–10 times the size of the plaque cells and show quite different ultrastructure. The plaque cells irregular apices are closely apposed to those of the cellular columnar uterine epithelium. In striking contrast the placentomal trophectodermal apices bear extensive long microvilli which are closely interdigitated with the uterine microvilli forming a wide microvillar band and the uterine epithelium is a syncytium, the only adequately illustrated example found in the reptiles so far.

Both plaque and placentome cells show ultrastructure indicating extensive histotrophic transfer. There is a further very anomalous type of small cell occasionally interrupting the microvillar junction of the placentome (not shown in Fig. 3.3). Their origin is uncertain but they do seem to have an intimate relationship with the uterine epithelial syncytium (Vieira et al. 2007).

Immunocytochemical, in situ hybridisation and molecular biological studies aimed at identifying the functions of these four very different chorioallantoic structures should provide a better understanding of how placental structure developed to support viviparity.

Types I to IV have almost certainly evolved separately, no evolutionary sequence has been suggested, but the range within the same order provides a unique opportunity to investigate placental function and evolution, as so ably demonstrated by the work of Blackburn, Stewart and Thompson and their colleagues.

3.4.4   Suborder Serpentiformes (Snakes)

In all snakes the shell membrane persists to term, and all have large amounts of yolk. The placental membranes of viviparous snakes are similar to those of Type I lizards in having a respiratory chorioallantoic placenta over the embryonic half and an avascular bilaminar omphalopleur showing ultrastructural evidence for considerable histiotrophic transfer (Blackburn and Lorenz 2003a,b; Stewart and Brasch 2003) over the abembryonic side throughout pregnancy. No specialised chorioallantoic placentomal areas have been reported and a major difference with the lizards is the growth of the allantois into the yolk sac cleft early in pregnancy to form what is referred to as an omphalallantoic placenta (omphaloplacenta plus allantois). Autoradiographic evidence indicates that the trophectoderm of the omphaloallantoic placenta is a major site of nutrient transfer (Hoffman 1970). However there is no evidence of significant organic uptake overall, since all snakes investigated so far show egg to neonate dry weight losses between 2 and 35% (Stewart and Thompson 2000).

3.4.5 Homologies Between Reptile, Fish and Amniote Placentas

The reptiles thus show many examples of the development of specialized fetomaternal extraembryonic membrane apposition as a result of reduction or loss of the
shell membrane in eggs retained in the uterus.

The yolk sac is important in many species, providing both respiratory and nutrient uptake. It is never as specialized as in some elasmobranchs (*Mustelus canis*) for
respiratory exchange because this function is provided by the chorioallantoic placenta and the two placentas coexist in snakes, lizards and other amniotes.

In lizards with Type IV placentas, *Mabuya sp.*, with the smallest, least yolky
eggs and earliest shell loss, the chorioallantoic placenta supersedes the early yolk
sac and develops four structurally diversified areas (see above). The placentome is
elaborately folded, increasing the surface area available for endocytotic uptake, and
much of the remainder is specialized for respiratory exchange. This arrangement is
functionally analogous, but structurally reversed, to that seen in the mammalian
chorioallantoic placenta, in which the folded area provides for respiration. The areolae and absorptive plaques also have structural parallels in the mammalian epitheliochorial placentas such as those in the pig.

Superficially the trophectodermal binucleate cells and the uterine epithelial syncytium in *Mabuya sp.* resemble the patterns typical of the ruminant placentomes in sheep
and goats (Wooding 1992). However in *Mabuya all* of the placentomal trophectodermal
cells are binucleate and of equivalent ultrastructure, not a minority population of unique
ultrastructure. In addition the syncytium is apparently formed solely from maternal
uterine epithelial cells and not by the migration and fusion of the fetal binucleate cells
to form a fetomaternal hybrid layer as in the ruminants. The use of the category "synepitheliochorial" (Wooding 1992) to describe this part of the *Mabuya* placenta (Ramirez-
Pinilla et al. 2006) is therefore inaccurate and misleading.

The results suggest that the chorioallantoic placenta is phenotypically more flexible than the yolk sac but also infer a basic limitation of that genetic repertoire
because such widely different phyla, Reptilia and Mammalia, both produce similar
structures from the basic uterine epithelium – trophoblast apposition. The 18 species of South American Mabuya have habitats ranging from the Andean mountains
(Ramirez-Pinilla papers) to tropical (*Mabuya heathii*, Blackburn and Vitt 2002) but
the details of the tiny eggs and placental complexity are very similar. They are considered monophyletic and there is no evidence that extremes of environment result
in any obvious changes. They all show the arolae and placentomal structures so
analogous to those in certain mammalian orders.

The structural similarities are the more surprising because the main nutrients
transferred, lipids in reptiles (Speake et al. 2004; Thompson and Speake 2006) and
glucose in mammals (Joost and Thorens 2001) are so different. Immunocytochemistry
of the various different transporters could provide useful structure/function correlates for the four Mabuya structural categories.

Chapter 4
Monotreme and Marsupial Placentation

Class Mammalia, Subclasses Prototheria (Monotremata, e.g. Platypus) and Metatheria (Marsupialia)

4.1 Introduction

Prototheria (e.g. Platypus) lay eggs and Metatheria (e.g. Kangaroo) produce shelled eggs in utero with the same reliance as birds on a bi- or trilaminar yolk sac and allantois for mobilization of yolk and respiratory exchange. However, they all have far less yolk and a much earlier hatching or birth than the birds and require subsequent mammary nutrition and in Metatheria, protection in a pouch (Luckett 1976, 1977; Tyndale-Biscoe and Renfree 1987). A few marsupials lose their shell during development in utero and show localized invasion of the uterine epithelium by the yolk sac trophectoderm, and other families develop an area of chorioallantoic placentation in the last quarter of the typically very short 12–14 day gestation.

4.2 Prototheria (Monotremata)

Two families only

Ornithorhyncus: Platypus
Tachyglossus: Echidna

The eggs are much smaller than those of equivalent sized reptiles or birds with a much greater increase in size by absorption of uterine secretions in utero prior to shell deposition. The blastocyst forms precociously and has 19 somites when the egg is laid after about 18 days of development. The mesoderm has spread to produce a vascularized trilaminar yolk sac over two-thirds of the blastocyst with the remainder bilaminar and avascular (Luckett 1977; Hughes 1993) (Fig. 4.1).

The trophectoderm of the trilaminar region is thin, minimizing the gas diffusion distance to the yolk sac capillaries, while the endodermal cells are columnar, actively absorbing yolk and transporting it to the same capillaries. In the avascular bilaminar portion the cell types are reversed, a columnar absorptive ectoderm and thin endoderm.

P. Wooding, G. Burton, *Comparative Placentation*,
© Springer-Verlag Berlin Heidelberg 2008

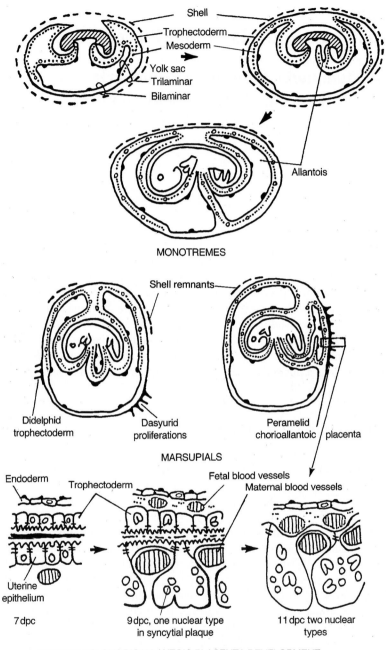

Fig. 4.1 Placental membranes in monotremes and marsupials

After the egg is laid the amnion folds fuse and the allantois develops as the usual hind gut outgrowth eventually, about halfway through incubation, forming a vascularised chorioallantoic membrane over half the inside of the egg. This has a much higher capillary density than the trilaminar yolk sac which persists over the other half; the bilaminar segment is finally a mere remnant (Luckett 1977; Hughes 1993) (Fig. 4.1).

The trophectoderm over the late-gestation trilaminar yolk sac and chorioallantois is said to be thin, which would facilitate haemotrophic exchange, but there are no detailed studies available to verify this.

4.3   Metatheria (Marsupialia)

Seven major families

Dasyuridae: insectivores and carnivores
Didelphidae: American opossums
Peramelidae: bandicoots
Macropodidae: kangaroos, wallabies
Phalangeridae: arboreal Australian opossums (Trichosurus)
Phascolarctidae: koala
Vombatidae: wombats

There is an excellent general account of marsupial reproduction (Tyndale-Biscoe and Renfree 1987) and a recent review (Renfree 2006).

Metatheria have even less yolk than Prototheria, cleavage is complete (holoblastic) and the hollow blastocyst is retained in the uterus initially bounded by a shell membrane.

Pregnancy is short (from 13 days in the 100 g *Monodelphis* to 36 days in the [largest] 30 Kg Kangaroo) and completed within the duration of one ovarian cycle; there is no system for prolonging the life of the corpus luteum. The shell membrane breaks down in mid pregnancy.

Most families have similar fetal membrane development to Prototheria, but the allantois is smaller and forms a chorioallantoic placenta in only a few species (Fig. 4.1).

The bilaminar yolk sac persists to the end of pregnancy absorbing uterine secretions by endocytosis before and after shell membrane rupture. Amino acids are taken up by this route and the yolk sac lumen can contain up to ten times the concentration in the maternal plasma. The trilaminar yolk sac occupies as much as two-thirds of the egg surface and shows a thin trophectodermal layer over the yolk sac capillaries, suggesting a predominantly respiratory function.

After the shell membrane disappears in mid- to late pregnancy the trophectoderm of the bi- or trilaminar yolk sac is apposed to the uterine epithelium in most

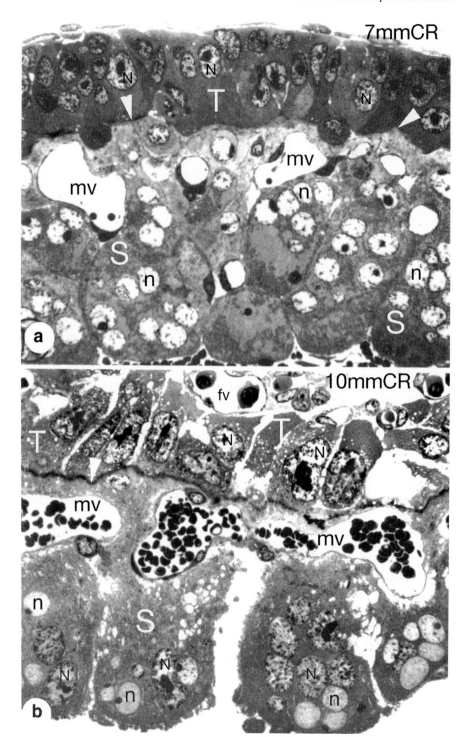

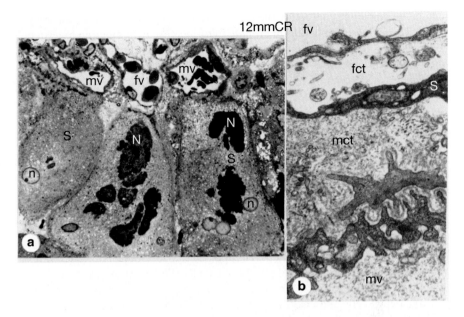

Fig. 4.3 Definitive structure of the chorioallantoic placenta in the marsupial family Peramelidae. (**a**) Near term, 12 CR: The fetal trophectoderm is no longer visible but the fetal blood vessels (FV) are separated from the maternal (MV) by only a thin layer of cytoplasm continuous with the syncytial plaques (S) which contain two sorts of nuclei (N and n) *Isoodon*, ×480. (**b**) 12 mm CR: Electron micrograph of the attenuated cytoplasmic layer (S) between fetal (FV) and maternal (MV) blood vessels. CT, maternal connective tissue. *Isoodon*, ×15,000. From Padykula and Taylor (1976)

families. There is no interdigitation of microvilli but there is an increase in the area of apposition of the trilaminar yolk sac by growth of folds in the uterine epithelium in the second half of pregnancy (Harder et al. 1993). Late in pregnancy (at 92% of term in *Monodelphis* and 97% in *Sminthopsis*, Freyer et al. 2003) some species show a trophectodermal invasion of the uterine epithelium from either the bi- (*Sminthopsis*) or the trilaminar (*Monodelphis*) portions of the yolk sac (Fig. 4.1). The trophectodermal invasion may be by "giant" cells (Didelphidae: *Philander*. Enders and Enders 1969; Dasyuridae: *Sminthopsis*. Roberts and Breed 1994) or by a syncytium (Didelphidae: *Monodelphis* Freyer et al. 2007). This brings trophectodermal processes up to and possibly across the uterine epithelium basement

Fig. 4.2 Development of chorioallantoic placenta in the marsupial family Peramelidae (compare with the diagram on Fig. 4.1). (**a**) 7 mm Crown Rump (CR) length; close apposition (*white arrowheads*) between the cellular fetal trophectoderm (T) (m, fetal mesoderm) and uterine epithelial syncytial plaques (S). Note that the plaques are deeply indented by the maternal capillaries (MV) *Isoodon*, ×400. (**b**) 10 mm CR. The fetal blood vessels (FV) do not indent the cellular fetal trophectoderm (T) in the way that the maternal vessels (MV) indent the uterine epithelial syncytial plaques (S). The plaques contain two sorts of nuclei: N, morphologically similar to those in the trophectoderm cells; and n, similar to those in the original syncytial plaques. CT maternal connective tissue. *Perameles*; ×400

membrane, apparently targeting the endometrial blood vessels (Zeller and Freyer 2001). These invasive plaques in *Monodelphis* increase in number up to term but never replace all of the uterine epithelium. The extent and consequent function of this epitheliochorial to endochorial structural transition at such a late stage of pregnancy is unknown.

The allantois is reported to form a chorioallantoic placenta in only three families, Phascolarctidae, Vombatidae and Peramelidae, and only the last has been investigated in detail. In the peramelids, *Perameles* and *Isoodon*, the allantois vascularizes the trophectoderm in the last one-third of a very short 12-day pregnancy (Padykula and Taylor 1976). It forms a chorioallantoic placenta in close apposition to the uterine epithelium occupying a far smaller area than the persistent bi- and trilaminar yolk sac placentas (Fig. 4.1).

In *Perameles* at 9 days post coitum or post-ovulation, the entire uterine epithelium apposed to the conceptus develops from a unicellular layer into individual plaques of syncytium containing uniformly small round nuclei. The uterine capillaries deeply indent these syncytial plaques. The apposed trophectoderm is undifferentiated and unicellular with large irregularly shaped nuclei (Fig. 4.2a). As the vascularization of the chorioallantoic trophectoderm increases, the cell membranes between the trophoblast cells and also those shared with the uterine syncytium disappear. The fetal blood vessels indent the fused epithelia and now two sorts of nuclei are recognizable in the uterine epithelial syncytial plaques, small round ones and others with large irregular profiles (Fig. 4.2b). Eventually the unicellular trophectoderm can no longer be distinguished and only very thin laminae continuous with the syncytial plaques separate maternal and fetal blood vessels (Padykula and Taylor 1976) (Fig. 4.3a, b).

From the time course of the trophectodermal disappearance and the evidence of the two types of nuclei in the syncytial plaques it seems that the trophectoderm cells fuse with the uterine epithelial syncytial plaques. Such fetomaternal cell fusion would significantly reduce the number of membrane barriers between fetal and maternal blood and greatly facilitate fetomaternal exchange. The trophectoderm does not vanish but forms part of a fetomaternal hybrid syncytium directly analogous to the similar system produced in sheep and goat chorioallantoic placentas by fusion of fetal binucleate cells with uterine epithelial cells (Chap. 6).

The chorioallantoic placenta in the peramelids supports considerable embryonic growth. Even though their gestation is the shortest of any marsupial, their neonates are relatively well developed at birth (Tyndale-Biscoe and Renfree 1987). This indicates that the chorioallantoic placenta may well be more efficient than the yolk sac type (Freyer et al. 2004).

The marsupials thus show several features, such as trophectodermal invasion and formation of chorioallantoic fetomaternal hybrid tissue, which have close parallels in eutherian Ruminant placentas (see Sect. 6.2.4). This seems to be another example of parallel evolution, the same functional demands producing similar structures from basically the same fetal membrane repertoire, rather than any indication of an evolutionary relationship between the groups.

Chapter 5
Eutheria: Epitheliochorial Placentation Pig and Horse

5.1   Introduction

In this form of placentation the uterine epithelium and trophectoderm come into contact, the fetal and maternal microvilli interdigitate and placental development essentially involves a vast increase in area with no loss of any layers between the fetal and maternal bloodstreams. The exclusively cellular layers may become exceedingly attenuated, reducing the diffusion distance as pregnancy proceeds, but all persist to term (Amoroso 1952; Steven 1975a, 1983; Ramsey 1982; Mossman 1987).

The uterine glands are numerous and actively secreting through pregnancy (Bazer and First 1983), and there are fetal expansions over the gland mouths, forming areolae consisting of cells specialized for histotrophic absorption. The domestic pig placenta will be described as the basic type, and variations on that pattern considered for the horse, American mole *(Scalopus)* and the primates (lemurs and *Galago*).

5.2   Pig, Sus scrofa

Oestrous cycle: 21 days
Ovulation: spontaneous

- Uterus: bicornuate
- Litter: 6–12
- Gestation: 114 days
- Implantation: superficial
- Amniogenesis: folding
- Yolk sac: initially large and vascularised, remnant persists, apparently functional to 50 dpc
- Chorioallantois: forms definitive placenta
- Shape: diffuse, simple villous
- No decidual reaction
- Interhaemal membrane: epitheliochorial

Accessory placental structures: areolae (Brambel 1933; Friess et al. 1981)

The major characteristics (•) are found in the families Suidae, Hippopotamidae and Camelidae (Mossman 1987).

5.2.1   Pig Fetal Membranes

At implantation (15 dpc, see below) the conceptus is a mostly trilaminar, filamentous blastocyst. The yolk sac forms what is probably a briefly functional bilaminar absorptive yolk sac placenta before its attachment to the trophectoderm is disrupted by growth of the allantois. The yolk sac lumen then collapses but still retains a good blood supply. Healthy yolk sac endodermal cells with enormous mitochondria and large amounts of smooth endoplasmic reticulum are found up to 50 days of gestation (Tiedemann and Minuth 1980) but they regress in the second half of pregnancy.

Between 22 and 28 dpc the allantois expands to contact and vascularize most of the inner surface of the trophectoderm. This forms the chorioallantoic placenta and the vascularization also spreads into the amniochorionic membrane over the back of the embryo (Amoroso 1952) (Fig. 1.1).

5.2.2   Development of the Pig Placenta

The pig conceptus passes into the uterus as a spherical 100 mm blastocyst within a zona pellucida. By 8–9 dpc it is correctly spaced in the uterus and out of the zona (see chapter 2.1.1). It then grows with rapid mitoses to a 700 μm sphere by 11 dpc. The mitotic rate slows but then with peak IL1β secretion by the blastocyst (Ross et al. 2003), remarkably rapid changes in cell shape and size remodel the sphere into a 100 mm thin filament by 12 dpc (Fig. 5.1). The rapid mitotic rate resumes and by 16 dpc the filament is up to 1 m long. Recent molecular genetic studies show a large number of genes are upregulated in the remodelling (Blomberg et al. 2006) and indicate more cell division at this time than the previous workers thought (Geisert et al. 1982a,b,c; Pusateri et al. 1990).

Between 11 and 12 dpc the blastocyst secretes estrogen, which switches the endometrial secretion of PGF2α from being directed into the uterine capillaries to release into the uterine lumen. This is the "maternal recognition of pregnancy" signal, preventing the luteolysis of the corpus luteum by the prostaglandin and thus maintaining the progesterone secretion essential for a successful pregnancy (Bazer et al. 1986; Spencer et al. 2004a).

A minimum of four conceptuses are necessary to produce enough estrogen to stimulate the uterus to modify its prostaglandin secretion so that the luteal progesterone production is maintained. Subsequent fetal death may result in litters of less than four (Polge et al. 1966). There is also evidence for an optimal length of uterus per conceptus at 20–25 dpc (Wu et al. 1989).

Further production of oestrogen by the blastocyst between 15 and 30 dpc stimulates secretion from the uterine epithelium and glands (Geisert et al. 1991; Spencer et al. 2004). This produces specific proteins which are thought to be involved in the growth of the conceptus and the subsequent immunological readjustments necessary for successful implantation. They also support continued luteal growth. Similar proteins are produced by the uterus of a non-pregnant pig treated at the correct stage of the cycle with artificial oestrogens (Geisert et al. 1991).

Changes in uterine oedema and increased fenestrations in the capillaries (Keys and King 1988) coincide with the adhesion of the filamentous blastocyst to the uterine epithelium at 13–14 dpc. The fetal and maternal epithelia come into close apposition, involving loss of microvilli and considerable glycocalyx erosion, along the strip region (Fig. 2.1; Dantzer 1985; Bowen et al. 1996). Molecules exposed by the modification of the glycocalyx (Bowen et al. 1996) together with newly expressed integrins on both surfaces zip the trophoblast and uterine epithelium close together at a constant separation of ~20 μm. Subsequently the fetal and maternal microvilli reform but now interdigitate (see Fig. 2.1). This increases the fetomaternal exchange area by 10–12 times over a flat apposition (Baur 1977). There is no good evidence for formation of any junctions between the apices of the fetal and maternal epithelia or for penetration of their tight junction apical seals by cellular processes from either side. Neither epithelium forms even a limited syncytium at any stage of gestation (Dantzer 1985). There are few differences in the fine structure of the uterine epithelium between the cycle and early pregnancy (Keys and King 1989, 1990). Since fetal trophectodermal syncytium formation has been reported from ectopic pig blastocyts transferred into the pregnant endometrium below the uterine epithelium (Samuel and Perry 1972), this layer may be instrumental in blocking fetal syncytium formation and possible subsequent invasion. No decidual changes have been reported in the endometrium at the implantation site.

In pig endometrium there is a conceptus dependent threefold increase in CD44+ lymphocytes at implantation sites between 10 and 20 dpc (Dimova et al. 2007; Croy et al. 2006), Isolated populations of such lymphocytes show NK lytic activity from 12 to 28 dpc, with 10% containing perforin and a similar percentage separating as large, possibly granulated cells on flow cytometry. All these indicators, as well as their probable CD16+ expression, suggest this is a population equivalent to the rodent and human uNK cells. However pig *blood* NK cells are quite different, being smaller and with few if any granules compared to their rodent blood counterparts, so it seems this uterine population could equally well be macrophages. No detailed morphological characterisation has been carried out as yet. An indication of a possible role for these pig endometrial cells comes from experiments with normal pregnancies where 35% of the conceptuses arrest implantation and degenerate between 15 and 25 dpc, providing an invaluable comparison between normal and adjacent arresting sites (Croy et al. 2006). Laser microdissected lymphocytes from normal sites produce angiogenic cytokines (VEGF, HIF) with virtually no inflammatory types (IFN γ, IL12), while arresting sites show the reverse. It is suggested (Croy et al. 2006) that both lymphocyte and trophoblast VEGF secretion are necessary for the initiation of the growth of the maternal capillary net which starts at 16 dpc (Dantzer and Leiser 1994). The lymphocytes have disappeared by 30 dpc, and subsequent maternal and fetal capillary growth presumably depends on the VEGF and other factors produced by both trophoblast and uterine epithelium (Winther et al. 1999).

At 17 dpc proliferation of fetal trophectoderm over the gland mouths has started to form the regular areolae, where the fetal and maternal epithelia become widely separated by glandular secretion (Amoroso 1952) (Figs. 5.1b, c and 5.2c).

About 7,000 regular areolae per conceptus are present, fairly evenly spaced, from early gestation to term (Fig. 5.1b, c). They triple in area between 30 and 114 dpc forming 10% of the total chorioallantoic surface area at mid-gestation and 4% at term (Brambel 1933). They show a characteristic fine structure

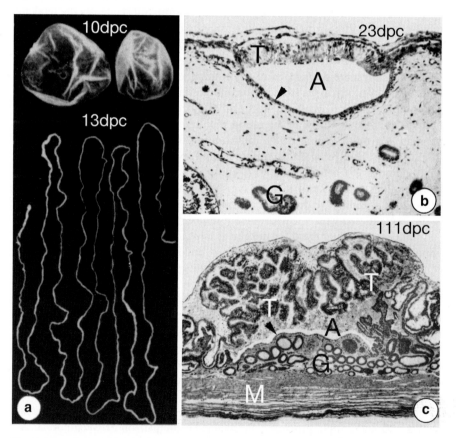

Fig. 5.1 Epitheliochorial placentation. (a) Note the remarkable change of the shape of the pig blastocyst from a 5 mm diameter sphere at 10 dpc to a 1 m long thread at 13 dpc; ×0.5. From Perry and Rowlands (1962). (b,c). Development of a regular areola in a pig placenta. (b) 23 dpc; ×80 (c) mature, 111 dpc; ×25 . A, areolus; G, gland; M, myometrium; T, trophoblast; arrowhead, uterine epithelium.

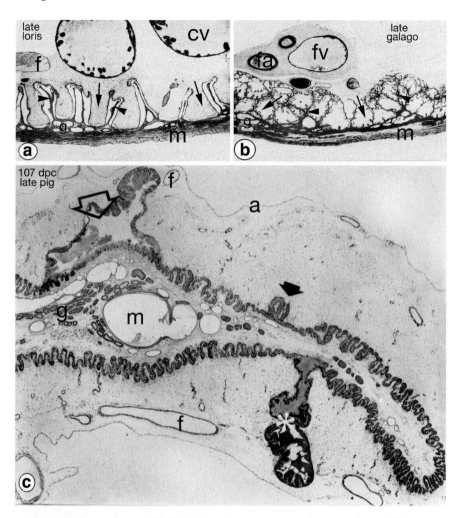

Fig. 5.2 Epitheliochorial placentas. Compare the structures of the Lemuroid placentas of (**a**) Loris and (**b**) Galago with that of (**c**) the Pig. The pig has mucosal folds whose sides consist of fetal villi inserted into endometrial crypts, seen at higher magnification in Fig. 5.3b. A chorionic vesicle (*asterisk*), irregular (*open arrow*) and regular (*closed arrow*) areolae can be seen. The section passes through the extreme edge of the regular areola, whose characteristic structure is better illustrated on Fig. 5.1c. Loris X15; Galago X 7; Pig X 12. Arrowheads, maternal villi; arrows, fetal villi; a, allantois; cv, chorionic vesicle; f, fetal vessel; fa, fetal artery; fv, fetal vein; g, glands; m, myometrium.

(Sinowatz and Friess 1983) and a broad spectrum of enzyme activities (Skolek-Winnisch et al. 1985; Firth et al. 1986a,b). Uteroferrin, an iron-containing glycoprotein, is among a variety of molecules synthesized in the gland cells (Chen et al. 1975; Stroband et al. 1986). This is released into the lumen, taken up by the areolar trophectodermal cells and transferred to the fetus, for which it represents an important iron source (Raub et al. 1985). The larger and less

frequent (about 1,500 per placenta) irregular areolae have significantly different fine structure from the regular type but nothing is known of their specific
functions (Fig. 5.2c) (Brambel 1933; Perry 1981).

Between 15 and 22 dpc the conceptus expands so that it touches the uterine epithelium round the entire circumference and establishes microvillar interdigitation
over the whole area of the trophectoderm (Dantzer 1985; Hasselager 1985). Further
growth in the area of the fetomaternal junction is accomplished by development
first of folds and then division of these into simple villi (Michael et al. 1985;
Wigmore and Strickland 1985) (Fig. 5.3).

The result is to maximize the surface area available for exchange between the
fetal and maternal capillaries and minimize the distance between them. This is

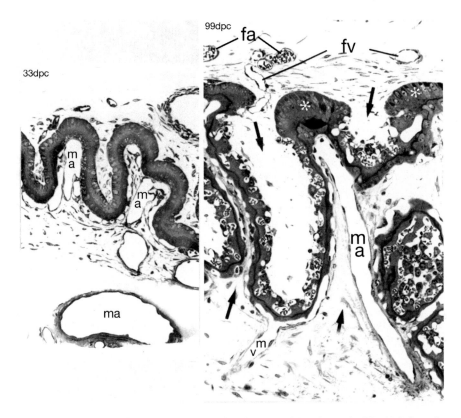

Fig. 5.3 Epitheliochorial placenta of the pig, development of the placental villi which form the
fetomaternal hemotrophic exchange area. Note the increase in regional differentiation. The uniform trophectoderm and uterine epithelial cell layers forming the folds at 33 dpc develop into very
attenuated layers deeply indented by fetal and maternal capillaries around the apex and sides of
the fetal villus by 99 dpc. At the base of the 99 dpc fetal villi the trophectoderm cells are considerably enlarged (*asterisks*) and are involved in pinocytotic, histotrophic uptake. Upward arrows,
maternal crypts; downward arrows, fetal villi; *fa* fetal arteriole, *fv* fetal venule, *ma* maternal arteriole, *mv* maternal venule (**a**) 33 dpc; ×130 (**b**) 99 dpc; ×170. (**b**) from Leiser and Dantzer (1988)

achieved in the pig by the growth and expansion of a meshwork of capillaries just under the epithelium of each fetal and maternal villus. Injection of latex or resin into the fetal or maternal vasculature and examination of the resulting casts of the blood vessel architecture shows how the maternal meshwork develops, and that the fetal and maternal villous meshworks are complementary. Each is supplied by a stem artery running up one side of the villus before dividing near the apex into a capillary meshwork which drains to veins around the base of the villus (Figs. 2.8, 5.3, 5.4, and 5.5; (Tsutsumi 1962; MacDonald 1975, 1981; Leiser and Dantzer 1988) (for the physiological exchange implication of this arrangement see Chap. 2).

Increase in size and branching of the villi can be accommodated by expansion of the capillary meshwork. As the conceptus grows the fetomaternal junctional area and the transport capacity of the placenta increase continuously right up to term (Michael et al. 1985). The villi and folds show local differentiation of structure. The trophectodermal cells at the bases of the fetal villi are tall, columnar and specialized for histotroph absorption; they are apposed to similar columnar maternal uterine epithelial cells, often with cell debris between (Dantzer et al.

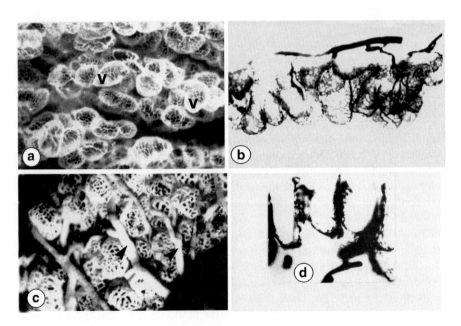

Fig. 5.4 Epitheliochorial placentation. Pig placenta vascular supply demonstrated with corrosion casts (**a,c**) and cleared sections of injected specimens (**b,d**) (**a**) Latex injection of fetal blood supply viewed from the maternal side showing the capillary network in each villus (v). 105 dpc; ×40. (**b**) Injection of the fetal blood supply shows the villus capillary net in cross section. 105 dpc; ×40. (**c**) Latex injection of the maternal vasculature viewed from the back showing the network of capillaries at the base of each crypt with their supplying or draining arterioles and venules (*arrowheads*). 105 dpc; ×40. (**d**) Injection of the maternal blood supply demonstrates on a section the maternal crypts into which the fetal villi fit. 75 dpc; ×50. All from Tsutsumi (1962)

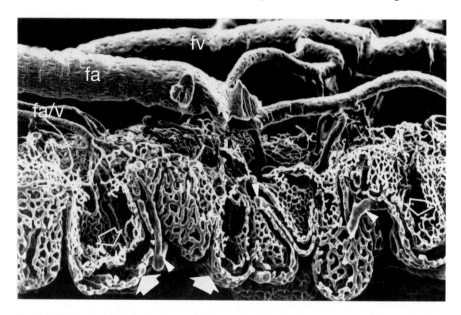

Fig. 5.5 Epitheliochorial placentation. Scanning electron micrograph of a corrosion cast from a pig placenta with both maternal and fetal circulations injected with acrylic resin in situ and then cross fractured. The complementary nature of the fetal villus (*white open arrows*) and maternal crypt (*solid white arrows*) capillary networks is elegantly demonstrated. *fa/v* fetal arteriole and venule; *fa* fetal artery; *fv* fetal vein; white arrowheads, maternal stem arteries. 99 dpc; ×95, from Leiser and Dantzer (1988)

1981). The fetal and maternal blood vessels are situated just below the basement membranes. The structure is analogous to a miniareola with material produced by degeneration of, or secretion by, the maternal epithelium rather than a gland. In contrast, at the sides and tips of the fetal villi (sides and base of the maternal villi) both epithelia are thin and deeply indented by blood vessels from each side (Figs. 5.3b and 5.6b) (Steven 1983).

This reduces the interhaemal diffusion distance to as little as 2 μm, although the same number of membrane barriers is present as at the base of the fetal villi where the interhaemal distance ranges from 20 to 100 μm (Figs. 5.3 and 5.6).

Indentation of the blood vessels into the epithelia increases as pregnancy proceeds (Figs. 5.3 and 5.6). There is no evidence for any active process of 'tunnelling into' or 'invasion' of the epithelia by the blood vessels. The endothelium is always separated from the epithelium by a basement membrane. It seems likely that indentation results from differential growth, the endothelium growing less quickly than the epithelium.

Breeds differ significantly in the vascular frequency per unit volume of placenta (Wilson et al. 1998) but the basic structure is very similar (see Chap. 2.3.1).

During pregnancy the fetal and maternal villi can be separated manually but there is always some maternal epithelium pulled away with the fetal side and *vice versa*.

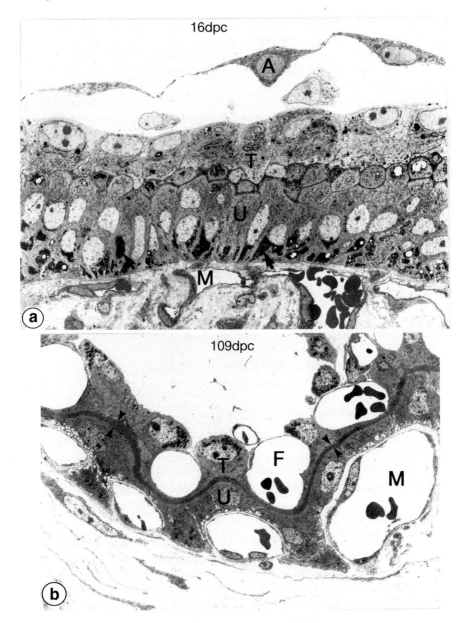

Fig. 5.6 Epitheliochorial pig placenta. (**a,b**) Electron micrographs at the same magnification illustrating the development of the extreme indentation of the blood vessels into the trophectoderm (T) and uterine epithelial layers (U). The depth of the microvillar interdigitation can also clearly be seen in (**b**) (*arrowheads*). *F* Fetal vessels; *M* maternal vessels; arrows, glycogen in uterine epithelium. (**a**) 16 dpc; ×400 (**b**) 109 dpc; ×400

At parturition the placenta separates cleanly at the microvillar fetomaternal junction and the incidence of retained placentas (see synepitheliochorial placentation, Chap. 6) is negligible. There is a massive release of relaxin from the corpora lutea as a result of the rise in prostaglandins which triggers parturition (First and Bosc 1979). Processes initiated by these hormonal changes (Bazer and First 1983) are no doubt involved in releasing the microvillar interdigitation between fetal and maternal epithelia to allow an easy placental separation.

5.3   Horse, Equus caballus

Oestrous cycle: 22 days

- Ovulation: spontaneous
- Uterus: bicornuate

Gestation: 336 (± 20) days
Litter: one, rarely two (Ginther 1989)

- Implantation: superficial
- Amniogenesis: folding
- Yolk sac: initially large and functional to 38 dpc
- Chorioallantois: forms definitive placenta
- Shape: diffuse, microcotyledonary, villous
- No decidual reaction
- Interhaemal membrane: epitheliochorial

Accessory placental structures: trophectodermal girdle producing endometrial cups; areolae
The major characteristics • are found in the Perissodactyla families.
Equidae, horse, donkey, zebra
Tapiridae
Rhinocerotidae

The unusual length of standing heat in horses (4–5 days; compare with sheep, 1 day; pig, 1 $^{1}/_{2}$ days) and the consequent uncertainty about the time of ovulation makes days post coitum a far less reliable indicator of gestational age than in the other domestic species. Use of the more accurate days post conception is only realistic if the preovulatory surge in luteinizing hormone is monitored in the maternal blood for a particular pregnancy.

Like the pig, the definitive horse placenta is a simple apposition of unchanged fetal and maternal epithelia with interdigitated microvilli for haemotrophic exchange with a large number of areolae for histiotrophic transfer (Steven 1982). Unlike the pig, the horse conceptus does not elongate and implantation, defined as full fetomaternal microvillar interdigitation, occurs much later at about 40 dpc (pig 15 dpc) (Ginther 1979). Prior to this, uniquely, the equid trophectoderm produces (between 30 and 40 dpc) a girdle around the conceptus of binucleate cells which migrate into the maternal endometrial stroma to form the gonadotropin-producing endometrial cups (Allen et al. 1973; Allen 1982). This pregnant mare serum gonadotropin (PMSG) is now known as equine chorionic gonadotropin (eCG).

5.3.1   Horse Fetal Membranes

At 20 dpc the 30–40 mm spherical conceptus has a fully formed amnion and the allantois is just visible as a small outgrowth of the hind gut (Fig. 5.7a).

About one-third of the conceptus is bilaminar and most of the rest forms a rapidly vascularizing trilaminar absorptive choriovitelline layer. By 25 dpc the allantois has formed a vascularized chorioallantoic cap over the embryo and its further expansion reduces the relative extent of the still absorptive choriovitelline layers. From 25 dpc a band of trophectodermal cells proliferates at the boundary between the chorioallantois and the trilaminar yolk sac to form the chorionic girdle (Figs. 5.7–5.11).

By 40 dpc these cells have migrated into the maternal endometrium to form a ring of endometrial cups (Figs. 5.7a, b and 5.9) and the yolk sac is considerably reduced. Subsequent growth of the conceptus is based entirely on the chorioallantois. Unlike the pig there is no amniochorion formation since the embryo is isolated from the chorion by growth of the allantois (Fig. 1.1). The umbilical cord is unusually long, having amniotic and allantoic portions linking the fetus to the membranes (Whitwell and Jeffcott 1975; Ginther 1979).

5.3.2   Development of the Horse Placenta

5.3.2.1   Initial Conceptus and Uterine Interaction

The conceptus moves freely between the two horns of the uterus between 9 and 16 dpc as it expands from 12 mm to a 28-mm sphere. The mobility has been demonstrated by ultrasonography and is presumably caused by contractions of the uterine musculature. Maternal recognition of pregnancy is suggested to require conceptus estrogen secretion between 14 and 16 dpc which blocks the endometrial production of prostaglandins, the normal cyclic luteolytic signal. The details of this vital process are still unresolved (Allen and Stewart 2001). At 17 dpc the expanding conceptus reaches a critical diameter and fixes in the caudal portion of the horn which offers most luminal resistance (Ginther 1983, 1984a,b; twin conceptuses, Ginther 1989). It can subsequently be detected by rectal palpation as a uterine swelling. Recent evidence suggests that there is little or no subsequent migration because of closure of the uterus around the conceptus with development of a considerable muscular tone in the myometrium. Cellular interaction is initially

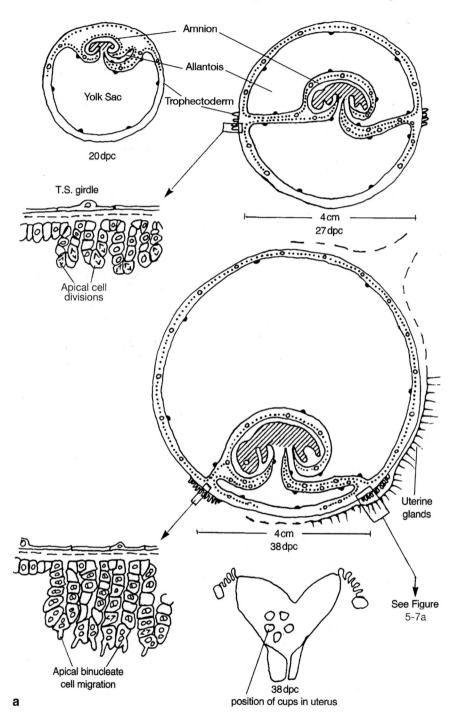

Fig. 5.7 (**a,b**) Diagrams of the cellular changes in the development of the endometrial cups in Equidae

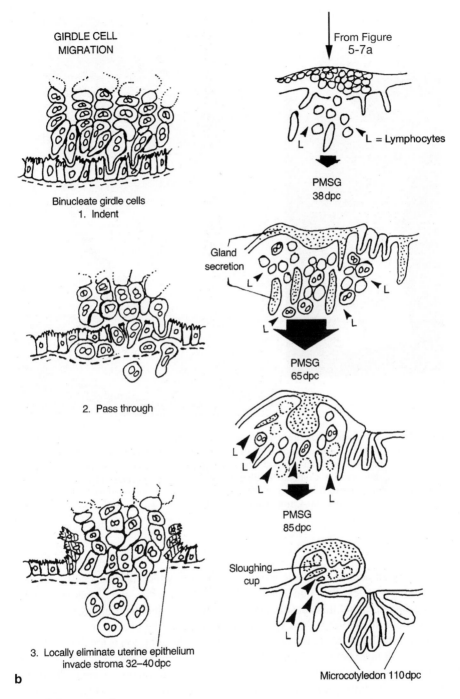

GIRDLE CELL
MIGRATION

Binucleate girdle cells
1. Indent

2. Pass through

3. Locally eliminate uterine epithelium
invade stroma 32–40 dpc

b

From Figure
5-7a

L = Lymphocytes

PMSG
38 dpc

Gland
secretion

PMSG
65 dpc

PMSG
85 dpc

Sloughing
cup

Microcotyledon 110 dpc

Fig. 5.7   b (continued)

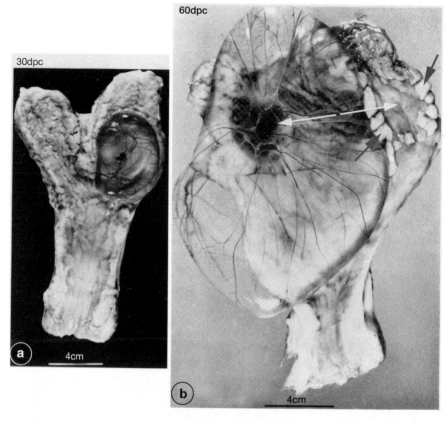

Fig. 5.8 Horse conceptus, chorionic girdle and endometrial cups. The chorionic girdle forms between 28 and 36 days post conception (dpc) around the conceptus in a band immediately above the junction of the allantois and yolk sac. The girdle remains the same size while the allantochorionic sac expands above it. Compare the diameter of the 30 dpc conceptus in (**a**) with the ring of cups (*black arrows*) in (**b**) from which the 60 dpc conceptus has been rolled back (*white arrows*). (a) 30 dpc; ×0.4, (b) 60 dpc; ×0.5, from Ginther (1979)

limited by an acellular capsule around the embryo which is produced by the tro-phectoderm after the conceptus reaches the uterus and loses the zona pellucida(9 dpc) and this capsule persists until 21–22 dpc then ruptures and disap-pears (Betteridge 1989). Subsequently, it is probably the bilaminar portion of the yolk sac which adheres to the uterine epithelium. The adhesion is tenuous and if the uterus is cut open between 22 and 40–50 dpc the conceptus is expelled by the inter-nal pressure in the uterus (Ginther 1979). There is a copious secretion from the uterine glands during this time (Allen and Stewart 2001), most of which is absorbed initially by the trilaminar yolk sac (from 16 to 28 dpc) and later by the chorioallantois.

5.3.2.2 Chorionic Girdle and Endometrial Cup Formation

From 27 to 35 dpc the avascular trophectoderm at the boundary between the chorioallantois and the trilaminar yolk sac proliferates to form a 'chorionic girdle' many cells thick (Figs. 5.7, 5.9, 5.10 and 5.11) (Allen 1982; Enders and Liu 1991a,b).

This proliferation is triggered by local production of growth factors (Stewart et al. 1995). Initial cell interaction (30–33 dpc) between mother and fetus is indicated by localized changes in lectin labelling of the uterine epithelium apposed to the developing girdle (Whyte and Allen 1985). By 35 dpc the apical cells of the girdle have become binucleate, and start to synthesise eCG (Fig. 5.12; Wooding et al. 2001) They also express MHC-1 antigens (Donaldson et al. 1990, 1992) which provoke the maternal immune system to produce an antipaternal antibody in pregnancy characteristic of, and unique to Equidae (see Chap. 9.1.5).

The girdle cells then invade the uterine epithelium against which they are pressed by expansion of the conceptus against the firm uterine muscular tone (Figs. 5.7, 5.11 and 5.12). The first girdle cells protrude pseudopodia into and are said to push through the body of the uterine epithelial cells, well away from their tight junction seal. They are then briefly delayed by the basement membrane of the uterine epithelium (along which they may migrate into uterine glands) and finally invade the endometrial stroma between the uterine glands (Allen et al. 1973; Enders and Liu 1991b) (Figs. 5.7, 5.11 and 5.12). The uterine epithelium is subsequently phagocy-

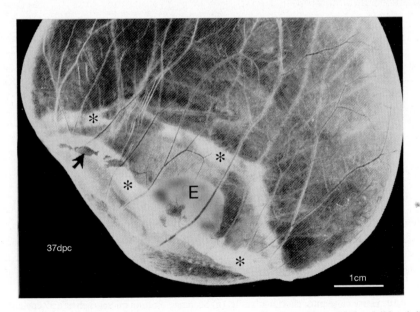

Fig. 5.9 Horse conceptus, chorionic girdle. A 37 dpc conceptus with a clearly visible girdle (*asterisks*) from which some girdle cells have been lost on removal from the uterus (*arrow*). E embryo. 37 dpc; ×2, from Ginther (1979)

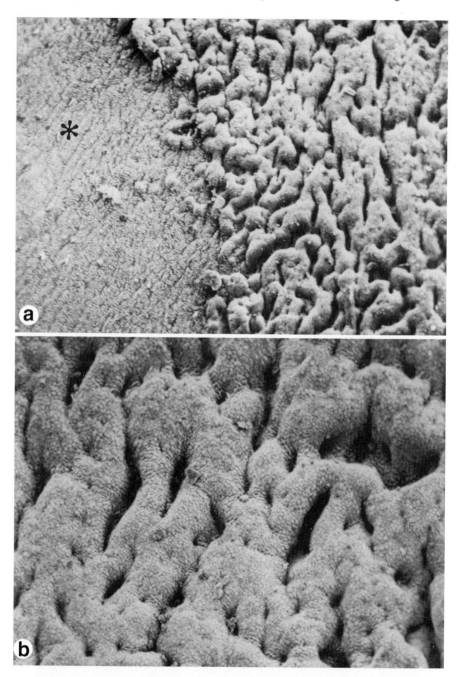

Fig. 5.10 Horse conceptus, chorionic girdle. (**a**) Scanning electron micrograph of a 32 dpc girdle. Note the very sudden transition from smooth trophectoderm (asterisk) to the convolutions of the girdle. (**b**) Detail from (**a**) showing how the girdle cell proliferation forms microfolds and crevices. (**a**) 32 dpc; ×60 (**b**) 32 dpc; ×170. Courtesy of Dr D.F. Antczak

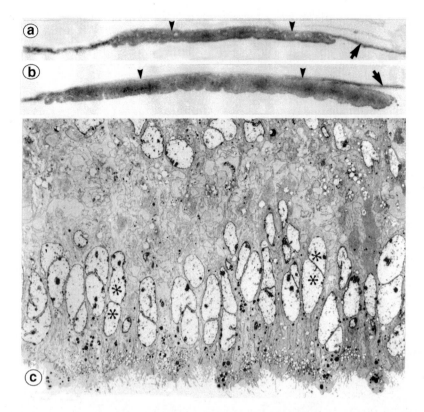

Fig. 5.11 Horse conceptus, chorionic girdle. Cross sections through (**a**) 32 and (**b**) 36 dpc girdles showing the considerable proliferation of the trophectodermal cells but the basement membrane (*arrowheads*) remains flat. Compare the thickness of the unmodified trophectoderm outside the girdle (*arrows*): (**c**) Electron micrograph of the apex of fetal chorionic girdle at 37 dp. Most of the cells are binucleate (e.g. asterisks). (**a**) 30 dpc; ×25 (**b**) 36 dpc; ×25 (**c**) 37 dpc; ×1,100 (**a**) and (**b**) courtesy of Prof W.R. Allen

tosed and removed by continuing binucleate cell migration (Fig. 5.12), which allows the whole chorionic girdle free access to the endometrial stroma (Fig. 5.7).

Once in the uterine stroma (36–40 dpc) the binucleate cells have not been seen to divide but develop into aggregates of very large cells forming a ring of 'endometrial cups' (Figs. 5.7 and 5.13). By 44 dpc they no longer express MHC (Donaldson et al. 1992).

The uterine epithelium grows in from the sides to cover the cup area. The cup cells secrete eCG which appears in the maternal (but never the fetal) circulation via the capillaries and lymphatic vessels in and around the cup. The mature cup cell has a cytoplasm full of mitochondria, rough endoplasmic reticulum cisternae and a large Golgi apparatus (Hamilton et al. 1973) (Figs. 5.14 and 5.15; Wooding et al. 2001).

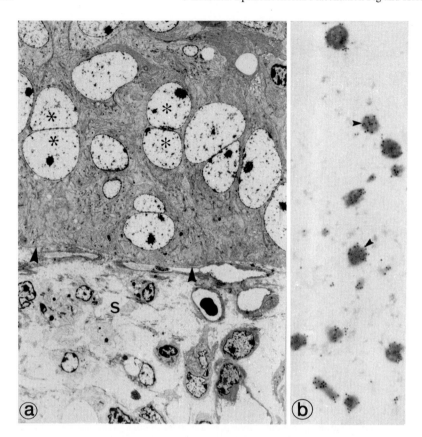

Fig. 5.12 Horse chorionic girdle. (**a**) The fetal chorionic girdle has invaded and eliminated the uterine epithelium but further migration into the maternal endometrial stroma (S) is delayed by the residual basement membrane (arrowheads) of the epithelium. Most girdle cells are binucleate (*asterisks*) 37 dpc, ×1,700. (**b**) Binucleate girdle cells in the position shown in (**a**) often contain groups of small granules. This section has been processed with immunocytochemical methods to demonstrate that the granules contain equine chorionic gonadotrophin indicated by the gold colloid marker particles (*arrowheads*). 37 dpc; ×30,000

Ultrastructural immunogold studies have shown that eCG is localized in the Golgi cisternae and in small dense granules similar to those found in the migrating girdle cell and present both in the Golgi region and at the peripheral plasmalemma (Wooding et al. 2001; Figs. 5.12 and 5.15).

Release of eCG would therefore seem to be by the usual exocytotic mechanism as found for other protein hormones. The amount of eCG produced depends on the size of the cup, which is determined by the number of fetal binucleate cell migrants. This last seems to depend upon the genotype of the conceptus. For example horse ♀x horse and horse ♀x donkey crosses produce large cups, while donkey ♀x donkey or donkey ♀x horse produce small cups (Allen 1982).

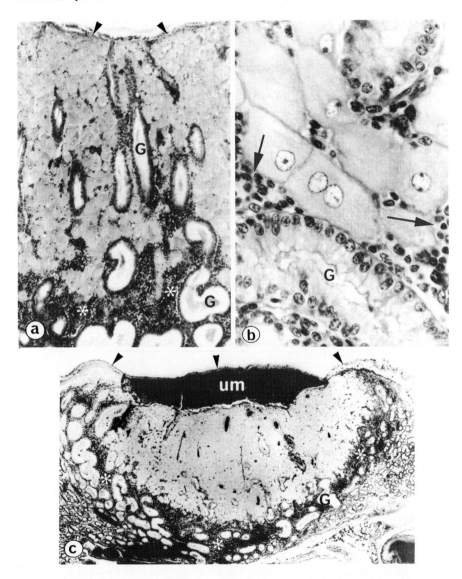

Fig. 5.13 Horse endometrial cup structure. (**a**) Section through a newly formed cup at 40 days post conception with cup cells closely packed around the uterine glands (G). The uterine epithelium (*arrowheads*) reforms rapidly after migration of the girdle cells is completed at 38–40 dpc There is an accumulation of lymphocytes around the base of the cup (*asterisks*). (**b**) Note the large size of the cup cells compared with the gland cells (G) and the small size of the invading lymphocytes (*arrows*). (**c**) Mature cup (60 dpc) with accumulated secretion (um) localised between the allantochorion (*arrowheads*) and the cup. Considerable numbers of lymphocytes have accumulated in a continuous zone (*asterisks*) around the cup. (a) 40 dpc; ×70. (b) 40 dpc; ×2,000. (c) 60 dpc; ×30. (**a,b,c**) from Ginther (1979)

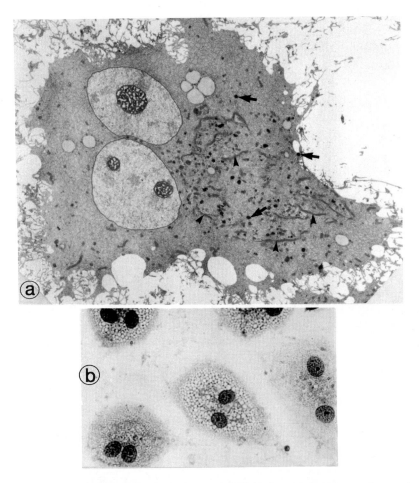

Fig. 5.14 Horse endometrial cup cells (**a**) in vivo and (**b**) in vitro. (**a**) An electron micrograph of a characteristic binucleate cell from a 50 dpc cup fixed by perfusion in situ. The material was processed without osmium and stained with phosphotungstic acid (Wooding 1980). This emphasizes the two nuclei and extensive golgi cisternae (arrowheads) and small granules (*arrows*) in a typical endometrial cup cell. 50 dpc; ×3,200. (**b**) Light micrograph of cells isolated from a 35 dpc girdle and grown in tissue culture for 50 days, they are very large, characteristically binucleate and produce considerable amounts of eCG. The cytoplasm accumulates numerous lipid droplets giving a stippled appearance ×1300. Courtesy Dr R. Moor

Cells scraped from the girdle between 36 and 37 dpc will grow in tissue culture to form large, usually binucleate, cells which produce considerable amounts of eCG for up to 200 days (Allen and Moor 1972; Fig. 5.14b). On the analogy of the ruminant binucleate cell it seems unlikely that binucleate girdle cells can divide, so any increase in cell numbers in culture would have to be based on division of uninucleate cells scraped from the girdle. In vivo, maternal lymphocytes and macrophages start to

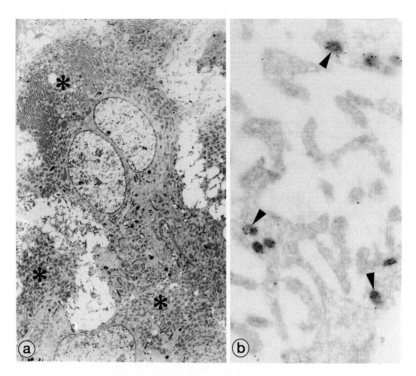

Fig. 5.15 Horse endometrial cup cells. (**a**) Electron micrograph of cells processed without osmium and stained with phosphotungstic acid followed by lead citrate which enhances the massive arrays (*asterisks*) of rough endoplasmic reticulum cisternae filling most of the cytoplasm of the cup cells. 50 dpc; ×3,200; (**b**) The small granules produced by the golgi apparatus of the cup cells (see Fig. 33a) contain eCG as indicated by the immunogold colloid marker particles (arrowheads). They are usually located peripherally but never seem to accumulate in any number. 50 dpc, ×44,000

accumulate at the edge of a cup as soon as it establishes itself (Fig. 5.13a, b) but coculture studies show that the invasive MHC positive girdle cells produce factors which inhibit the lymphocyte activation and proliferation (Flaminio and Antczak 2005).

Maximum secretion of eCG from the cup occurs between 50 and 80 dpc, stimulating the ovaries to produce the secondary corpora lutea, necessary to maintain progesterone levels. As the eCG (and presumably the lymphocyte inhibitory factors) decline the lymphocytes start to invade and kill the cup cells. The uterine glands in the cup also produce a considerable amount of secretion at this time (Fig. 5.13c), presumably stimulated by the cup cells. This glandular secretion contains very large amounts of eCG secreted by the cup cells only, but surprisingly this does not reach the fetal circulation.

The variation of cup size with genotype, the lymphocyte accumulation and the ability of the cup cells to secrete eCG in tissue culture (Fig. 5.14) for up to 200 days suggests that the cytolysis and regression of the cup cells is caused by an immunological reaction rather than a programmed cell death (Allen 1982). Subsequently, the whole mass of degenerating cup tissue is everted from the endometrium, encapsulated by the overlying allantochorion, and remains as a necrotic residue in the allantoic cavity (Whitwell and Jeffcott 1975). The uterine epithelium grows over the scar from the edges (Ginther 1979).

Interspecies embryo transfer experiments (e.g. zebra conceptus into horse, Summers et al. 1987) have shown that as long as there is some cup development, however small, normal gestation results and the usual antipaternal antibody is elicited in the maternal blood. However, a donkey conceptus transferred to a horse produces neither a cup reaction nor antipaternal antibody and abortion usually follows between 80 and 100 dpc in the presence of a massive lymphocyte infiltration throughout the entire endometrium, ostensibly similar to the infiltration round a normal size (horse in horse) cup. In some cases this abortion can be blocked by infusing pregnant mare's serum (but not eCG alone) or paternal lymphocytes into the mare between 20 and 80 dpc (Allen et al. 1987).

Once a horse has carried one donkey pregnancy successfully (and this is as yet entirely unpredictable) with or without treatment, then most subsequent donkey conceptuses are carried to term without further therapy – the system displays immunological memory.

The girdle and the early cup cells are the only allantochorionic cells to express MHC antigens (Crump et al. 1987; Donaldson et al. 1990, 1992) and this stimulates the maternal immune system to produce the antipaternal antibodies. How this antibody and the establishment of the cup ensures that the uterine environment is immunologically tolerised to accept placental growth is as yet not understood.

5.3.2.3   Chorioallantoic Placenta Formation

At 35–38 dpc the allantois has vascularized more than 90% of the total trophectoderm, and this is the earliest stage at which microvillar interdigitation between uterine epithelium and trophectoderm has been reported (Steven 1982). The fetal trophectoderm and apposed maternal uterine epithelial maternal cell sheets then produce villi by a process of complementary growth controlled by steroid hormones and the growth factors EGF and IGF II (Allen and Stewart 2001) Fetal villi are formed over the entire surface of the conceptus except over the openings of the numerous endometrial glands. Here the trophectoderm develops a dome of highly absorptive columnar cells over the gland mouth and its secretion, forming areolae exactly as in the pig, if a little smaller in size (Fig. 5.16).

By 60 dpc the fetal villi are longer and have started to group together between the areolae. Further extension of the villi with secondary branching plus the development of a bounding maternal connective tissue capsule produces the characteristic microcotyledon between the glands, each 1–2 mm in diameter at term (Samuel et al. 1974; MacDonald et al. 2000) (Figs. 5.16 and 5.17).

After girdle cell migration has finished (about 40 dpc) there is no published evidence for any specialized cells nor any syncytium in either the trophectoderm or the uterine epithelium. Like the pig, the trophectoderm is uniformly uninucleate and its microvilli are closely interdigitated with the uniformly uninucleate uterine epithelium. This fetomaternal junctional area increases (as does fetal weight)

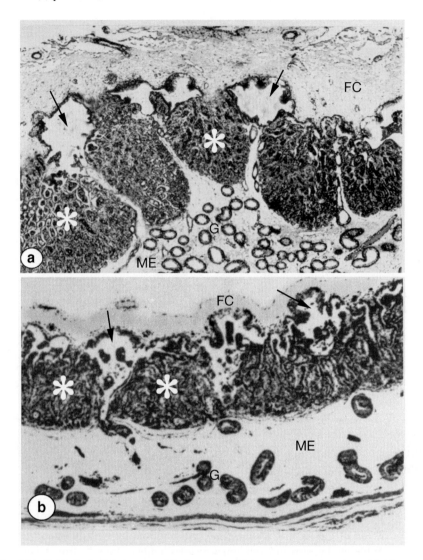

Fig. 5.16 Epitheliochorial placentas. The horse (**a**) with its microcotyledons (*asterisks*) between areolae (*arrows*) between has a very similar placental architecture to the American Mole, Scalopus (**b**). In (**b**) the areolae are again indicated by arrows, the microcotyledons by asterisks. In the endometrium (ME) of both there are many glands (G) which empty into the areolae. *FC* fetal connective tissue. (**a**) Midpregnant, from Bjorkman (1970) (**b**) Late gestation; ×37, from Prasad et al. (1979)

throughout pregnancy by mutual growth of the fetal and maternal epithelia (Baur 1977), and both are increasingly indented by their initially subepithelial capillaries (Samuel et al. 1976). Indentation is most marked in the fetal trophectoderm but is not so extreme as observed for the pig.

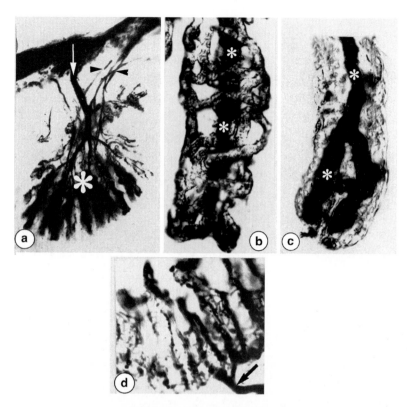

Fig. 5.17 Horse epitheliochorial placenta. (**a–d**) Vascular injections illustrate the tuft of fetal vessels forming a microcotyledon (*large asterisk*), and partly filled villus tips show that there is a central arteriolar supply (*small asterisks*). (**d**) Shows the capillary network forming the maternal cotyledonary blood supply with a large draining venule (arrow). (**a**) 210 dpc; ×40. (**b**) 195 dpc; ×400. (**c**) 195 dpc; ×400. (**d**)150 dpc; ×70. All from Tsutsumi (1962)

Restriction of the area of fetomaternal proliferation, by twin pregnancy, prior damage to the uterine structure (endometriosis) or embryo transfer from a larger to a smaller species reduce growth of the fetus and produce small foals with poor growth prospects (Allen and Stewart 2001). The fetomaternal area at term per gram of neonate peaks at the second gestation but other factors ensure that later normal pregnancies produce equivalent foal weights (Wilsher and Allen 2003).

There are no extravasations of maternal blood at the microvillar junction (as found in ruminants and carnivores). Iron transport is based on a similar carrier system to the pig protein uteroferrin in the glandular secretions (Fig. 2.8). The 9 kDCBP protein marker of calcium transport is also localised only to the gland and areolar epithelium as in the ruminants (Fig. 2.8). Glucose transporters 1 and 3 are used in sequence for maternofetal transfer and are found only in the microcotyledonary epithelia (for details see Fig. 2.6, Chap. 2.2.1 and Wooding et al. 2000).

The trophectodermal cell fine structure changes during pregnancy. At 200 dpc there is a considerable amount of smooth endoplasmic reticulum and this persists to term. Rough endoplasmic reticulum is found from 180 dpc to term, when the number of dense bodies (probably lysosomal) has also increased considerably (Samuel et al. 1976).

The endocrine profile of the horse during pregnancy is complex (Allen 1975, 1982) (see Fig. 5.18).

A peak level of 10 ng ml^{-1} serum progesterone is initially produced at 8 dpc by the primary corpus luteum, but this secretion gradually decreases. The eCG produced by the endometrial cup cells starting at 38 dpc then stimulates ovulation and the production of the secondary corpora lutea, which increase the progesterone levels considerably (up to 14 ng ml^{-1}). As the cups and secondary corpora lutea regress (100–150 dpc) the placenta takes over production of a much reduced level of progesterone (3–4 ng ml^{-1}), presumably related to the considerable amount of smooth endoplasmic reticulum in the trophectoderm of the microcotyledons. Oestrogens

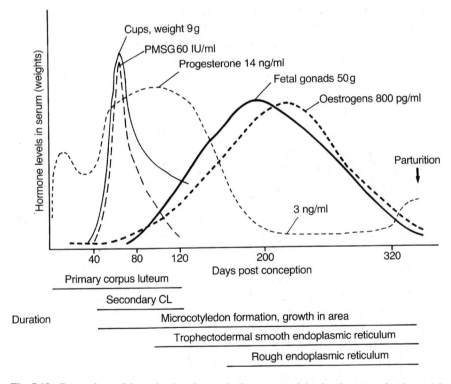

Fig. 5.18 Comparison of the endocrine changes in the serum and the development of endometrial cups and fetal gonads during pregnancy in the horse. Note that the lifespan of the corpora lutea correlates with high progesterone levels; the levels necessary to support pregnancy in second half of gestation are much lower and are supplied by placental synthesis (Data from Ginther 1979; Samuel et al. 1974, 1976.)

are also produced, mainly oestrone and the unique equid β-ring unsaturated oestrogens equilin and equilenin (Allen and Stewart 2001). Precursors for the oestrogens are produced by the hypertrophied interstitial cells in the fetal gonads (Fig. 5.18).

The pattern of blood vessels is very different in the horse fetal villus when compared with that in the pig. There is a central arteriole which passes directly to the villus tip before dividing into a peripheral capillary network draining back down the villus to its base (Figs. 2.4 and 5.17). The maternal arteries run first to the tips of the maternal villi or septa which are inserted between the bases of the fetal villi. The blood then drains down through a network of peripheral capillaries to the collecting veins at the base of the maternal villus (septum) (Tsutsumi 1962; Steven 1982; Abd-Elnaeim et al. 2006) (Figs. 2.4 and 5.17).

This arrangement should allow classical countercurrent flow between fetal and maternal capillaries assuming they run alongside one another for a sufficient distance. At parturition the interdigitated microvilli of the fetal and maternal cells separate by a process probably requiring changes in adhesion at the molecular level. The short fetal villi and maternal septa (villi) then disengage, probably as a result of decrease in blood pressure and/or volume. There is rarely any problem with separation and expulsion of the membranes at partus (Steven et al. 1979). Details of the changes in progestagens (Ousey et al. 2003) and increases in prostaglandin and oxytocin at this time are given in Haluska and Currie (1988).

5.3.3   Other Epitheliochorial Types

Scalopus aquaticus, the American mole, is an insectivore with a bicornuate uterus, producing an average litter of four after 35 days of gestation. Implantation is superficial with no apparent decidual formation, the amnion forms by folding and there is a functional yolk sac until term. The definitive chorioallantoic placenta covers the antimesometrial half of the uterus, is villous and epitheliochorial with numerous areolae (Prasad et al. 1979). *Scalopus* is the only insectivore known with an epitheliochorial placenta, most are haemochorial with a few endotheliochorial examples (Mossman 1987). However, they all have a similar yolk sac development, with a mesometrial bilaminar yolk sac which persists to term and a functional trilaminar choriovitelline placenta on the lateral and antimesometrial, glandular uterine walls. The vitelline capillaries are displaced by development of the allantoic circulation to produce the simple villi. In *Scalopus* the chorioallantoic villi are collected into groups similar to the microcotyledons of the equid type but they lack the connective tissue capsule (Fig. 5.16b). The uterine epithelium is extremely thin (<1 μm) in places and required EM studies were required to prove that it persists (Prasad et al. 1979). Areolae develop over the mouths of the uterine glands separating the villus groups. At parturition, separation occurs at the fetomaternal microvillar junction.

Scalopus with its epitheliochorial placenta has obvious skeletal and developmental affinities with insectivores such as *Talpa*, the European mole (discoid endotheliochorial placenta, Malassine and Leiser 1984), and with the shrews and the European hedgehog (haemochorial placentas). This emphasises the point that animals of very similar phyletic origins, size, habitat and reproductive behaviour can have widely differing definitive placental structures. On current evolutionary considerations (see Chap. 9), the epitheliochorial *Scalopus* shows an unusual, derived state for an Insectivore. Carter (2005) has raised some doubts about the interpretation of the results in the Prasad et al. (1979) paper and further study with unequivocal immuno-cytochemical identification of the layers is needed to settle the point.

Another very different order, Pholidota (scaly anteater or pangolin) has a similar epitheliochorial placenta to Equidae with grouped short villi between frequent areolae (Mossman 1987).

Cetacea, the whales, porpoises and dolphins, have diffuse epitheliochorial placentas like that of the pig with short villi or folds and numerous areolae opposite the mouths of uterine glands. None of these have yet been investigated in detail with the EM (Mossman 1987).

The strepsirhine primates, lemurs, lorises and bushbabies (*Galago*), all have diffuse epitheliochorial placentation based on narrow slender or short fat individual villi with large bulbous areolae between (Fig. 5.2a, b; compare with Fig. 5.2c, pig). At implantation *Galago senegalensis* is reported to produce a specialized area of giant trophoblast cells, which remove the uterine epithelium and act as an anchorage point for the conceptus (Butler 1967). This is reminiscent of the trophectodermal protrusions into the uterine epithelium produced by some of the marsupials (Enders and Enders 1969). In both cases these giant cells are transient and do not contribute to the definitive placental development. Whether the implantation is produced by delamination intrusion or fusion is not yet clear. In the definitive placenta of the lemur *Microcebus* a discoid structure with no uterine epithelium (Grosser's syndesmochorial) has been claimed (Reng 1977), but insufficient detail can be seen to verify this; and a giant, possibly double, cell layer has been suggested for *Galago demidovii* (Mossman 1987). Both these structures would be very atypical and it needs EM resolution to unravel the details. More recent ultrastructural studies of the definitive placenta of *Galago* emphasized the simplicity of the basic epitheliochorial villus pattern (King 1984; Njogu et al. 2006).

The Camelidae have large branched folds and villi forming diffuse epitheliochorial placentas (Morton 1961). Recent work has shown that among the uninucleate trophectodermal cells there are occasional characteristic giant cells, several times the size of their neighbours. These giants have large polyploid nuclei of bizarre shape (Klisch et al. 2005) and are steroid synthesising cells present from about 35 dpc to term with cytoplasmic structure quite different from that of their neighbours. Between implantation and 35 dpc steroids are produced by all cells of the uniformly unincleate trophectoderm, but this function is confined to the giant cells once they differentiate (Skidmore et al. 1996; Wooding et al. 2003).

Chapter 6
Synepitheliochorial Placentation: Ruminants (Ewe and Cow)

6.1   Introduction

In Grosser's syndesmochorial category (exclusively ruminants) the uterine epithelium was said to be lost so the trophectoderm was apposed to the maternal connective tissue directly. It has now been shown by a variety of techniques that throughout pregnancy characteristic ruminant fetal trophectodermal binucleate cells (BNCs) modify the uterine epithelium by apical fusion to form fetomaternal hybrid syncytial plaques at the junction of the fetal and maternal tissue (Wooding 1982a,b, 1984) (Figs. 6.1 and 6.2) . This fusion delivers fetal hormones and effectors to the maternal circulation.

This is so different from the epitheliochorial category that it amply justifies a separate grouping.

'Syndesmochorial' is of obscure derivation and synepitheliochorial is now accepted as a much more descriptive and easily understood category (Sect. 1.4.4). All ruminant cotyledons so far investigated have 15–20% BNCs in the trophectoderm which produce the fetomaternal hybrid syncytium with uterine epithelial cells but the degree to which the maternal epithelium is permanently altered does vary between Families. The 'syn-' indicates the contribution of the BNC-derived syncytium and the retention of 'epitheliochorial' emphasizes that large areas of simple cellular fetomaternal apposition persist in the definitive placenta (Wooding 1992).

The synepitheliochorial placenta has three identifying characteristics: the BNC in the fetal trophectoderm (Chap. 1.6.5) which form a structurally separate population among the uninucleate trophectodermal cells from which they are derived; the fetomaternal syncytium formed by BNC fetal migration and fusion with some or all of the maternal uterine epithelial cells; and the placentomal [or cotyledonary] chorioallantoic placental organization, with the restriction of fetal and maternal villus elaboration to discrete regions (placentomes) based on pre-existing non-glandular caruncular areas on the uterine epithelium (Amoroso 1952) which are well developed in the fetal uterus by mid-pregnancy (lamb, Wiley et al. 1987). A 'placentome' is formed by enmeshed fetal villi and maternal crypts; the tuft of fetal villi is correctly referred to as a 'cotyledon' and the corresponding maternal structure is a 'caruncle' (Fig. 6.3).

1. Immobilization by papillae, and elongation

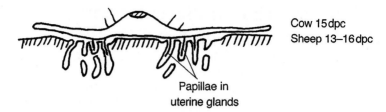

Cow 15 dpc
Sheep 13–16 dpc

Papillae in
uterine glands

2. Cellular apposition and subsequent interdigitation of microvilli,
   binucleate cell initiation (1-4)

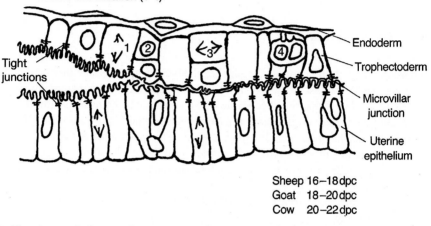

Endoderm

Trophectoderm

Microvillar
junction

Uterine
epithelium

Tight
junctions

Sheep 16–18 dpc
Goat   18–20 dpc
Cow   20–22 dpc

3. Binucleate cell differentiation and migration

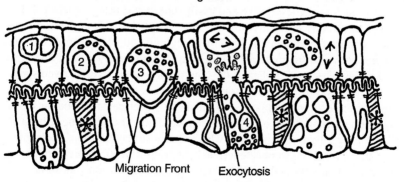

Migration Front   Exocytosis

The uterine epithelium is greatly modified by migration and
fusion (1-4) of the apices of the fetal binucleate cells with
some uterine epithelial cells and the death of others (∗).

Fig. 6.1 The cellular changes at implantation in the ruminants

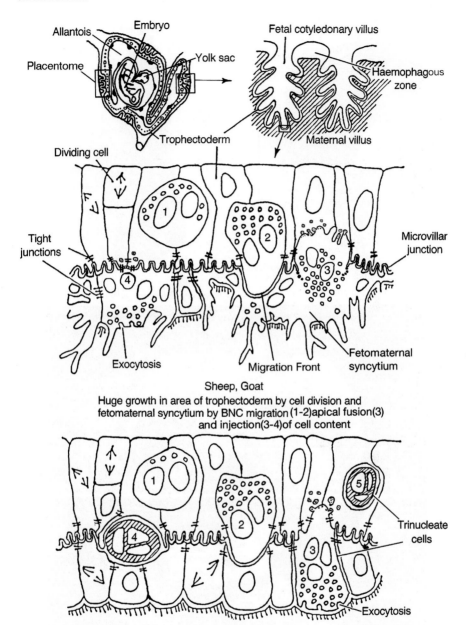

Fig. 6.2 The binucleate cell contribution to the definitive ruminant placenta

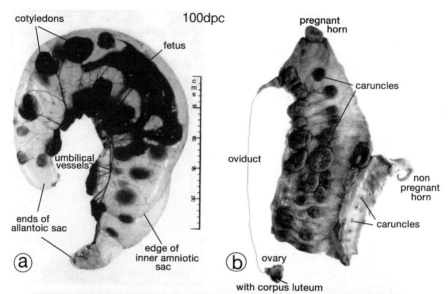

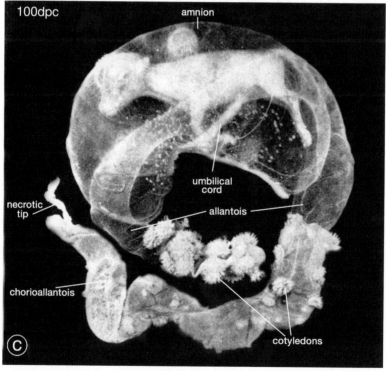

However, cotyledon is also frequently used for the whole structure, so that a 'maternal cotyledon' would correspond to a caruncle. Ruminant placentas may be oligocotyledonary with 3–8 large placentomes, as in deer, or polycotyledonary with numerous (20–150) but smaller placentomes as in cows, goats and sheep (Fig. 2.4). Since there are a finite number of caruncles per uterus, although a singleton conceptus may not utilize all the sites it may have more placentomes than twins, which can only have about half the total number each. However, the twins will have larger individual placentomes and their combined mass will be greater than that of the singleton. There is a close correlation between conceptus placentomal mass and birthweight of the fetus (Bell 1984, 1991; Bell et al. 1999), and recent studies indicate a pivotal role for IGF-1 and -2 in placental growth (Bassett 1991; Geisert et al. 1991; Owens 1991). Placentomes also differ in gross structure, being concave to the fetal side in sheep and goats, but convex in the cow and flat in antelopes (Amoroso 1952). If the sheep (normally concave) has some of its caruncles removed surgically from the non-pregnant uterus then the usually adequate resulting placenta has fewer, larger, but now convex, placentomes (Harding et al. 1985). During the first two thirds of a normal pregnancy in the ewe all placentomes are concave but later in pregnancy a variable, usually small, proportion become convex with intermediate types. Recent work indicates that maternal stress (carunclectomy, nutrient imbalance, cortisol injection, high temperature) can influence the change (Ward et al. 2006), and the convex type is said to be more efficient since it has greater maternal and fetal vascular development per gram (Reynolds et al. 2005).

A logical sequence of mature villus evolution can be assembled starting with a completely diffuse epitheliochorial placenta with short stumpy individual villi as in the pig. The microcotyledonary epitheliochorial placenta of the horse is slightly more complex and has small tufts of longer more slender villi but the pregnancy depends on a transient population of hormone producing trophectodermal BNCs invading the maternal endometrial stroma to establish a progesterone secreting immunotolerant environment.

The most primitive ruminant, Tragulus has much longer, uniformly distributed villi and the BNC are less invasive than in the horse girdle but are present throughout pregnancy. Like the equid BNC they deliver fetal hormones and effectors to the maternal circulation, but this is accomplished by fetomaternal hybrid tissue formation at the maternofetal interface. Aggregation of these villi into discrete placentomes produces the typical ruminant placenta. Hradecky et al. (1988) could find no consistent correlation between the number or size of placentomes with the length or degree of branching of the villi in a wide variety of ruminant placentas.

Fig. 6.3 Synepitheliochorial placentation. (**a**) Cow fetus within intact extraembryonic membrane sac(s). Unusually this conceptus was contained within one horn only of the uterus (**b**), which has been opened out flat. Note the increase in size of the caruncles in the pregnant horn which in vivo were enmeshed with the fetal cotyledons. (**a** and **b**) 90 dpc, ×0.25. (**c**) A similar cow conceptus to that in (**a**) but the chorion has been dissected away over the central region to reveal the amnion and the allantois. The villi on the fetal cotyledons on the non-dissected part are well demonstrated; they are clumped into apparently solid masses in (**a**). 100 dpc, ×0.66

The advantage of concentrating the villi into placentomes is not obvious but the change is accompanied by a considerable increase in the fetomaternal exchange area per gram of delivered fetus. Such "compact" placentas have 5–10 times as much fetomaternal exchange area per gram of fetus as "diffuse" placentas (Fig. 2.9; Baur 1977). It is, however, unclear why the ruminant should need so much more exchange area than the horse or if there are any advantages in a few large rather than many small placentomes.

The fetal trophectoderm has microvilli closely interdigitated with those of the uterine epithelium or derived syncytium, forming the characteristic microvillar junction. In the placentome local separation at this junction plus a local intermittent vascular leak forms haemophagous zones for transport of iron to the fetus. These are large and localised in sheep but small and scattered in the cow and in them stagnant maternal blood is apposed to tall columnar trophectoderm which actively phagocytoses the red blood cells (Burton 1982).

There are two major types of synepitheliochorial placenta: the sheep and goat type, with more than 90% of the original caruncular uterine epithelium modified to feto-maternal syncytial plaques; and the cow and deer type, with a similar proportion of (reformed) cellular uterine epithelium but with continuous formation of transient tri-nucleate minisyncytia (Fig. 6.2). In the interplacentomal areas, which have ~5% BNC, both types have a cellular uterine epithelium containing characteristic intraepi-thelial lymphocytes after the first third of pregnancy, (Lee et al. 1992) but with only occasional minisyncytia produced by BNC migration and fusion (King and Atkinson 1987). There are small areolae over the mouths of the glands. The percentage of feto-maternal exchange surface contributed by the interplacentomal area decreases rapidly until at term it is less than 2% of the total in the cow (Baur 1972).

Development of the sheep and cow placenta will be described as the basic rumi-nant types.

6.2   Sheep, *Ovis aries*; Cow, *Bos taurus*

Oestrous cycle: 16 days, sheep; 21 days, cow
Ovulation: spontaneous
Litter: 1–4
Gestation: about 150 days, sheep, goat; about 280 days, cow, depending on breed
Implantation: superficial; chorionic papillae into glands to immobilize conceptus in
    sheep and cattle
Amniogenesis: folding
Yolk sac: vascularized by 16 dpc (sheep), 20 dpc (cow) reduced to insignificance by
    30 dpc by growth of the allantoic sac
Chorioallantois: Forms definitive placenta, allantoic sac never surrounds the
    amnion as in the horse but the allantoic mesodermal vessels vascularize the
    entire chorion (Fig. 6.3)
Shape:Placentomal with complex villous structure. The placentome number varies
    from 3 to 150 depending on species

Decidual reaction: very limited, so far described only for the ewe (Johnson et al. 2003)

Interhaemal membrane: synepitheliochorial; migratory trophectodermal binucleate cells form fetomaternal hybrid tissue from implantation to term

Accessory placental structures: areolae in interplacentomal areas; haemophagous zones at bases of fetal cotyledonary villi, larger in sheep than cattle

Most ruminant families share the characteristics above.

Cervidae: deer, moose

Giraffidae

Antilocapridae: pronghorns

Bovidae: cattle, sheep, goats, antelopes, buffalo, bison, gnu

Tragulidae, chevrotains or mouse deer from East Africa and Malaysia. These have the smallest ruminant adults, about 2 kg, and from both molecular and skeletal evidence are considered the most primitive ruminant family. Surprisingly, they have recently been shown to have a non placental placenta with the villi diffusely distributed (Kimura et al. 2004). However the trophectoderm contains typical ruminant BNCs at the usual frequency which migrate and fuse with uterine epithelium to form the characteristic ruminant fetomaternal syncytium – two of the synepitheliochorial identifiers (Wooding et al. 2007a).

6.2.1 Ruminant Fetal Membranes

The ruminant morula enters the uterus about 4–5 dpc, transforms to a blastocyst, and loses its zona pellucida as it expands two or three times (Rowson and Moor 1966). One genus (roe deer, Aitken 1974) arrests the development of the blastocyst (diapause) at this stage prior to implantation. The physiological mechanism of arrest and reactivation is not understood but it does not seem to be hormonal (Lambert et al. 2001). Excellent summaries of the details of uterine secretions, the maternal hormonal levels and structural changes from oocyte to blastocyst in ruminants can be found in Bazer and First (1983) and Spencer et al. (2007). Subsequent hormonal changes are detailed in Heap et al. (1983).

In the sheep and cow the spherical conceptus comes to rest in a predictable region in the uterus (Lee et al. 1977), presumably as a result of myometrial contractions, and immobilizes itself by extending trophectodermal papillae, cellular protrusions, down the glands (Fig. 6.1) (Wooding and Staples 1981; Guillomot and Guay 1982; Wooding et al. 1982). The conceptus then expands and elongates to fill the uterine lumen (Wintenberger-Torres and Flechon 1974; Wales and Cuneo 1989). In the sheep or cow and many other ruminants, if there are two conceptuses each implants about one-third of the length of the uterus up from the cervix, one in each horn of the bicornuate uterus (Lee et al. 1977). Single conceptuses implant in a similar position, not always on the side they first enter the uterus, and may eventually expand into both horns. The location of the implantation sites are characterized by local edema which can be monitored with i.v. injections of pontamine blue (Boshier 1970).

The yolk sac is fully vascularized by 16 dpc in the sheep but has only a transient existence and is soon displaced by the rapidly elongating allantoic sac. This forms a tubular structure along the full mesometrial aspect of the conceptus by 26 dpc having started as a tiny diverticulum from the hind gut at 16 dpc (Figs. 1.1, 6.1, 6.2 and 6.3).

Fusion of the allantois with the chorion starts at ~30 dpc in the embryonic area and the allantoic vascularization rapidly spreads over the entire chorion including the extensive amniochorion (Bryden et al. 1972).

6.2.2   *Ruminant Implantation*

As the blastocyst elongates in vivo or in vitro the trophectoderm secretes Interferon τ(IFN τ) between 12 and 16 dpc. This downregulates the Oestrogen receptor in the uterine gland and surface epithelia and, consequently the Oxytocin receptor (OTR) also. Loss of the OTR blocks the OTR induced Prostaglandin pulses from the endometrium (Spencer et al. 2007). It is these pulses, transferred from the uterus by lymphatic and venous countercurrent exchange to the ovarian artery (Heap et al. 1985) which induce luteolysis. The IFN τ thus acts as the maternal recognition of pregnancy signal in the ruminants, maintaining the progesterone secretion.

The progesterone secretion acts on the surface uterine epithelium to downregulate the expression of the bulky adhesion resistant MUC-1 mucin and on the glands to produce proteins such as galectin, osteopontin and other integrins (Spencer et al. 2007) which act together with interleukins, TGFα and TNFβ (Aminoor-Rahman et al. 2004) to promote the adhesive phase of implantation. The gland secretions are essential. If the development of the uterine glands is blocked by prior hormonal treatment of the neonate, the glandless uterus cannot support pregnancy (Spencer et al. 2004b).

The changes in uterine epithelium from cellular to syncytial (Wooding 1984) have recently been shown to be coincident with what is considered to be a decidual type of reaction but only in the ewe endometrium so far (Johnson et al. 2003). Endometrial fibroblasts enlarge to form a compact subepithelial zone by 35 dpc and express proteins (smooth muscle actin, desmin and osteopontin) which are characteristic of rodent decidua.

At 16 dpc the trophectoderm and the uterine epithelium are closely adherent over the central region of the conceptus and have developed microvillar interdigitation, just as in the pig. Unique to the ruminant, from ~14 dpc in ewe, the trophectoderm produces binucleate cells (BNC). After maturation these BNC will migrate and fuse with the uterine epithelium (from ~16 dpc ewe, 19 dpc goat and 20 dpc cow; Figs. 6.1, 6.4 and 6.5) forming maternofetal syncytia (see later for detail).

The uterus also grows and the blood flow through the uterine vasculature increases considerably between 11 and 30 dpc (Reynolds and Redmer 1992). Although the sheep/goat and cow/deer types have significantly different definitive synepitheliochorial placental structure, at implantation BNC migration and fusion

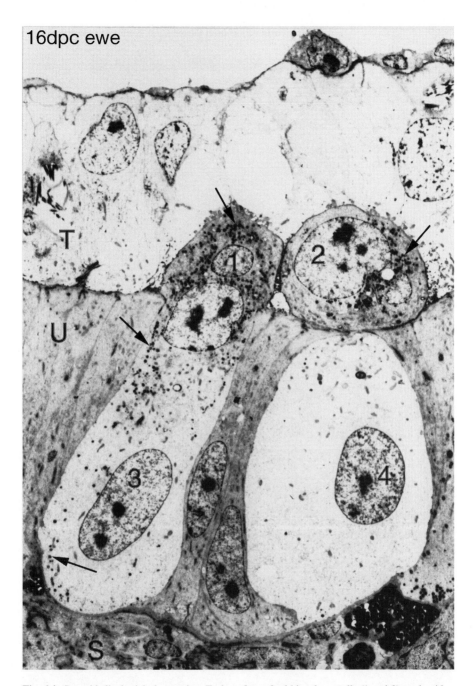

Fig. 6.4 Synepitheliochorial placentation. Fusion of two fetal binucleate cells (1 and 2), each with a uterine epithelial cell (3 and 4), at implantation (16 dpc) in the sheep. Cells 2 and 4 were continuous on a different plane of section (Wooding 1984). The material is non-osmicated and stained with phospho-tungstic acid to emphasize the binucleate cell granules (*arrows*), nuclei (1,2,3,4) and microvillar junction (*midway between T and U*), the cytoplasm appears empty. However, after osmium, conventional uranyl acetate and lead section staining produces from the same material micrographs equivalent to those in Figs. 6.11 and 6.12. *T* fetal trophectoderm, *U* uterine epithelium. 16 dpc, ×2,500

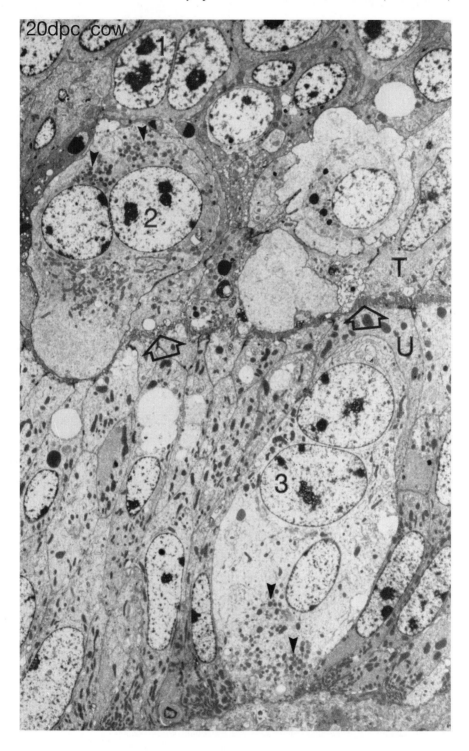

produces the same conversion of the caruncular epithelium from cellular to fetomaternal syncytial plaques (Wathes and Wooding 1980; Wooding 1984; Wango et al.
1990a,b). At the earliest stage of BNC migration in the sheep, evidence from serial
section counting of the number of nuclei within each cellular boundary in the uterine epithelium indicates that possible fusions are restricted. A BNC can fuse with
the apex of a single uterine epithelial cell to form a trinucleate cell (TNC) and further BNCs can fuse apically to extend this minisyncytium to five or seven nuclei or
beyond. No evidence was found for any uterine epithelial minisyncytia with two or
four nuclei (Wooding 1984; Wango et al. 1990a,b). Thus, the syncytia in the uterine
epithelium do not result from random lateral fusion of uterine epithelial cells which
would produce a continuous range of nuclear numbers per minisyncytium. There is
no technical problem in recognizing cells with two nuclei since the same studies
clearly confirmed that the trophectoderm consisted of uninucleate cells and BNCs
(Wooding 1984; Wango et al. 1990a,b).

By 20 dpc (sheep), 22 dpc (goat) and 24 dpc (cow) continued BNC migration
and fusion coupled with cell death produce syncytial plaques which cover the flat
caruncular surface (Leiser 1975; King et al. 1979; Wathes and Wooding 1980;
Wooding 1984; Wango et al. 1990a,b). The maximum number of nuclei per
plaque in sheep was estimated to be 24, and the plaques are bounded by tight
junctions significantly different from those sealing the uterine epithelial cells
(Morgan and Wooding 1983). The occasional uterine epithelial cell persists also.
There are more frequent uninucleate cells remaining in sheep intercaruncular
areas but here too, at this early stage, in the sheep and cow syncytial plaques
replace the uterine epithelium to a considerable extent (King et al. 1981; King
and Atkinson 1987). The intercaruncular uterine epithelium is re-established after
2–3 weeks by rapid division of residual epithelial cells, mostly from gland
orifices.

6.2.3   *Ruminant Placental Development*

Villus development in the caruncular areas starts at 24–26 dpc (sheep, goat) and
28–30 dpc (cow) (King et al. 1979). This is the time when the allantois is fusing
with the chorion, and the regularity of the pattern of chorioallantoic villus initiation

Fig. 6.5 Synepitheliochorial placentation. Implantation (20 dpc) in the cow. Glutaraldehyde and
osmium fixation. Binucleate cells (1 and 2) with characteristic granules (arrowheads) are migrating up to the microvillar junction (open arrows) between trophectoderm (T) and uterine epithelium
(U). The uterine epithelium includes a trinucleate cell (3) with granules (arrowheads) and two
round nuclei very similar to those in the binucleate cell. This trinucleate is probably a fetomatemal
hybrid cell produced by fetal binucleate cell fusion with a uterine epithelial cell (see Wathes and
Wooding 1980). 20 dpc, ×2,800

could well be determined by expression of genes analogous to the Gcm-1 type responsible for fetal villus initiation in the rodents (see Chap. 8.2.1; Anson-Cartwright et al. 2000).

Clusters of BNCs in the ruminant trophectodermal epithelium are the first indications and this patterning could originate from the trophectoderm or the allantoic mesoderm. At the same time, between the fetal villus initiations, subepithelial maternal capillaries are extending in the opposite direction pushing their overlying maternofetal syncytium and trophectoderm back towards the chorion. The processes would be like gradually inflating the fingers of a rubber glove (fetal villi) into a swelling mound of jelly (Figs. 6.6a, b). Once the patterns are established, the trophectoderm and maternal and fetal vascular systems produce growth (IGF, EGF) and angiogenic (VEGF, KDR) factors and the interactions between the two systems will drive the mutual growth of the placentomal villi. There is no evidence for any preformed maternal "crypts" into which fetal villi grow. Remodelling of the caruncle by matrix metalloproteinases (Salamonsen et al. 1995) facilitates the upward growth of the maternal villi and possibly a limited amount of invasive growth of the fetal villi but most of the growth is above the original flat implantation level.

Autoradiographic studies with [³H]thymidine later in pregnancy have produced no evidence of tip growth of villi (Wooding et al. 1981), but in the first half of gestation BNCs are undoubtedly concentrated at the tips of the fetal villi (Lee et al. 1985; Morgan et al. 1987). In all ruminants the main function of the BNCs is to deliver fetal hormones (Placental lactogens) and effectors (pregnancy associated glycoproteins [PAGs]; Prolactin related protein [PRP]) to modify the maternal environment to favour the fetus both metabolically and immunologically. In the ewe and goat they also contribute the bulk of the syncytial plaques, essential for the growth of the maternal villi. These plaques are very flexible structures with an amplified microvillar apical border. Basally they surround the maternal capillaries with a meshwork of fine elongations of their plasmalemma interspersed with basement membrane material. The apical and basal membrane elaborations enormously increase the surface area for membrane transport and are the location for the glucose transporters (Wooding et al. 2005a; see Fig. 2.6). The syncytium can also simplify and thin down locally with the apposed trophectoderm to minimise the diffusion distance between maternal and fetal capillaries to less than 1–2 μm. Autoradiographic evidence (Wooding et al. 1993) indicates that the plaques have a finite functional life with nuclear compaction and aggregation (as in the syncytial "knots" in the human placental syncytium) as a prelude to phagocytosis by the trophectoderm and replacement by new BNC migration.

In the sheep and goat, occasionally, between fetal syncytium and maternal vasculature there are large individual cells, probably residual endometrial fibroblasts or pericytes. They have a very uniform structure with a cytoplasm filled with much glycogen, tracts of fibrils and a large deeply divided nucleus (Fig. 7.8b). They are equipped as if capable of considerable synthesis and secretion having numerous dilated cisterna of rough endoplasmic reticulum, and a sizeable Golgi body with a complement of small dense vesicles. They are present throughout pregnancy. In position and structure they closely resemble a cell type seen in the endotheliochorial placenta which is generally considered 'decidual' (Figs. 7.8a, b).

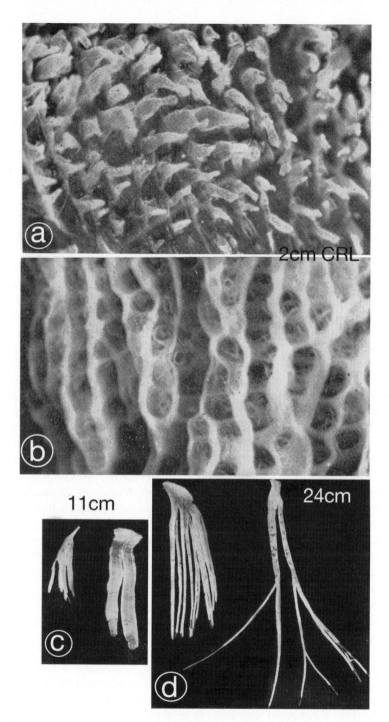

Fig. 6.6 Fallow deer placentomal development. Photographs of separated cotyledon and caruncle (**a, b**) and of dissected fetal villi (**c, d**). The regular pattern of the early villus development is clear on (**a**), and the corresponding crypts in the edematous endometrium on (**b**). The growth of the villi is one of elongation and branching. Villus branching in the definitive placenta is very variable between ruminant species (Hradecky et al. 1988) with this deer having much less than the ewe or the cow. (**a,b**) 2 cm CRL, ×23; (**c**) 11 cm CRL and (**d**) 24 cm CRL, both ×6 (Harrison and Hyett 1954)

In the sheep when the villi reach their maximum length and the placentomes their maximum weight – between 70–90 dpc – there is a considerable volume of fetal connective tissue core (Lawn et al. 1969; Steven 1983). Subsequently, the surface area of the villi (the fetomaternal hemotrophic exchange surface) grows by ever-increasing subdivision of the initial villi until at term there is very little connective tissue in either fetal or maternal villi. These now consist largely of the two blood vessel systems (Fig. 44) (see Leiser 1987) with very little separation because of indentation by the fetal vessels of the trophectodermal epithelium and attenuation of both that and the apposing syncytial plaques.

This is best appreciated on tissue fixed by perfusion via both maternal and fetal arteries (Fig. 6.7).

Corrosion cast studies of the goat villus vasculature suggest a gross countercurrent flow (Fig. 6.8) in the cotyledons.

The approximate 250-μm-long capillary meshworks in the terminal villi (Tsutsumi 1962) may have a cross-current or a mixed cross and countercurrent microflow (see Figs. 2.8, 2.9 and 6.8e).

The villus vasculature in the cow is more complex (Fig. 6.9 and 6.10) than in the goat but the gross and microflow systems are probably very similar.

In the ewe and goat the increase in the area of the syncytium must be generated solely by BNC migration for no nuclear division has been observed in the syncytium. Quantitative investigations (Wooding 1983) indicate that BNC migration is maintained at the same high level from implantation to term in all ruminants examined so far, and this is corroborated by the autoradiographic evidence in sheep (Wooding et al. 1981, 1993). These results agree well with Baur's morphometric investigations which show a continuous increase in fetomaternal exchange area virtually to term in the cow and in each of the wide variety of other genera he examined (Baur 1977). However the ewe shows no placentomal weight increase from mid pregnancy. Morphometric studies indicate that the surface area of the microvillar junction increases up to 120–125 dpc and the exponential growth of the fetus is then accommodated by an increase in fetal capillary proliferation (Reynolds et al. 2005; Borowicz et al. 2007).

The cow and deer synepitheliochorial placental type produces villi in a similar way but the syncytial plaques formed at implantation are overgrown by rapid division of the residual uterine epithelium (King et al. 1979). There is still a concentration of BNCs at the forming/advancing tips of the villi (Lee et al. 1986a) but they form only transient trinucleate cells (TNCs or 'minisyncytia') with the uterine epithelium (Fig. 6.16) (Wooding and Wathes 1980); there is no production of the extensive areas of syncytial plaques seen in the sheep (Wooding 1984). Plainly the syncytial transformation is not essential for villus growth. Rapid cell division in the uterine epithelium plus the contribution from BNC migration and TNC formation is sufficient to keep pace with the trophectodermal expansion in the cow. This produces an interhaemal membrane which is cellular on both sides as in a pig or horse placenta (Fig. 6.2, bottom diagram). In deer placentas there have been several reports of more extensive minisyncytia than trinucleate cells in the uterine epithelium, but only at the tips of the maternal villi (Hamilton et al. 1960; Kellas 1966; Sinha et al. 1969). Our observations on five deer genera (Chinese water, fallow,

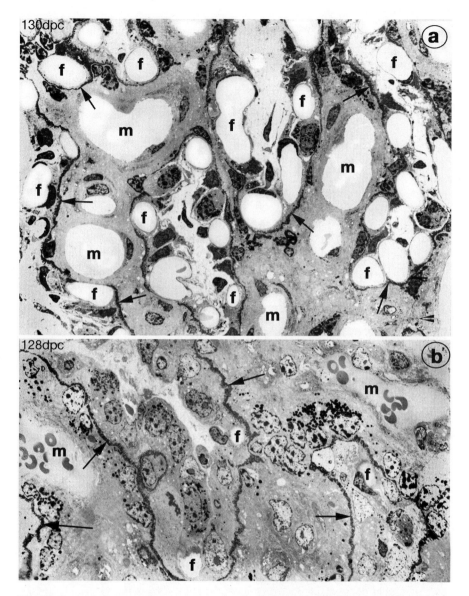

Fig. 6.7 Synepitheliochorial sheep placenta. Vascular relationships. Electron micrographs of sections of placentomes at a similar gestational age (**a**) perfused with fixative through both maternal and fetal blood vessels or (**b**) fixed by dicing up a fresh placentome in the same fixative used in (**a**). Only by the use of double perfusion can an accurate idea be obtained of the relative sizes and dispositions of the blood vessels in vivo. The result emphasizes the concept that the placenta should he considered primarily as a close apposition of blood circulations, not a mass of tissue through which blood circulates. Phosphotungstic acid staining on non-osmicated araldite embedded tissue. Arrows, microvillar junction (which is not the fetomaternal junction, the arrowheads are all on top of the fetomaternal hybrid syncytial layer); f, fetal capillaries; m, maternal capillaries. (**a**) 130 dpc, ×1,300. (**b**) 128 dpc, ×1,300

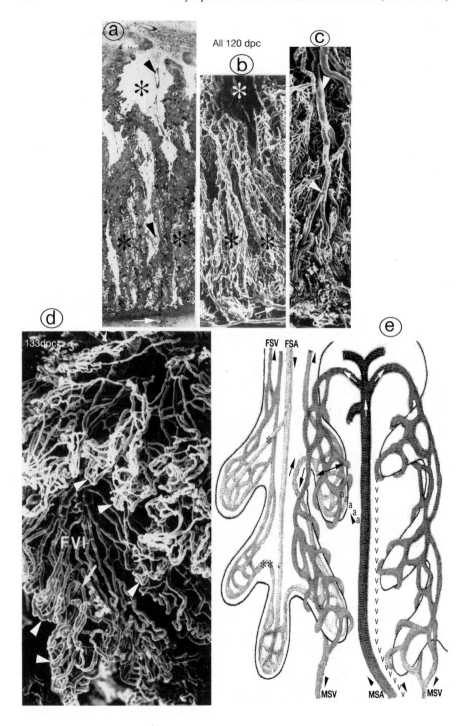

muntjac, red and roe) indicate that TNC formation is the usual outcome of BNC migration (Wooding et al. 1997).The limited life span of the syncytial plaque and TNC, with the latter clearly ending in "classical" apoptopic blebbing in the deer, have interesting parallels with the cycle of renewal suggested for the human syncytiotrophoblast. The same apoptopic cascade genes are expressed in cow BNC development (Ushizawa et al. 2006) as are implicated in the human syncytium (Huppertz et al. 2006).

Parturition in the sheep is heralded by a fall in progesterone levels. This is induced by cortisol production and secretion by the fetal adrenal. This and other hormonal changes (Bazer and First 1983; Thorburn 1991) possibly including relaxin are probably instrumental in causing a rapid but specific degeneration of the uninucleate trophectodermal layer (which the BNCs survive) (Perry et al. 1975; Steven 1975b). There is also an increase in the strength and synchrony of the myometrial cell contractions due to their considerable hormone-dependent development of gap junctions at this time (Garfield et al. 1979, 1988; Cole and Garfield 1989) and the cervix softens as a result of increased prostaglandin synthesis (Ledger et al. 1983; Magness et al. 1990). Together with the sudden drop at birth in the fetal placental blood pressure these processes are probably sufficient to produce an efficient means of placental separation within 1–4 h in the sheep and goat. The role in partus of the increase in number of the intraepithelial large granular lymphocytes in the interplacentomal uterine epithelium and their degranulation at term (Aminoor-Rahman et al. 2002) is at present unclear. The fetomaternal syncytium remains in situ on the maternal villi. As soon as the placenta separates there is a massive infiltration of the maternal caruncular residues by polymorphs and macrophages, which produce a rapid degeneration of the tissue, and the whole mass is sloughed at a level just above the myometrium (Van Wyk et al. 1972). The uterine epithelium is regenerated by cellular overgrowth from the glands at the edges of the caruncular areas (Wagner and Hansel 1969).

In the cow the progesterone level falls and cortisol rises in a similar manner at parturition and relaxin may also be involved (Musah et al. 1987) as well as oxytocin (Fuchs et al. 1992), but there is no degeneration of the trophectodermal layer as in the sheep. Release of the placenta depends upon changes in the molecular forces

---

Fig. 6.8 Goat placentomal vasculature at 120 dpc. Histological section (**a**) and corrosion casts of maternal vasculature, (**b**), and fetal vasculature, (**c**). All ×40 The fetal stem arteries (*arrowheads* in **a** and **c**) pass direct to the maternal side before dividing into capillaries to flow back. Upper asterisks on (**a**) and (**c**) indicate equivalent positions in a fetal villus. Lower asterisks on (**a**) and (**b**) show maternal villus positions. (**d**) Detail of the terminal villus capillary meshworks (white arrowheads). One terminal meshwork may drain into the next one in line down the intermediate villus (*double asterisk* on (**e**)), or each terminal villus may be independently supplied (*single asterisk* on (**e**)). 133 dpc, ×170. (**e**) Schematic reconstruction of the vascular system in an intermediate villus. The gross flow is countercurrent – MSA versus FSA – but the microflows (*small arrows*) are both cross- and counter-current – network flow, see Fig. 2.3. The main cascade down the maternal capillary meshworks shown on the right may be short circuited by capillaries following the aaaa or vvv pathways. Modified from Leiser (1987)

maintaining the microvillar interdigitation, as in the pig or horse placenta. However, the villi are very much longer and more branched than those of the ewe, pig or horse and this may be the cause of the significant percentage of cow placentas that do not separate easily (Grunert 1986). BNCs may also have a role at partus (Gross et al. 1991). A 'retained placenta' is one which has not been delivered 12 h after birth. Recent work indicates there may be an immunological component of placental release in the cow. Increasing expression of MHC1 restricted to the trophoblast of the intercotyledonary areas is reported in late pregnancy and the greater the immunological disparity between mother and father the less frequent the incidence of retained placenta (Davies et al. 2004).

Methods for isolating pure populations of BNCs have been reported in several laboratories. BNCs in short-term primary culture have been reported to produce prostaglandins, progesterone (Hamon et al. 1985; Reimers et al. 1985; Ullman and Reimers 1989; Wango et al. 1991) placental lactogens (Rhodes et al. 1986; Morgan et al. 1990) and Pregnancy associated glycoproteins (PAGs Nakano et al. 2001). However, all isolated BNC populations contain the complete developmental sequence of BNCs, in our experience usually biased towards the young BNCs with few or no granules. This is quite unlike other isolated exocrine cells such as mammotrophs or somatotrophs from the pituitary, which consist almost entirely of mature fully granulated cells. The activities of the isolated BNC population could therefore be characteristic of only one developmental stage, and it is difficult to assess their physiological significance without figures for synthesis of these hormones by the other cell types in the placenta (Shemesh 1990; Ben David and Shemesh 1990) which are not yet available. BNCs can be isolated because only they are robust enough to survive the procedure; other, gentler methods will be necessary to separate viable populations of the more fragile uninucleate trophectodermal cells or the syncytial plaques.

BNCs will survive in viable condition in primary culture for up to 10 days if grown on collagen (Nakano et al. 2001) before rapid deterioration and cell lysis. Most studies of isolated BNC have demonstrated release, not synthesis, of progesterone, prostaglandins or placental lactogen, usually at a declining rate (Reimers et al. 1985; Rhodes et al. 1986; Ullman and Reimers 1989). Progressive death of

Fig. 6.9 Corrosion casts of the vascular systems in cow placentomes. Stem arteries are indicated by arrows, stem veins by arrowheads. (**a**) Early gestation corrosion cast of the maternal vasculature showing long supply and collection vessels linked by a capillary network at the apex. 80 dpc, ×50 (**b**). Longitudinal section through a mid gestation placentome with fetal vessel ink injection showing comprehensive detail of the fetal capillary meshworks in the terminal villi. 150 dpc, ×80. (**c**) Cross section of an ink injected villus, showing the three stem vessels in the centre. 105 dpc, ×70. (**d**) Detail from (**b**) with a stem artery (*arrow*) and vein (*arrowhead*) indicated. 150 dpc, ×200 (**e**) Corrosion cast of a mid gestation primary fetal villus showing the three dimensional organisation of the capillaries in the terminal villus. Part of the centre of the villus exposed, showing the central stem artery (*arrow*). 220 dpc, ×100 (**f**) Light Microscopy of a section corresponding to (**d**) with conventional H and E stain, the intermediate villi arise at right angles to the primary. 150 dpc, ×140 (From Tsutsumi 1962; Leiser et al. 1997 and Pfarrer et al. 2001)

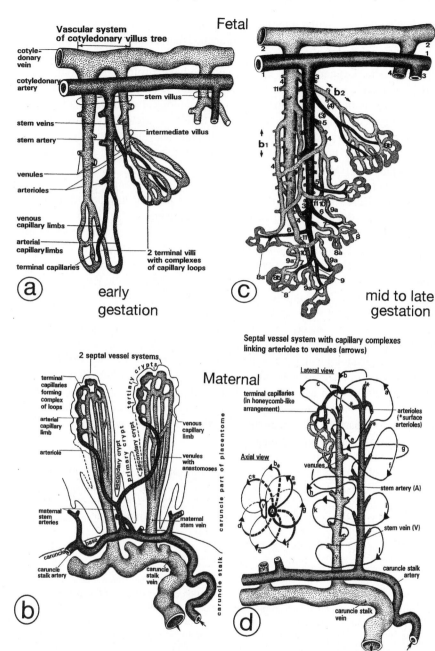

Fig. 6.10 Schematic reconstructions of the placentomal vasculature at early and late stages; (**b**) corresponds to Fig. 6.9a; (**c**) and (**d**) to 6.9 (**c**), (**d**), (**e**), and (**f**). In (**c**), b1 represents a full grown, and b2 a budding, villus; 1 and 2, uterine artery and vein; 3 and 4, stem artery and vein; 5, arterioles of intermediate villi, 9,10 and 11 the venous return. (From Leiser et al. 1997 and Pfarrer et al. 2001)

the cells and passive release of material seems to be as likely an explanation as normal secretion. This is especially true for placental lactogen since normal BNC development includes migration and fetomaternal fusion prior to release of secretory granules by exocytosis. No one has succeeded in reproducing this behaviour in cultures as yet, but isolated BNCs can synthesize progesterone from pregnenolone (Wango et al. 1991) and placental lactogen from amino acids (Morgan et al. 1990). In vivo the BNC population has considerable synthetic potential: bPL has fetal growth promoting activity and biases the maternal metabolism to favour fetal nutrient uptake and PAGs are suggested to play an important immunoregulatory role in the cow (Wooding et al. 2005a–c). Immature BNC express Platelet Activating Factor receptor (Bucher et al. 2006) and VEGF (Pfarrer et al. 2006a) and mature BNC show FGF and a variety of Integrins (Pfarrer et al. 2003; 2006b) which may be involved in BNC migration. Goat BNC express Matrix metalloproteinases and their inhibitors (Uekita et al. 2004).

Nakano et al. (2002, 2005) have established a uninucleate cell (UNC) line (BT-1) from the cow blastocyst which grown on collagen gel eventually produces "clusters" of bPL producing BNC (the UNC do not produce bPL). A goat cell line (HTS-1: Miyazaki et al. 2002) derived from late pregnant cotyledons also produces bPL after some time in culture but the bPL is expressed in both UNC and BNC, which does not occur in vivo. What controls the differentiation in these two cases and how equivalent developmentally, ultrastructurally and synthetically these BNC are to those in vivo remains to be established.

6.2.4 Structure and Function of Ruminant Placental Binucleate Cells

Five to ten percent of any dividing cell population will be binucleate [BNC] with an ultrastructure identical to their uninucleate [UNC] neighbours. The nuclei are diploid [2N] and cytoplasmic division will usually follow. In contrast, trophoblast BNC (or "giant cells") have significantly different [ultra]structure, size and shape to their adjacent UNC, are frequently polyploid and form hormone producing populations with specific functions. In rodents they are usually a transient stage in the development of highly polyploid giant cells (Pijnenborg et al. 1981; Ilgren 1983). In Equid girdle and cup cells they do have two diploid nuclei (Wooding et al. 2001) but in ruminants the trophoblast BNC have two 4N nuclei at maturity prior to migration (see refs in Klisch et al. 2004).

One of the characteristic properties of the binucleate cell seems to be a capacity for migration from its site of formation. In ruminants it can move into tight junctions, in the horse pass through the uterine epithelium and in hamsters and some other rodents the giants can invade the uterine arteries as deeply as the myometrial layer.

BNCs are most consistently found and in the greatest number throughout pregnancy, in the trophectoderm of all the ruminants so far studied (Wimsatt 1951;

Wooding 1982b, 1983). They first appear in the trophectoderm of the blastocyst during the third week of pregnancy (day 15, sheep, Boshier 1969; Wooding 1984; day 17, cow, Greenstein et al. 1958; day 18, goat, Wango et al. 1990b). They are found in the fetal trophectoderm apposed to the uterine epithelium at the earliest stages of implantation (day 16, sheep; day 18, goat; day 20, cow), and constitute approximately one-fifth of the total trophectodermal cells for almost the whole of the remainder of pregnancy. There is a decrease in the last week of pregnancy in the sheep and goat (Table 6.1).

These generalizations apply to both singleton and multiple pregnancies (Wooding 1983).

There is no evidence for any specialized stem cell; binucleate cells can originate by division of any uninucleate cell in the trophectodermal epithelium. The division produces two cells, one of which has no apical tight junction. The nucleus of this cell divides again without subsequent cytokinesis, thereby forming a binucleate cell. This binucleate cell rapidly releases any contact with the basement membrane. It therefore starts and usually completes its maturation process with only rare tiny desmosomal-like contacts with the uninucleate cells of the epithelium; no tight or gap junctions have been observed (Lawn et al. 1969; Boshier and Holloway 1977;

Table 6.1 Binucleate cell numbers throughout pregnancy in ruminants[a]

Stage of gestation days post coitum	Percentage of BNC in trophectoderm	Percentage of BNC population migrating[b]
Sheep		
12–15 (isolated preimplantation conceptuses)	0–3	28
16–29	18	24
41–67	20	21
70–105	18	26
114–139	17	18
142–147 (term = 145)	8	20
Goats		
27–90	21	21
95–130	21	23
135–148	16	25
Cows		
18–28	20	25
37–260	21	12
Deer		
Implantation	16	12
Implantation to 60	18	15

[a]For details of breeds, counting methods, statistical limits see Wooding (1983)

[b]A migrating BNC is defined as one which is part of the trophectodermal apical tight junction

Wooding 1992). The young binucleate cell has a few mitochondria in a dense, ribosome-filled cytoplasm with no glycogen (Fig. 6.11a).

It rapidly develops an extensive system of swollen rough endoplasmic reticulum cisternae and a very large Golgi apparatus. The Golgi produces large numbers of dense granules containing microvesicles (Fig. 6.11b) and the cell grows very large.

In the first third of pregnancy in sheep and goat a fully granulated 'mature' binucleate cell frequently also contains a large lipid droplet and/or a membrane-bound crystalline inclusion which are typical of the surrounding uninucleate cells, emphasizing the derivation of one from the other. All the binucleate cells in the trophectoderm can be arranged in a single developmental sequence (Wooding 1982a,b, 1983, 1984); there is no clear structural evidence for two populations as suggested by Boshier and Holloway (1977).

BNCs from all species so far examined contain a unique structural element, the double lamellar body (DLB). This is a discrete small area of double membranes of characteristic branching structure intermittently continuous with endoplasmic reticulum cisternae and usually located near the Golgi body (Fig. 6.12).

In a recently divided BNC the double membranes have numerous associated small clear vesicles (Fig. 6.12a, b). DLBs in mature migrating BNCs lack these clear vesicles as do DLBs in the fetomaternal syncytia, which are derived from the migration (Fig. 6.12c).

The DLB thus shows a clear developmental sequence but as yet its function is unknown. Autoradiographic (Wooding et al. 1981), histochemical (Wooding 1980) and immunogold EM (Lee et al. 1986b–d) studies show no involvement of the DLB in protein synthesis, placental lactogen storage, phosphatase localization or BNC granule production.

The maturing binucleate cell lies very close to the microvillar junction but normally does not form part of the tight junction between the apices of the uninucleate trophectodermal cells. The mature BNC inserts a pseudopodium into the tight junction, separating the UNC but maintaining the structure of the junction with the UNC. The BNC then expands the area of the plasmalemma above the junction (Figs. 6.1, 6.2 and 6.13) with a specialised population of small vesicles (Fig. 6.14) probably from the golgi body (Wooding et al. 1994).

This modifies the microvillar junction to form a unique area of of flat apical BNC plasmalemma, the 'Migration Front', closely apposed to and/or bulging into the flattened plasmalemma of the uterine epithelial cell or derivative (Figs. 6.2, 6.13 and 6.14). There is no further migration through the tight junction. The newly formed apical BNC plasmalemma is designed to fuse with the apposed uterine epithelial plasmalemma and the BNC cell content is injected into the uterine cell/syncytium. All of the original fetal BNC plasmalemma remains in the trophoblast, vesiculates and is resorbed by the trophoblast UNC (Fig. 6.2).

Throughout pregnancy 15–20% of the total population of binucleate cells are found inserted into the tight junction in all four genera enumerated so far (cow, deer, goat and sheep) (Wooding 1983; Table 6.1).

Only mature binucleate cells are found with tight junction structures and electron micrographs of these cells show all stages of insertion into the tight junction, injection

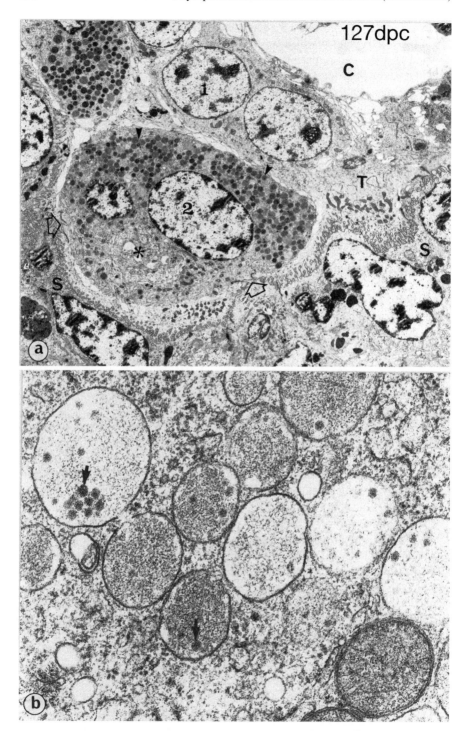

of BNC content and membrane resorption (Figs. 6.13, 6.15–6.17). Autoradiographic (Wooding et al. 1981, 1994) freeze-fracture (Morgan and Wooding 1983) and immunological (Wooding et al. 1994) evidence provides a clear confirmation of this remarkable process (subsequently referred to as "migration").

This BNC "migration" starts at implantation and in the goat, sheep and cow produces a syncytium bordering the maternal tissue by fusion of the inserted binucleate cells with some individual uterine epithelial cells plus an elimination of the remainder (Wathes and Wooding 1980; Wooding 1984) (Figs. 6.1, 6.2, 6.4 and 6.5). In the sheep and goat placentome this syncytium is present throughout pregnancy (Lawn et al. 1969) (Fig. 6.15a, b and 6.17).

In the cow at implantation there is a transient formation of syncytium by binucleate cells fusing with uterine epithelial cells but only a partial loss of the uterine epithelium. This bovine syncytium is then eliminated by displacement by continuing division of the remaining uterine epithelial cells (King et al. 1979; Wathes and Wooding 1980). Subsequently, during the remainder of pregnancy in cow and deer, the "migrated" binucleate cell fuses with an individual cell derived from the original uterine epithelium to form a transient trinucleate cell (Fig. 6.16) which dies after the granules from the original binucleate cell have been released (Wooding and Wathes 1980; Wooding 1987) (Fig. 6.2).

This "migration" was suggested by the earliest workers on the ruminant placenta (Assheton 1906; Wimsatt 1951), but subsequent electron microscopic investigations were unable to find any supporting evidence (Bjorkman and Bloom 1957; Lawn et al. 1969; Bjorkman 1968, 1970). A study of sheep binucleate cells in 1977, for example, concluded that there were 'no published, well resolved pictures of binucleate cells in contact with the syncytium' (Boshier and Holloway 1977). By developing a method for selective staining of non-osmicated material with phosphotungstic acid to emphasize only the binucleate cell granules and the microvillar junction (Figs. 6.13–6.17) (Wooding 1980) and by subsequent use of ultrastructural (Wooding et al. 1980) and immunocytochemical (Fig. 6.13) (Wooding 1981; Lee et al. 1986b–d) techniques it has now been possible to demonstrate clearly both the fact and extent of the insertion and fusion process.

The frequency of "migration" in the sheep and goat needs to be sufficient to maintain the syncytium, in which no nuclear division has ever been demonstrated, throughout the enormous expansion in area which occurs during formation of the

---

Fig. 6.11 (a) Synepithelio-chorial placentation. Glutaraldehyde/osmium fixation. Development of fetal binucleate cells (1, young; 2, mature) in the trophectoderm (T) of the definitive placenta of the goat. Note the numerous characteristic granules (*arrowheads*) and large Golgi body (*asterisk*) in the mature binucleate cell, which has started to migrate up to the microvillar junction at two points (*open arrows*). *S* fetomaternal syncytial layer; *C* fetal connective tissue. 127 dpc, ×2,500. (b) Cow binucleate cell granules containing characteristic microvesicles (*arrows*). 49 dpc, ×46,000

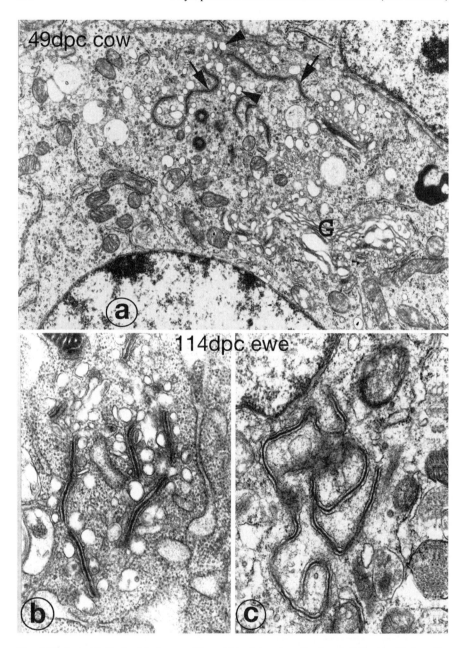

Fig. 6.12 Synepitheliochorial placenta Glutaraldehyde/osmium fixation. (**a**) The 'double lamellar body' (DLB, at *arrows*), characteristic of ruminant binucleate cells, is closely associated with the golgi body (G) and has electron lucent vesicles (*arrowheads*) associated with it in the young non granulated binucleate cell in a cow. 114 dpc, ×50,000. (**b**) The DLB is very similar in structure in BNC in the ewe. 114 dpc, ×50,000. (**c**) In the ewe the mature DLB in the fetomaternal sycytium which is formed mostly from migrated BNC shows no associated vesicles such as those seen in the immature form in (**b**). 114 dpc, ×40,000

cotyledonary villi. It is clear that static micrographs can never provide direct evidence for a dynamic process. However, injection of radioactive thymidine into fetal sheep or goats and subsequent autoradiography of the fetomaternal interface layers shows that labelled nuclei are initially found exclusively in the uninucleate trophectoderm, subsequently in mature binucleate cells and finally in the syncytium (Wooding et al. 1981, 1993). There was no evidence for a population of binucleate cells which became labelled but did not migrate.

In the goat and sheep cotyledon, maintenance of the substance of the syncytium would seem to be as important a function for "migration" as granule delivery (Fig. 6.17).

In the cow and deer the primary function of the "migration" is to deliver the binucleate cell granules to the plasmalemma bounding the maternal connective tissue (Steven et al. 1978). Exocytosis would then deliver the granule contents close to the maternal circulation, and there is good ultrastructural evidence for such a process from sheep and goat syncytium and cow trinucleate cells (Wooding 1987; Wango et al. 1990a). This is the ideal position for the fetal products to influence the maternal metabolism and immunological reactions to favour the fetus, evading the tight junction barriers in the placenta.

Ovine and bovine placental lactogens (oPL and bPL) have been localized exclusively in the BNC granules and Golgi body and the granules in the syncytium by electron microscopic immunocytochemistry (Fig. 6.13b; see refs in Wooding et al. 1992 for sheep; Wooding and Beckers 1987 for cow).

No label was found over the Golgi bodies in the syncytium, indicating that synthesis and packaging of BNC granule contents occurs before BNC "migration" (Wooding 1981; Lee et al. 1986b–d; Wooding and Beckers 1987). Biochemical organelle isolation techniques have also shown that placental lactogens co-purify with a granule fraction (oPL, Rice and Thorburn 1986; bPL, Byatt et al. 1986, 1992). More recently, hybridization histochemical studies (Milosavljevic et al. 1989; Kappes et al. 1992) have shown that PL mRNA is found only in ovine and bovine BNCs.

The Pregnancy Associated Glycoproteins (PAGs) are another important BNC granule constituent. Molecular studies have shown there are many distinct PAGs – 21 in the cow for example- and they belong to the aspartic proteinase family. The PAGs are probably enzymically inactive but may retain specific peptide/protein binding ability. There are two subgroups, one evolving >87 million years ago (mya) and a second >52 mya. The older types are present in ALL trophoblast cells (though concentrated in the BNC), the younger ONLY in the BNC granules (Hughes et al. 2000). Recent immunological studies indicate important functional roles in maternofetal adhesion and as immunomodulatory effectors in the maternal connective tissue (Wooding et al. 2005b).

There is evidence that, although all BNCs appear similar ultrastructurally, the content of their granules may depend on the location of the BNCs or the age of the conceptus. Ovine interplacentomal BNCs do not express oPL or the PAG known as SBU-3 (Lee et al. 1985, 1986c,d), and although oPL is present in placentomal BNC in conceptuses from 16 dpc, SBU-3 is not found until 24 dpc (Morgan et al. 1987).

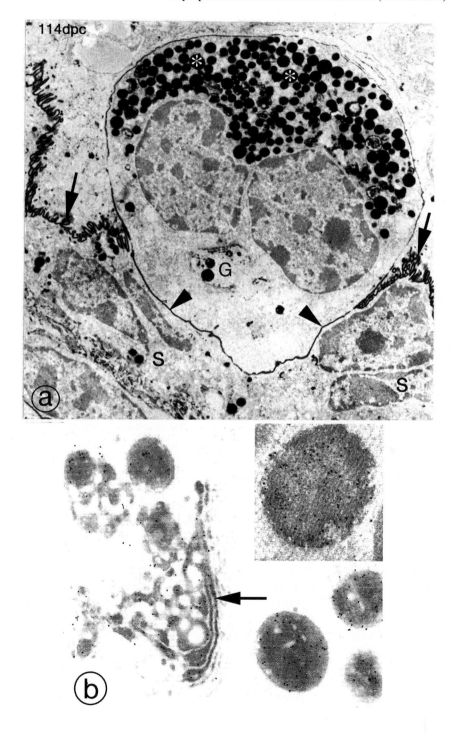

Binucleate cell "migration", therefore, provides the fetus with a general method of delivering large molecules to the maternal circulation. However, in the sheep, oPL is present in both maternal and fetal circulations and if all binucleate cells migrate to the maternal side it is difficult to understand the origin of the fetal serum oPL (Chan et al. 1978; Martal and Lacroix 1978). What little evidence there is indicates that there is no transfer of oPL from maternal to fetal circulation (Grandis and Handwerger 1983) or from fetus to mother (Schonecht et al. 1992). This problem is more extreme in the cow, in which the bPL concentration in the maternal circulation is always very much lower than on the fetal side (Beckers et al. 1982; Wallace et al. 1985). Despite this, all the bPL-containing BNCs appear to migrate and bPL-containing granules are found only in BNCs and their derivatives (Wooding and Beckers 1987). Even allowing for the vastly greater maternal blood volume, the amount reaching the fetal circulation is very significant from the earliest stage measured.

There is no evidence in cow or sheep for a subpopulation of BNCs which degranulate on the fetal side, but there may be a constant low level of release of granules prior to BNC migration, which would be very difficult to detect ultrastructurally. Alternatively, there may be a constitutive pathway for oPL and bPL secretion independent of the conventional regulated exocytotic mechanism (Burgess and Kelly 1987). This is an important point because exogenously administered oPL elicits biological responses in the fetal sheep (Battista et al. 1990) consistent with earlier suggestions (Bell 1984; Gluckman et al. 1987) that oPL may serve as an important somatotrophic hormone in the fetus.

This presence of placental lactogens in both fetal and maternal circulations is in sharp contrast with the restriction of equine chorionic gonadotropin (eCG) produced by the horse endometrial cup BNCs entirely to the maternal circulation (Ginther 1989). In this case the BNCs do not mature (i.e. produce eCG-containing granules) until they are in transit to the cup site in the maternal endometrium (Chap. 5).

It has been suggested that binucleate cell migration in the sheep can be increased by fetal pituitary stalk section or adrenalectomy, but this was based on the assumption that migration was insignificant in normal pregnancy (Barnes et al. 1976; Bass et al. 1977; Steven et al. 1978; Lowe et al. 1979). Quantitative

---

Fig. 6.13 (**a**) Synepitheliochorial sheep placenta. Phosphotungstic add staining. Binucleate cells migrate across the microvillar junction (*arrows*) by forming the Migration Front (MF; *arrowheads*). Migration occurs continuously throughout pregnancy and fusion by vesiculation of the MF with the apposed fetomaternal syncytium plasmalemma releases the characteristic BNC granules (*asterisks*), formed before migration in the Golgi body (G), into the fetomaternal syncytium (S). 114 dpc, ×5,000. (**b**) Binucleate cell Golgi body immunostained for ovine placental lactogen. The gold particles localize the lactogen exclusively to the trans cisternae of the Golgi stack (*arrow*) and the granules. 114 dpc, ×38,000. (Inset on b) Colocalization of ovine placental lactogen (5 nm gold) and the SBU-3 antigen (10 nm gold) on the binucleate cell-derived granules in the placental syncytium. *N* nucleus (From Lee et al. 1986c). 114 dpc, ×40,000

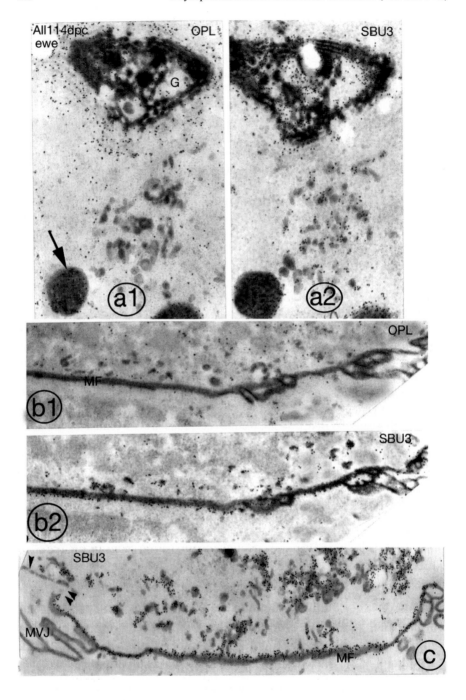

studies have shown in the sheep that there is no significant difference between the BNC migration level in normal pregnancy and that found after a variety of experimental procedures including pituitary stalk section, adrenalectomy, carunclectomy, hypophysectomy and administration of progesterone synthesis inhibitors and EGF (Wooding et al. 1986). The production, maturation and migration rates of the BNCs seem to be totally unaffected by administration of this wide variety of compounds and procedures, as are the oPL blood concentrations (Taylor et al. 1982, 1983). In the last week of pregnancy there is a fall in binucleate cell numbers (Wooding 1982a,b, 1983) and this correlates with a similar fall in oPL concentration in both fetal and maternal circulations (Martal et al. 1977). Fetal cortisol from the adrenal is the trigger for parturition with its accompanying fall in BNC number. Injection of cortisol into the fetus in late pregnancy (130 dpc), long before normal partus (145 dpc) has recently (Ward et al. 2002) been shown to produce a similar reduction in BNC numbers. If betamethasone, a glucocorticoid analogue is injected into the fetus earlier in pregnancy (104 dpc) a 30% fall in BNC number and maternal oPL concentration result 5 days later but the BNC numbers and oPL concentrations recover to normal levels near term (Braun et al. 2007). Corticoids can therefore stop BNC production but what initiates and maintains it remains to be demonstrated.

During implantation the earliest evidence for binucleate cell formation in sheep (Boshier 1969) is found at the same time as the first demonstration of placental lactogen and PAGs in the blastocyst of the sheep (Martal et al. 1977; Morgan et al. 1987) or cow (Flint et al. 1979; Wooding et al. 2005). The detection of either oPL or PAGs in the maternal circulation have been used as the earliest indication of pregnancy (Vandaele et al. 2005). However, until it is possible to predictably manipulate the oPL and PAG concentrations, there is insufficient evidence to assume that the blood concentrations are directly related to binucleate cell numbers. It could be that it is the rate of migration of the cells or exocytosis of the granules which is the primary locus of control of release of oPL into the maternal circulation. Secretion from placentomal slices (said to involve protein kinase C. Battista et al.

Fig. 6.14 Differential labelling of the Binucleate Cell Migration Front (MF) which is the new plasmalemma the migrating BNC forms past the tight junction (Figs. 6.1, 6.2 and 6.13). This new plasmalemma will fuse and break down into vesicles with the apical membrane of the maternofetal syncytium to which it is apposed. The fusion releases the BNC cytoplasmic content into the syncytium. Serial sections show that Ovine Placental Lactogen (OPL) antibody labels (a1,b1) the golgi body (G) and granules (arrow) but not the group of tiny vesicles (v) near the golgi nor (b1,2), the migration front (MF) whereas SBU3 antibody labels all three (a2,b2 and c) Quantitation of the label shows the specificity of the SBU3 label with the MF at 19 grains per μm and none of the other membranes, including the BNC basolateral plasmalemma (single arrowhead on (**c**)) and the adjacent microvillar junction (MVJ on (**c**)) showing levels significantly above background. (From Wooding et al. 1994). 114 dpc, (a1,a2), ×24,000; (b1,b2), ×19,000; (**c**), ×13,000

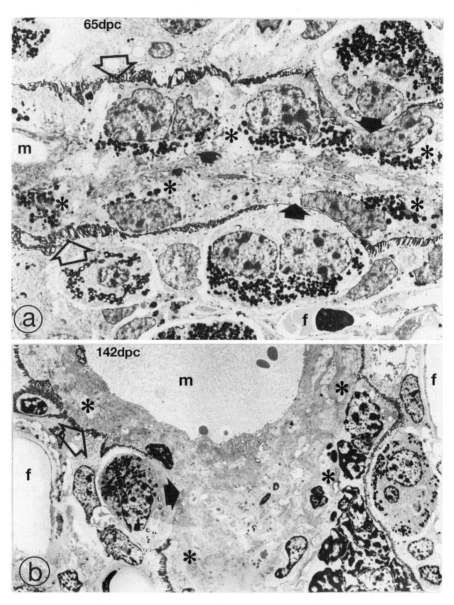

Fig. 6.15 (a) Synepitheliochorial sheep placenta. Binucleate cells migrate (*solid arrows*) across the microvillar junction (*open arrows*) throughout pregnancy ((a) 65 dpc; (b) 142 dpc) delivering their granules to the base of the fetomaternal syncytium (asterisks). There are usually many granules in the syncytium at early stages (Figs. 6.15a and 6.17) of pregnancy, few are found near term (6.15b) when BNC still migrate (**b**, *solid arrow*) but are far fewer. *m* maternal *f* fetal, blood vessels (Wooding 1981). (**a**) 65 dpc, ×2,600; (**b**) 142 dpc, ×1,400

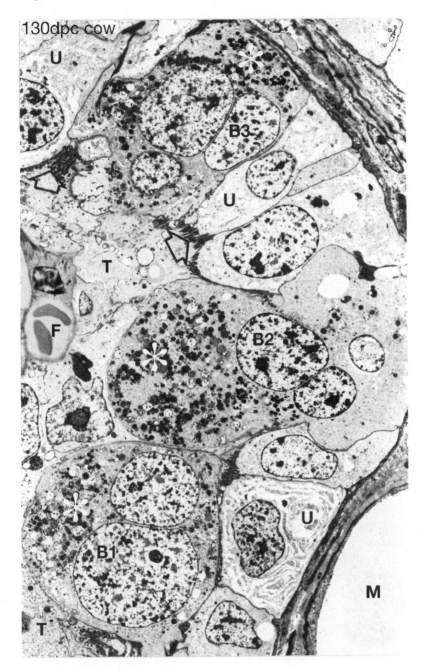

Fig. 6.16 Synepitheliochorial cow placenta. Throughout pregnancy mature fully granulated (*white asterisks*) binucleate cells (B1) migrate from the fetal trophectoderm (T) across the microvillar junction (*open arrow*) to fuse with uterine epithelial cells (U) producing trinucleate cells (B2). These eventually release their granules (B3, *asterisk*) close to the maternal blood vessels (m), die and are resorbed by the trophectoderm (Wooding and Wathes 1980) *F* fetal blood vessel. 130 dpc, ×2,700

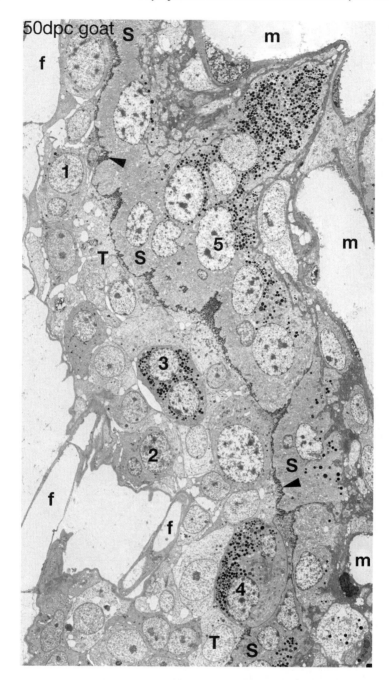

Fig. 6.17 Synepitheliochorial placenta in the goat. Numerous BNC are present in the trophectoderm, a sequence of development 1–4 is shown, with BNC4 migrating to form the syncytium (S). This contains numerous granules, all derived from the BNC and many nuclei (e.g. 5) most of which are from BNC. 50 dpc, ×1,500

1990) will include oPL secreted to the fetal side as well as from granules delivered to the syncytium by BNC migration – the normal route into the maternal circulation. The latter is an order of magnitude greater than the former in vivo. Isolated trophectodermal uninucleate and binucleate cells (Burke et al. 1989) and tissue culture monolayers (Nakano et al. 2002, 2005) will presumably be restricted to only the fetally directed secretion type.

There is an increasing amount of evidence that the placental lactogens can modify the maternal metabolism and placental angiogenesis to favour fetal nutrition and growth (Freemark and Handwerger 1986; Freemark et al. 1987; Chene et al. 1988; Forsyth 1991; Byatt et al. 1992; Corbacho et al. 2002) but the details of the mechanisms and control of release of the hormone remain to be elucidated.

Chapter 7
Endotheliochorial Placentation: Cat, Dog, Bat

7.1 Introduction

In endotheliochorial placentation no uterine epithelium survives implantation (Figs. 1.5 and 7.1). The cellular chorioallantoic trophectodermal cells produce, by division and fusion, a fetal syncytium apposed to an amorphous layer around the persisting maternal endothelium. The amorphous layer (also referred to as the interstitial layer or, misleadingly, 'membrane') is probably produced by the syncytium and is equivalent, ultrastructurally and in molecular constituents, to the basement membrane of the maternal endothelium. It is neither uniformly thick nor continuous; processes of the fetal syncytium penetrate through it to reach a close relationship with the maternal endothelium in places, and it includes occasional large residual endometrial cells. This structure is characteristic of the early and definitive endotheliochorial placentas which are zonary (Figs. 1.4 and 7.1).

The zones are made up of grossly parallel fetal villi, which are labyrinthine in architecture. Usually, but not always, there is a large haemophagous zone at the edge or in the centre of the zonary structure. Outside the main zone of the placenta a cellular fetal trophectoderm is loosely apposed to a persistent uterine epithelium and there are small areolae over the gland mouths. Endotheliochorial placentation is characteristic of carnivores and many bats but is also found in individual genera in other orders – Proboscidea (elephant), Tubulidentata (aardvarks), Rodentia (*Dipodomys*, the kangaroo rat), Insectivora (*Talpa*, the European mole, and Blarina, the shrew) and also Bradypodida (the sloths) (Mossman 1987).

The cat and dog will be taken as the characteristic examples of this placental type.

7.2 Dog, *Canis familiaris*; Cat, *Felis domestica*

Oestrous cycle

Dog, one 21-day cycle during the single yearly breeding period.
Cat, 15–21 day cycles; several during each of the breeding periods which occur 2–3
    times per year.

P. Wooding, G. Burton, *Comparative Placentation*,
© Springer-Verlag Berlin Heidelberg 2008

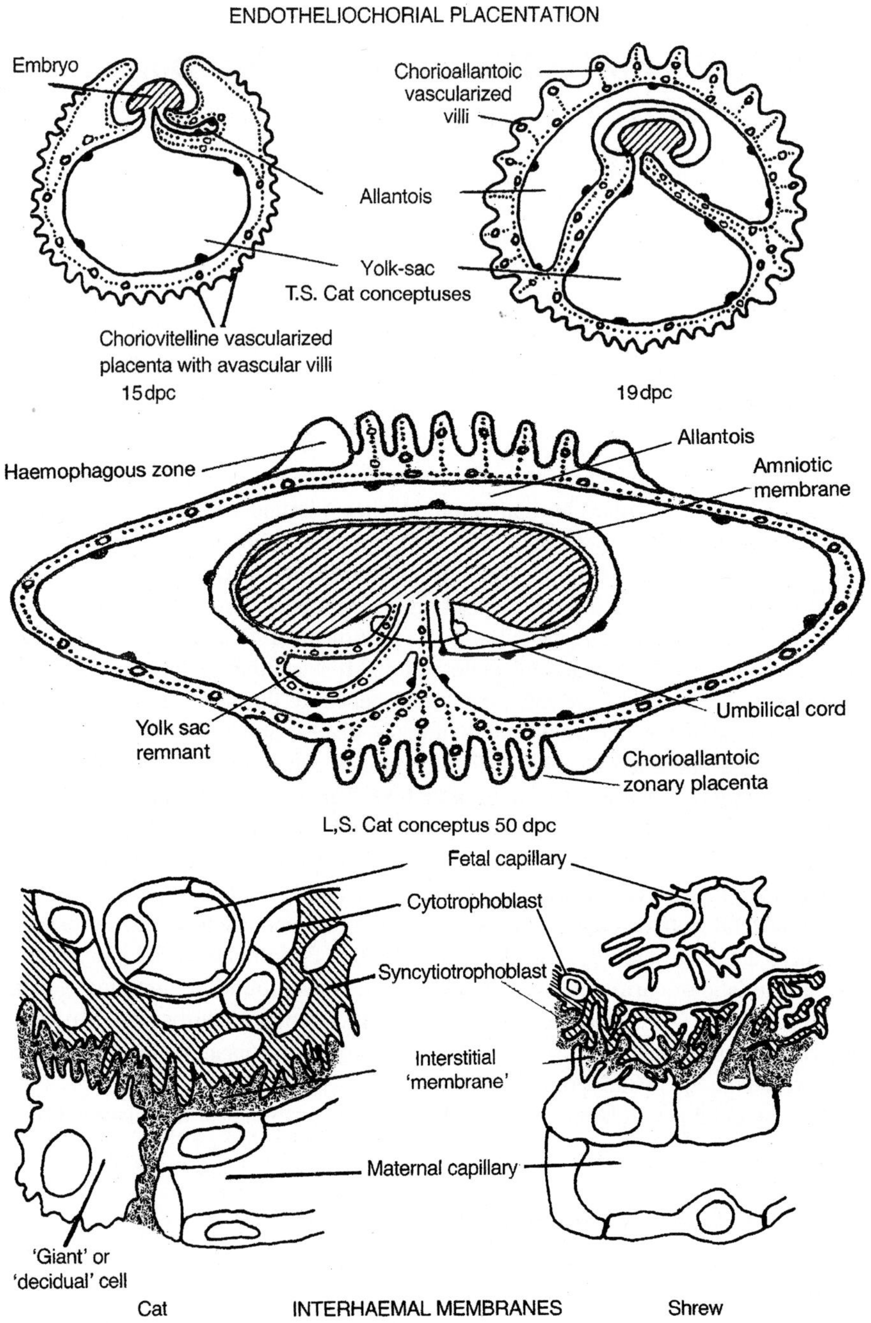

ENDOTHELIOCHORIAL PLACENTATION

Fig. 7.1 Endotheliochorial placental membrane development, mostly Carnivora

Ovulation: Spontaneous in the dog, induced by coitus in the cat. For endocrinological details during pregnancy see Concannon et al. (1989), Swanson et al. (1995).

Litter: 3–6.

Gestation: 55–65 days. Relaxin assay provides a possible pregnancy test at 28 dpc (Van Dorsser et al. 2006)

Implantation: superficial; central, antimesometrial.

Amniogenesis: folding.

Yolk sac: well-vascularized choriovitelline placenta up to 24 dpc with non-vascularized short villi of endotheliochorial structure (Fig. 7.1). The yolk sac is then separated from the chorion by growth of the allantoic sac but persists to term (Fig. 7.1) as a highly vascularized collapsed sac with a characteristic ultrastructure (Lee et al. 1983).

Chorioallantois: forms definitive placenta, the allantoic sac surrounds and vascularizes the amnion and chorion.

Shape: zonary.

Decidua: very infrequent modification of stromal cells in the endometrium just below the invading tips of the villi.

Cats, and to a lesser extent dogs, have large cells, probably of endometrial stromal origin, throughout the placental lamellae between the trophectodermal syncytium and the maternal endothelial cells. Most authors refer to them as decidual cells but nothing is known of their function (Fig. 7.8a) (Malassine 1974). Ruminants have similar cells in an equivalent position (see Fig. 7.8b).

Interhaemal membrane: endotheliochorial in the placental zone.

Accessory placental structures: areolae in the paraplacental region which has cellular trophectoderm loosely apposed to the cellular uterine epithelium. Haemophagous zones at the margin or centre of the zonary placenta (Figs. 1.4 and 7.5).

7.2.1 Fetal Membranes and Placental Development in Cat, Dog and Mink

The newly pregnant bicornuate uterus has a uniform development of numerous glands around the lumen (Wislocki and Dempsey 1946; Barrau et al. 1975). In the cat and dog, blastocysts enter the uterus at 5 dpc and establish an even spacing by 8 dpc (see chapter 2.1.1), at which time they are sufficiently large to distend the uterus. If only two blastocysts are present one implants in each horn about half-way up (Amoroso 1952). There is evidence for up to 40% transuterine migration between horns in cat and dog (Shimizu et al. 1990).

In several animals with endotheliochorial placentas, such as bears, badgers, mink and seals, the blastocyst can be held in diapause (Enders and Given 1977; Mead 1993). Implantation can be delayed for several months and reactivated by physiological mechanisms usually involving photoperiod and in mink, release of hypophysial Prolactin (Desmarais et al. 2004). Once the blastocyst has expanded to distend the uterus (central implantation) the yolk sac is vascularized over most of its mesometrial half, and a functional trilaminar choriovitelline placenta is formed (Figs. 7.1 and 7.2a).

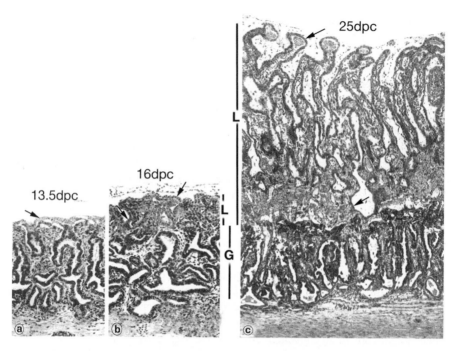

Fig. 7.2 Endotheliochorial cat placenta (see Fig. 7.1 for diagrams). Arrows indicate the tropho-blast. Developmental sequence, all at the same magnification, showing (**a**) early apposition (13 $^1/_2$ dpc), (**b**) short yolk sac placental villi, non-vascularized (about 16 dpc) and (**c**) vascularized fetal chorioallantoic villi or lamellae (25 dpc). Note the increase in length of both the fetal villi and maternal crypt sides (L) but the relative constancy of the depth of the glands plus junctional zone (G). (**a**) 13 $^1/_2$ dpc, ×120. (**b**) 16 dpc, ×120. (**c**) 25 dpc, ×120

In the cat, cellular contact and subsequent interdigitation of microvilli between trophectoderm and uterine epithelium starts on 13 dpc. Around this time full api-cal junction complexes have been observed (Leiser 1979) between cells from the two epithelia, but it is not clear whether these are formed as a result of fetal trophectoderm intrusion between, or fusion with, the cells of the uterine epithelium. The uterine epithelium undergoes degeneration with sporadic lateral fusions to form symplasm and a considerable amount of individual cell death. At this stage the cellular trophectoderm (cytotrophoblast) shows numerous divisions and pro-duces a syncytium inextricably enmeshed with remnants of the cellular and newly symplasmic remnants of the uterine epithelium (Leiser 1979). The fetal syncytio-trophoblast is actively phagocytic, capable of displacing and replacing the uterine epithelium and its basement membrane (Barrau et al. 1975; Leiser 1979). It then flows around the maternal capillaries and establishes the endotheliochorial pattern of the interhaemal membrane.

Initial penetration of the endometrium by the syncytium backed by the rapidly dividing cytotrophoblast is down the gland lumina, eroding the gland epithelium and basement membrane. Corrosion casts of the maternal endometrial vasculature

60dpc

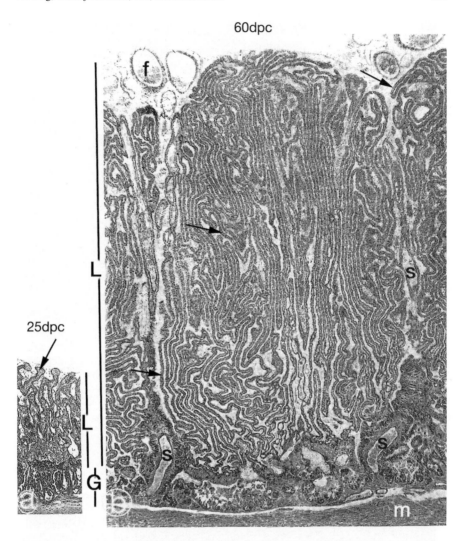

Fig. 7.3 Endotheliochorial cat placenta. Placental development showing the enormous increase in length and proliferation of the labyrinthine fetal and maternal lamellae (arrows) between 25 dpc (**a**) and about 60 dpc (**b**). Term is 61–63 dpc. The gland region depth (G) does not change much. Both (**a**) and (**b**) are at the same magnification. *f* fetal artery; *m* maternal connective tissue, *s* maternal stem arteries. (**a**) 25 dpc (**b**) 60 dpc, both ×24

at this stage (Pfarrer et al. 1999) show a network of capillaries around each regularly spaced gland. These networks are connected to a stem artery running up to a subepithelial position before branching to supply the capillaries. The initial *pattern* of the fetal villi is therefore dictated by the pre-existing gland distribution and their *growth* by the close proximity to the vascularised yolk sac. The shallow chorivitelline villi form over most of the spherical conceptus between 12 and 18 dpc but are not vascularised. The yolk sac is then pushed away from the chorion by the rapid

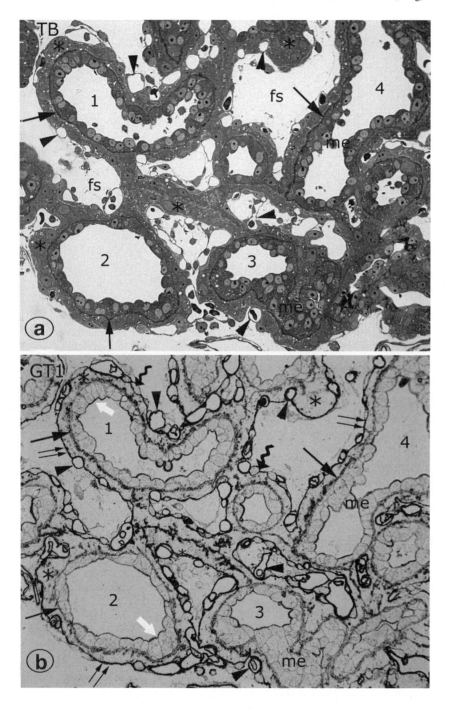

growth of the allantois. This membrane sac is much more aggressively vascularised and initiates blood vessel growth into each villus, initiating the chorioallantoic placenta at about 20 dpc (Fig. 7.1).

At the same time the maternal villi are formed by upward growth of the stem arteries. In the cat, the maternal villi are well established before there is any fetal vascularisation, but in mink the growth and vascularisation of both proceed in parallel (Wooding et al. 2007b). Growth of both villi depends upon remodelling of the connective tissue (Walter and Schonkypi 2006) and the effect of angiogenic stimuli provided by VEGF and its receptors expressed by the trophoblastic syncytium and the endothelia of maternal and fetal capillaries (Winther et al. 2001; Lopes et al. 2003).

As Mossman (1987) has pointed out, formation of the zonary placental villi cannot just be based on erosion by the fetal syncytium and its advance into the endometrium, the final length of the villi is much too long for this. There has to be considerable expansion of the maternal blood capillary system. This point can be best appreciated by examination of a developmental series of transverse sections through the cat placenta (Figs. 7.2 and 7.3) .

In the cat the structure of the initial interhaemal membrane is endotheliochorial, as it is at term (Figs. 7.8a and 7.9a). Indentation of the fetal cyto- or syncytiotrophoblast by the fetal capillaries is considerable towards the end of pregnancy (7.4a) and there is usually a definite attenuation of the amorphous interstitial layer (Anderson 1969; Barrau et al. 1975). The maternal endothelium varies considerably in thickness depending upon the species. In the bear and mink it is typically two to three times thicker than in seals and cats, but the functional significance of the differences is obscure (Amoroso 1952; Sinha and Erickson 1974; Wimsatt 1974).

The area of the interhaemal membrane increases and its thickness decreases as the fetus grows, thus continuously improving the transport potential of the placenta (Baur 1977). Only one glucose transporter isoform, GT1, is found on the apical and basal syncytiotrophoblast plasmalemma in cat, dog and mink (Fig. 7.4b, Wooding et al. 2007b). Amino acid transporters are present on the basal plasmalemma of the cat syncytiotrophoblast (Champion et al. 2004). There is also some ultrastructural evidence for immunoglobulin transport across the dog placenta to the fetus (Stoffel et al. 2000).

When the chorioallantoic membrane is first forming, the whole conceptus is expanding and elongating to become sausage shaped, with paraplacental regions

◄──

Fig. 7.4 Glucose transporter 1 localisation in the mink placenta. (**a, b**) Adjacent sections showing (**a**) toluidine blue (TB) stained resin section and (**b**) GT1 immunogold labeled section, no osmium, no counterstaining. (**b**) The endothelium (me) of the maternal capillaries (1–4) expresses a high level of GT1 on the apical plasmalemma (*solid white arrows*), with some on the lateral plasmalemmas also. The strongest GT1 immunoreactivity (IRT) is shown by the fetal capillaries (*arrowheads*) that indent the trophoblast and the basal plasmalemma (*double arrows*) of the syncytiotrophoblast (*asterisk*), which is occasionally interrupted by the cytotrophoblast (*wiggly arrows*). The apical syncytiotrophoblast displays a more diffuse GT1 IRT (**b**, *arrow*) and is apposed to the 'intermediate membrane', the thickened basement membrane of the maternal endothelial cells which is selectively stained with TB (**a**, *arrows*). 1.2 cm CRL, ×200

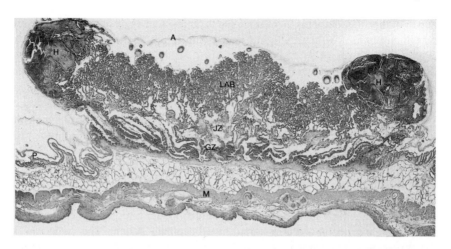

Fig. 7.5 Endotheliochorial dog placenta. Cross section through the zonary placenta of the dog at mid pregnancy. Note the large marginal hemophagous zones (H). *A* Allantoic membrane, *LAB* labyrinthine chorioallantoic placenta, *JZ* junctional zone, *GZ* glandular zone, *M* muscle in uterine wall, *P* paraplacental chorioallantoic placenta. About 35 dpc ×15 (from Bjorkman 1970)

outside the central zone of the placental villi. These regions have unchanged cellular trophectoderm and uterine epithelium loosely apposed with small areolae over the mouths of the numerous glands which are actively secreting throughout pregnancy (Leiser and Enders 1980a). The area of this paraplacental region (Figs. 7.1 and 7.5) is macroscopically larger than the zone of the placental villi but probably less than 5% of the total interhaemal area at term.

A characteristically large haemophagous zone develops at the edge or in the middle of the zonary endotheliochorial placenta of carnivores (Fig. 7.5).

It is lined on the fetal side by high columnar trophectoderm showing all stages of very active phagocytosis and digestion of the red blood cells from the stagnant pool of blood at the fetomaternal interface. On the maternal side the uterine epithelium may be either cellular or syncytial. Transient rupture of the occasional maternal blood vessel is probably instigated by the fetal trophectoderm at the edge of the haemophagous zone, but the details of the mechanism are unclear. The endothelial break is soon repaired and the uterine epithelium reforms over the top (Leiser and Enders 1980b).

Fig. 7.6 Endotheliochorial cat placenta. (**a,b**) Corrosion casts of resin injected fetal blood vessels. Note the lamellae formed from the fetal capillary meshwork in the labyrinthine zone in (**a**) and (**b**). The casting resin filled fetal arterioles (AI) and capillaries but not the venules (VI) whose lumina are indicated by the dashed arrows in (**b**). (**a**) 57 dpc; ×25. (**b**) 57 dpc; ×100, from Leiser and Kohler (1984). (**c**) Corrosion cast of maternal blood vessels. The maternal stem artery (SA) runs through the full depth of the placenta before supplying (arrowheads) the capillaries of the labyrinthine zone (LZ). The capillaries collect into broad venules (SV) in the junctional zone (JZ) which drain into veins (white arrow) in the endometrial muscle layer (M). Compare this figure with the histological cross section of the dog placenta on Fig. 7.5.From Leiser and Kohler 1983

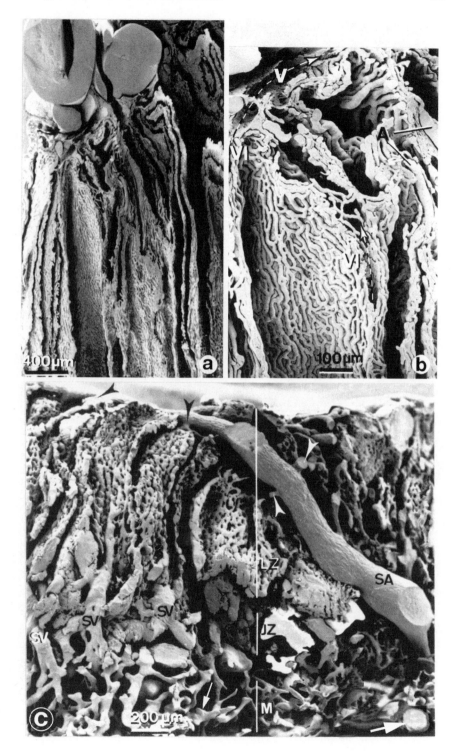

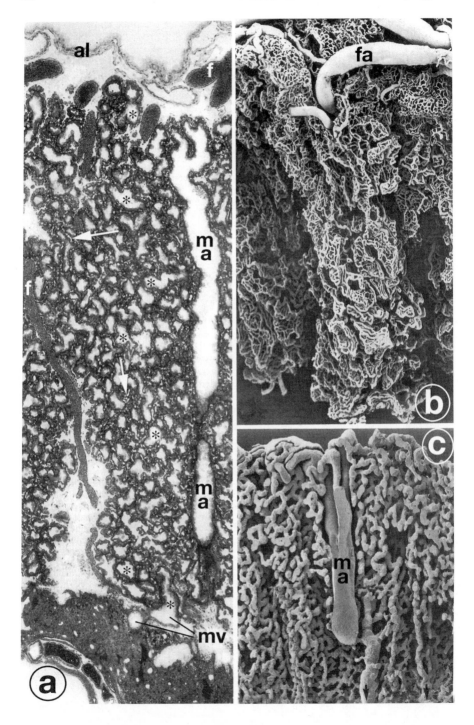

The gross form of the haemophagous zone or organ is very characteristic of the species. Cats and dogs (Fig. 7.5) have green or brown borders to their zonary placentas coloured by blood pigment residues, and in the sea otter there is a large central pedunculate sac. There is an excellent review of haemophagous zones by Burton (1982).

The intimate relationship of the fetal and maternal blood vessels in the carnivores and the relative volumetric insignificance of the tissue in between is elegantly demonstrated by perfusion of both fetal and maternal vasculatures prior to TEM (Fig. 7.4a). The overall vascular architecture is best demonstrated by histological and SEM studies of corrosion casts of the two vasculatures (Figs. 7.6, cat; Fig. 7.7, mink, Leiser and Kohler 1983; Krebs et al. 1997).

These clearly show that maternal blood passes directly to the fetal side of the placenta and fetal blood to the maternal side prior to flowing back through their capillary systems. The arrangement of the arteriolar supply to the capillary networks is such that the flow is considered to be predominantly cross-current (Fig. 2.6a) (Leiser and Kohler 1983).

Erosion of the original endometrial cells in the dog and cat is practically complete except for a population of cells which are distributed fairly evenly throughout the placenta in the amorphous (interstitial) material between fetal syncytium and maternal endothelium (Fig. 7.1 and 7.8).

Wislocki and Dempsey (1946) suggested that these cells originate from fibroblasts. They are typically large, 30–50 μm with one, or often two, large nuclei. They resemble fibroblasts in their massive fibrillar tracts containing actin and vimentin (Winther et al. 1999) but also have swollen rough endoplasmic reticulum and a large Golgi apparatus which suggests they may have considerable synthetic potential. In the cat they express TGFα, EGF and EGFR (Boomsma et al. 1997) amino acid transporters (Champion et al. 2004), and matrix metalloproteinases (Walter and Schonkypi 2006). Once the placenta is established these cells can divide and maintain their frequency as the placenta grows. They are very similar in ultrastructure and position to the residual endometrial cells described above for the sheep and goat synepitheliochorial placentas, and their function is equally enigmatic (Fig. 7.8).

Descriptions of endotheliochorial placental development refer to a zone of necrosis in the endometrium at the tips of the advancing fetal villi. In this region the gland epithelium becomes syncytial, the connective tissue oedematous, and both degenerate. From transverse sections of the placenta the impression given is that the necrotic

---

Fig. 7.7 Endotheliochorial mink placenta. (**a**) LM section through the full depth of the near term placenta. Perfusion fixation has emptied the maternal capillaries (*asterisks*) which are much larger than the meshwork of fetal capillaries (*white arrows*) with which they are surrounded. The maternal stem arteries (ma) pass through the full depth of the placenta before supplying the capillaries. This is also elegantly shown by the corrosion cast of the maternal blood supply on (**c**). (**b**) Corrosion cast of a fetal primary villous complex corresponding to the meshwork of capillaries indicated by the white arrows on (**a**). The stem artery (fa) enters, and vein (fv) exits at the top of the placental unit shown on (**a**). (**a**) ×65; (**b**) ×50; (**c**) ×75. From Krebs et al. 1997

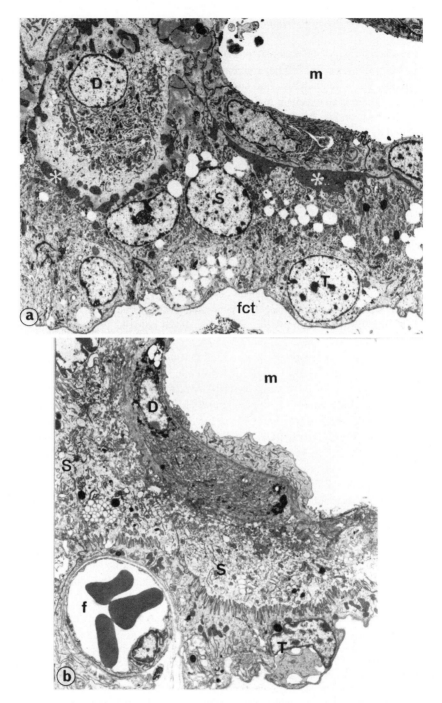

Fig. 7.8 Endotheliochorial placentation. Electron micrographs of the definitive interhemallayers in a cat (**a**), and a sheep (**b**), placenta. "Decidual" cells (D) in the maternal connective tissue are consistently present in the same position adjacent to the maternal endothelial cells. They have extensive rough endoplasmic reticulum, large golgi bodies and glycogen deposits. *fct* fetal connective tissue; *T* cellular trophectoderm; *S* syncytial trophectoderm; asterisks, 'interstitial membrane'; *f* fetal and *m* maternal, blood vessels (**a**) Cat. 35 dpc; ×3,000. (**b**) Ewe 140 dpc; ×4,000

zone represents an area of total tissue breakdown induced by the fetal syncytium, which then phagocytoses the resulting debris (Amoroso 1952; Barrau et al. 1975). The process is in fact very selective since the maternal blood vessels pass through this zone completely unchanged and they probably represent 30–50% of the volume of the structures initially present (Leiser and Kohler 1983). The closest analogy might be a building on extensible stilts, which are sufficiently robust to support and supply the building of the upper floors. These would correspond to the initial choriovitelline placenta. The blood vessel 'stilts' would eventually be incorporated into the final structure by growth of the chorioallantoic lamellar placenta.

The characteristics of carnivore placentation are remarkably consistent. The initially weakly vascularised endotheliodichorial villous choriovitelline placenta is colonized by the rapidly proliferating allantoic vasculature to produce the endotheliodichorial chorioallantoic placenta. This is always associated with a haemophagous zone, a large allantoic sac and a significant yolk sac remnant. The only exception so far reported is the hyen a, which has the typical large allantois and persistent yolk sac but a haemomonochorial definitive placenta which still has a marginal sinus similar in size to that of the cat haemophagous zone but filled with circulating rather than static maternal blood (Wynn et al. 1990; Enders et al. 2006).

7.2.2 Other Endotheliochorial Placentas

Definitive bat placentas show a considerable range of trophoblast forms with a few endotheliochorial but mostly (more than 90% of species) haemochorial in structure. However, uniquely, many definitively haemochorial bats show a sequence of two different chorioallantoic placental types. There is an initial diffuse endotheliodichorial area which is functional until just after midgestation when it regresses. Just before midgestation the definitive, separate discoidal hemochorial placenta develops which is functional for the second half of pregnancy when the exponential growth of the fetus requires maximum nutrient transport (Rasweiler 1993; Badwaik and Rasweiler 2000). It is the families with the more primitive anatomical characteristics which have definitive endotheliochorial placentas, but only one (Emballonuridae) has a clearly defined equivalent of a haemophagous zone. The trophoblast barrier may be cellular (Rhinopomatidae), syncytial (Emballonuridae) or dichorial with continuous cellular and syncytial layers *(Rhinolophus)*, (Mossman 1987; Badwaik and Rasweiler 2000). For details of the haemochorial bat placentas, see Chap. 8.6.2.1.

It was only by a study of the ultrastructure that the placenta of the insectivore *Blarina* (common shrew) was shown to be endotheliochorial (Wimsatt et al. 1973). The syncytial trophoblastic layer is uniquely attenuated and profusely fenestrated, and both maternal and fetal capillary endothelial cells extend slender processes from their bases to or through this lace-like trophectoderm (Fig. 7.9b). It certainly would be no barrier to macromolecules but may have some residual role in regulation or prevention of maternofetal cell traffic. A recent suggestion that in the final week of pregnancy it forms no significant barrier at all (Kiso et al. 1990) requires confirmation. *Blarina* (Soricinae) also has an unusual annular haemophagous zone

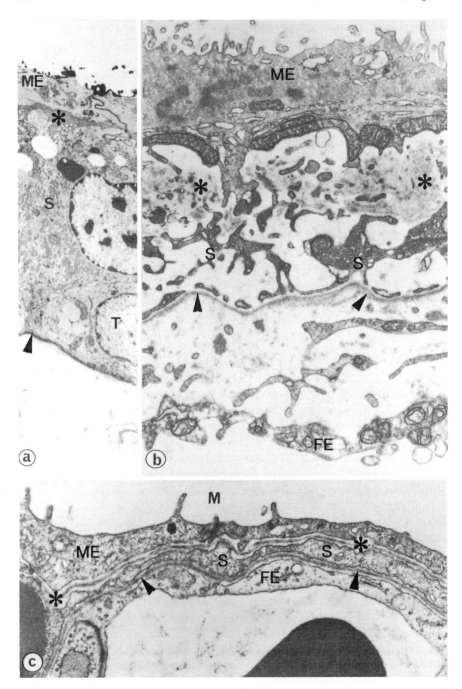

with absorptive trophectoderm underlain by vitelline (yolk sac) capillaries (King et al. 1978) rather than the usual allantoic vascular system as in carnivores or bats. This emphasizes again the plasticity of the fetal membranes with quite different parts adapting to serve the same function.

Fenestrated syncytiotrophoblast has been found in only one other order, Bradypodidae, the sloths, again with an endotheliochorial placenta. Here the layer is more substantial than in the shrews but still essentially porous. There is no haemophagous zone and the rest of the fetal membranes are, according to Mossman (1987), more reminiscent of anthropoids with a very early vascularization of the eventually discoid chorioallantoic placenta and the small non-persistent yolk sac (King et al. 1982).

The African otter shrews are endotheliomonosyncytiochorial plus the obligatory discontinuous cytotrophoblast layer to allow growth (Carter et al. 2006), with a central haemophagous sac. Neither of these features is found in the European shrews which are haemochorial (Mossman 1987).

Other examples of endotheliochorial placentation similar to the carnivore interhaemal membrane in structure are found in widely separated orders. Most [the aardvark (Orycteropodidae), two insectivores, *Talpa* (Malassine and Leiser 1984) and *Tupaia* (Luckhardt et al. 1985)] lack a haemophagous zone.

In the Elephantidae the trophoblast is a monocellular layer, no syncytium is formed. There is less invasion of the endometrium, the chorioallantoic lamellae form above the plane of the initial implantation. For iron transport there are many small haemophagous zones at the tips of the maternal villi, while glucose transport requires two different isoforms, GT1 on the basal and GT3 on the apical trophoblast plasmalemma (Wooding et al. 2005c).

Dipodomys, the kangaroo rat (King and Tibbitts 1969), also has cytotrophoblast as the only layer between the maternal and fetal capillaries (Fig. 7.9c), but there is no haemophagous zone.

The endotheliochorial placenta thus appears to be a successful solution for a variety of animals in a wide range of habitats with very different reproductive habits and maternal sizes. A haemophagous zone is a sufficiently frequent accompaniment to endotheliochorial placentation to indicate one advantage in developing a fully haemochorial placenta. Most orders with endotheliochorial placentas also have members with the haemochorial type.

Fig. 7.9 Most endotheliochorial placentas display a fairly uniform interhemal membrane (see Fig. 7.8), as typified by the cat (**a**) with a discontinuous cellular cytotrophoblast (T) which generates the thick syncytial layer (S) separated from the maternal endothelium (M) by a dense acellular interstitial layer (*asterisk*). However the Insectivore, *Blarina* (**b**) with its exiguous meshwork of discontinuous syncytiotrophoblast and the Rodent, *Dipodomys* (**c**), with its extremely thin syncytiotrophoblast layer (S) clearly demonstrate the variability possible within the endotheliochorial framework. The interstitial layer (asterisks) is also very different in these examples. Arrowheads, trophoblast basement membrane; *FE* fetal capillary endothelium; *M* maternal capillary; *ME* maternal endothelium. (**a**) 35 dpc; ×8,000 (**b**) Near term placenta; ×18,100 from Wimsatt et al. (1973). (**c**) Near term placenta; ×18,000, from King and Tibbits (1969)

Chapter 8
Haemochorial Placentation: Mouse, Rabbit, Man, Apes, Monkeys

8.1 Introduction

The epitheliochorial-synepitheliochorial-endotheliochorial sequence represents a gradual loss of interhaemal layers. The haemochorial is an apposite end to such a sequence; with the loss of all the maternal tissue, maternal blood irrigates the chorioallantoic fetal trophectoderm directly (Fig. 1.5).

Previous suggestions that in some species the fetal trophectoderm is also lost have not been substantiated by subsequent EM studies. The trophectoderm may be very thin but no placenta yet accurately described lacks a trophectoderm or its derivative as a barrier for the maternal blood, although pores in the trophectoderm have been reported (King 1992). Depending on the species the trophectoderm may have one, two, or three layers, each of which may be cellular or syncytial. The species-specific pattern is established soon after implantation and is maintained to term.

Fetal placental tissue is usually aggregated in a discrete area, discoid or limited zonary, they are all 'compact' placentas. The structure of the paraplacental area varies, with the yolk sac placenta playing an important role throughout pregnancy in rodents, lagomorphs and insectivores (Figs. 1.2 and 1.3) but probably only in early pregnancy in primates. No areolar development has been reported; the glands are now thought to play an important role in the first third of human pregnancy, in other species their secretions are probably significant only very early in pregnancy.

Typically only the species with haemochorial placentation induce decidual tissue transformation in the endometrium prior to placental formation. It is this newly expanded and decidualised tissue that is invaded and eroded to produce the definitive placenta which is 99% fetal tissue. The variation in decidualisation is enormous, with massive amounts in the rodent, moderate in the human and negligible initially in the rhesus monkey, where the placenta forms above the plane of implantation, once the maternal blood supply has been accessed.

It must be emphasized that, although most maternal tissue is lost in deeply invasive types, major maternal arterial channels persist throughout pregnancy through the full depth of the placenta in most haemochorial placentas, supplying fully oxygenated maternal blood direct to the fetal surface (subchorial lakes) of the placenta.

P. Wooding, G. Burton, *Comparative Placentation*,
© Springer-Verlag Berlin Heidelberg 2008

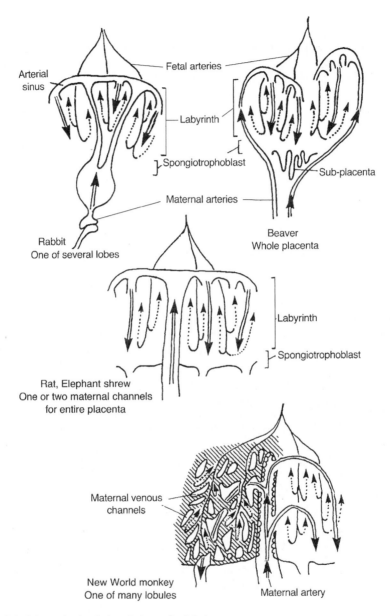

Fig. 8.1 Schematic circulations in hemochorial placentas

Usually several channels persist (rabbit, see below), sometimes only one [in the insectivorean elephant shrew (Starck 1949)] (Fig. 8.1).

The most notable exceptions to this are man, apes and the Old World monkeys where blood spurts up into fetal villi from below from arterial openings in the basal plate.

Haemochorial placentas are rather more varied in gross architecture than previous types with villous or labyrinthine structure and a variable but specific number of up to three interhaemal layers (see Chap. 8.6 and the excellent diagrams in Bhiwgade 1990; Enders et al. 1998).

As examples of the type with haemotrichorial structure the rat and mouse will be discussed in detail with rabbit (labyrinthine,haemodichorial) and man (villous; haemomonochorial), for comparison. Monkeys and other anthropoids, bats and other haemochorial examples are also briefly reviewed.

8.2 Haemotrichorial: Mouse, Mus musculus; Rat, Rattus norvegicus

Oestrous cycle: 4–7 day/s
Ovulation: spontaneous
Uterus: duplex
Litter: 8–9
Gestation: 19 day/s (mouse) 22 days (rat)
Implantation: interstitial
Amniogenesis: cavitation

Yolk sac: Initial bilaminar yolk sac which is the main conduit for nutrients for the first half of pregnancy but ruptures and disappears by late pregnancy. No vascularised antimesometrial choriovitelline placenta formed. The completely or incompletely inverted well-vascularized villous mesometrial (splanchnopleuric) half of the yolk sac placenta transports the nutrients diffusing in across the bilaminar yolk sac. The splanchnopleuric sac persists until term (Fig. 1.2).

Chorioallantois: Forms definitive placenta, vascularized by a solid allantoic mesenchymal bud (Figs. 1.2 and 8.3). The allantoic sac is vestigial or absent.

Shape: discoid, labyrinthine structure.

Decidua: extensive development throughout pregnancy in the endometrium around the conceptus (see Fig. 1.9).

Interhaemal membrane: haemotrichorial, two syncytia next to the fetal tissue and a cellular layer lining the maternal blood space. The fetal blood vessels probably have a villous organisation, the maternal blood space is labyrinthine (see below).

Accessory placental structure: completely inverted yolk sac.

8.2.1 Fetal Membranes and Placental Development in Mouse and Rat

In these rodents the blastocysts reach the uterine lumen on 3–4 dpc, and can go into embryonic diapause if the ovarian estrogen secretion required on 4 dpc for implantation is blocked, for example by a suckling stimulus (Lopes et al. 2004). Given the

estrogen stimulus, the blastocyst will implant after prostaglandin induced uterine contractions, influenced by the individual blastocysts, have ensured an even spacing
(Hama et al. 2007). Closure of the uterine lumen then immobilizes the blastocysts and
ensures an intimate apposition between blastocyst and uterine epithelium (Enders et al.
1980). This involves specific changes in several transmembrane proteins (for example:
osteopontin, White et al. 2006; HBEGF, Das et al. 1997) and glycocalyx carbohydrate
expression (Kimber and Lindenberg 1990; DeSouza et al. 1999) probably controlled
in part by LIF (Kimber 2005) and requiring 9CBP or 28CBP in the uterine epithelium
(Luu et al. 2004). There is no appreciable swelling of the blastocyst prior to closure,
and the blastocyst invariably implants at the antimesometrial side of the uterine lumen,
with the embryonic disc mesometrial (Amoroso 1952; Kaufmann 1983).

   Cellular interactions start at 5 dpc with trophectodermal penetration between and
phagocytosis of some uterine epithelial cells (Fig. 8.2) (Bevilacqua and Abrahamsohn
1988) followed by complete death and delamination of the whole uterine epithelium (Enders et al. 1980; Schlafke et al. 1985; Blankenship et al. 1992 (hamster);
Welsh and Enders 1991a; Blankenship and Given 1992).

   At this stage the blastocyst is bilaminar but mesoderm is starting to form at the
edge of the mesometrial embryonic rudiment underneath the trophectodermal cells

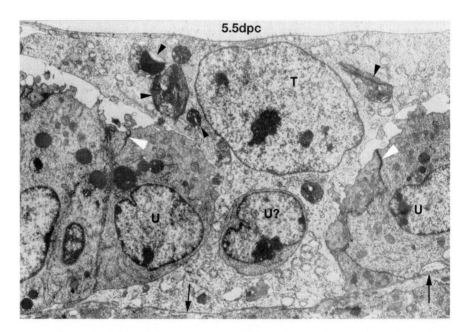

Fig. 8.2 Hemochorial placentation. Early displacement implantation in the mouse. A trophectoderm cell (T) containing several lysosomes (*arrowheads*) has completely penetrated through to the
basement membrane (*arrows*) of the uterine epithelium. It is phagocytosing one uterine epithelial
cell (U?) and undermining and delaminating the adjacent epithelial cells. No junctions are formed
between trophoblast and uterine epithelial cells equivalent to those between uterine epithelial cells
(*white arrowheads*). 5.5 dpc; ×3,450, from Kaufmann (1983)

surrounding the proamniotic cavity. Above the cavity, an aggregate of trophectodermal cells is produced, referred to as the ectoplacental cone.

Also at this stage (6 dpc), the antimesometrial decidual cells form an avascular, closely packed zone around the conceptus, isolating it from maternal blood (Parr and Parr 1986; Parr et al. 1986) Synthesis of the tight junction proteins sealing the decidual cells together requires a trophoblast signal, since oil droplet deciduoma formation does not trigger such a zone (Wang et al. 2004). However there is no ultrastructural evidence for such tight junctions, only for gap and adherens junctions.

The peripheral trophectoderm of the ectoplacental cone will differentiate into the polar or secondary giant cells which invade the uterine endometrium, by now considerably decidualized (Parr et al. 1986) (Fig. 1.9). Over the antimesometrial two-thirds of the blastocyst, away from the ectoplacental cone, the endodermal cells secrete material which thickens the basement membrane separating the endoderm from the mural trophectoderm. This has by now differentiated into the primary giant cells (Fatemi 1987; Mazariegos et al. 1987). This thickened basement membrane is then referred to as Reichert's membrane (Fig. 1.9). As the blastocyst expands, the decidual exclusion zone regresses. This allows the mural or primary giant cells to rupture the maternal blood vessels and form lace-like extensions through which the blood circulates but does not penetrate past Reichert's membrane (Kaufmann 1983; Welsh and Enders 1987). The endoderm and primary giant cells then form a non-vascularized bilaminar yolk sac placenta which is the only conduit for embryonic nutrients for the first half of pregnancy. As the blastocyst develops and increases in size, this bilaminar region maintains a similar structure but becomes ever thinner, eroding the decidua capsularis between the giant cells and the myometrium but finally completely rupturing on 17–18 dpc in the rat (term is 22–23 dpc). This exposes the vascularized endodermal layer (referred to as the inverted visceral yolk sac placenta) to the reformed uterine epithelium which is completely restored by 15 dpc in the rat (Welsh and Enders 1983) (Figs. 1.3 and 1.9). Absorption of nutrients and maternal immunoglobulins across this yolk sac placenta throughout pregnancy is of considerable importance to embryonic growth and neonatal immunoregulation (Bainter 1986).

After the initial differentiation of the primary giant cells the yolk sac is inverted below the embryo as it grows; above, the allantoic outgrowth from the fetus reaches the base of the ectoplacental cone on 8 dpc (mouse) and 10 dpc (rat) (Fig. 1.2). This has been beautifully demonstrated by SEM in the rat (Fig. 8.3) (Ellington 1985, 1987).

The cone appears to have three layers: basal or inner, outer and a shell of secondary giant cells invading the mesometrial decidua. As soon as the allantoic outgrowth reaches the bottom of the ectoplacental cone it erodes the mesothelium and spreads out into a mesodermal pad closely underlying the basal ectodermal layer of the cone. An excellent pictorial sequence of placental membrane and embryonic development can be found in Rugh (1975) which illustrates the growth of the labyrinth and spongiotrophoblast areas of the mouse chorioallantoic placenta (see Fig. 1.9).

Recent molecular biological work, notably Cross's group (Cross 2006), has elucidated some of the genes involved in growth of the fetal vasculature. Gcm-1, for example, at 9 dpc is only expressed in a pattern of small groups of basal ectoplacental

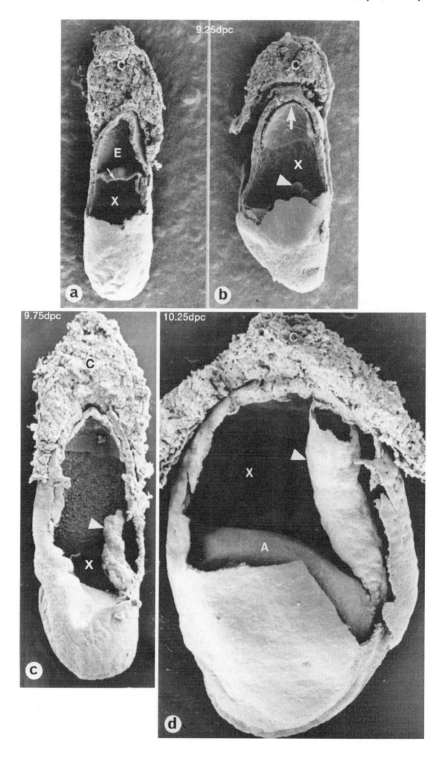

cone cells immediately above the mesodermal pad (Anson-Cartwright et al. 2000). These are the cells which are the initiators of the mouse fetal villi and transform into the three chorioallantoic placenta trophoblast layers covering the villi as they invade the expanded maternal blood sinusoids (Fig. 8.4).

Gcm-1 KO mice do not form villi and die soon after 10 dpc. The three layers express different molecular markers, are self-generating and rapidly proliferate, driven by the growth of the fetal blood vessels that the villi contain. At the tips of the elongating villi the outermost cellular layer is continuous with the spongiotrophoblast layer whose cells have replaced the endothelium of the maternal capillary sinusoids draining the maternal blood (Fig. 8.4). Outside the spongiotrophoblast the secondary giants face the mesometrial decidua which has expanded enormously. The endometrial spiral arteries, after crossing the uterine muscle layers have elongated into this expansion and feed the dilated sinusoidal capillaries throughout the decidua (Welsh and Enders 1991b). This tissue expresses many remodelling enzymes and their modifiers (Reponen et al. 1995; Leco et al. 1996) and will be eroded by apoptosis and the secondary giant cells to provide space for placental growth. This growth is based on the elongation of the fetal villi below the spongiotrophoblast. Elegant corrosion cast studies of the vascular system of the developing mouse placenta (Adamson et al. 2002; Rennie et al. 2007) clearly show that the labyrinth consists of individual fetal capillary units each of which has a central arteriole running to its tip before dividing into capillaries to drain the blood back to the embryo (Fig. 8.5). Initially each villus presumably consists of a single loop of allantoic arteriolar supply and venular drainage both enveloped by the three layers of trophoblast (Fig. 8.4). The vein then grows more slowly because the arteriole as it elongates is constantly proliferating a meshwork of trophoblast covered capillaries forming a capillary unit (see Fig. 8.5) as they drain back down to the vein. This growth pattern ensures that in the fetal capillary units blood always flows back (down) towards the fetus.

Development of the trophoblast-enveloped fetal vasculature (Metz 1980 [rat]; Carpenter 1972, 1975 [hamster, which is also haemotrichorial]) depends upon a variety of growth factors. VEGF (Steingrímsson et al. 1998), IGF-2 (Sibley et al. 2004) and PDGF (Ohlsson et al. 1999) have been implicated from knockout studies. The fetal blood flow is always in the opposite direction to the maternal blood. The maternal blood is carried to a sinus around the bases of the fetal villi by 2–3 persistent maternal arterial channels or conduits and then flows back to the mother through the expanding labyrinth region (Fig. 8.5d, e, g). These channels are developments of the spiral arteries which have been invaded by trophoblast cells from the ectoplacental cone as early as 7.5 dpc in the mouse (Hemberger et al. 2003). At this stage they feed into the ectoplacental-cone-cell-surrounded dilated subepithelial

---

Fig. 8.3 Scanning electron micrographs of the rat conceptus after removal of the parietal yolk sac, all at the same magnification, dissected to show the internal membrane development. (**a, b**) Expansion of the exocelom (X) inverts the chorion (*small arrow*) which bounds the ectoplacental cavity to form a double layer of trophectoderm lined with mesoderm (*large arrow*) under the ectoplacental cone (Fig. 8.3a). This forms the basis for development of the allantochorionic placenta when vascularised by growth (Figs. 8.7c, **d**) of the solid mesodermal allantois (*arrowheads*) which originates from the hind end of the embryo here covered by the amnion (A). (**a**) 9.25 dpc; ×220. (**a**) 9.25 dpc; ×220, (**c**) 9.75 dpc; ×220. (**d**) 10.25 dpc; ×220, from Ellington (1985, 1987)

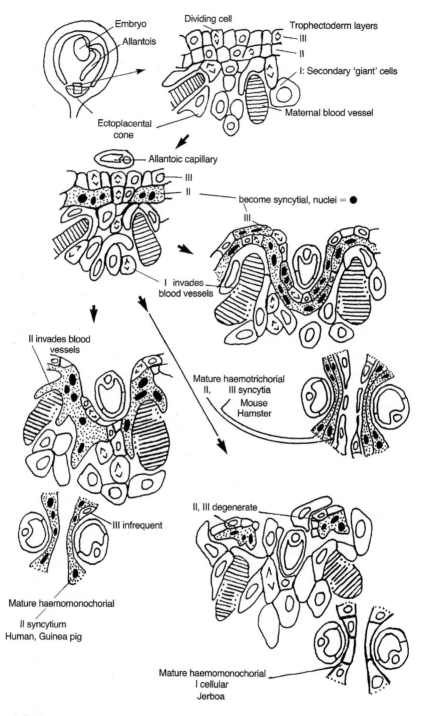

Fig. 8.4 Placental development in multilayered hemochorial placentas

sinusoidal maternal vessels into which the fetal vasculature will grow. The channels will continue expansion and elongation up to 15 dpc. The trophoblast cells replace the endothelium of the channels and spiral arteries and in the decidua erode the muscle sheath of the spiral arteries down as far as the myometrial layers in the mouse but much further in other species (Pijnenborg et al. 1981 (man); Carpenter 1982 (hamster); Hees et al. 1987 (guinea pig); Vercruysse et al. 2006 [rat]). These vascular modifications are dependent upon uterine natural killer (uNK) cells as well as trophoblast cells, the uNK cells form cuffs around the vessels at 11 dpc and start to degenerate from 15 dpc (Croy et al. 2002, 2006).

From the start of growth of the fetal vessels, fetal and maternal blood flows are separated by three layers of trophoblast, two syncytia derived from the inner ecto-placental cone layers and the outer cellular layer continuous with the spongiotro-phoblast cells. These are the basic interhaemal layers in the labyrinth through the rest of gestation and the usual nomenclature for this maternofetal 'barrier' is mater-nal blood/layer I/layer II/layer III/fetal endothelial basement membrane/fetal endothelium/fetal blood (Fig. 8.4).

It has generally been considered that the individual capillaries in the capillary units (see above and Fig. 8.5a) of the fetal villi become confluent with one another as they increase in length forming the labyrinth at the expense of the mesometrial decidual tissue but there is no evidence for this from the 3D corrosion cast studies of the mouse fetal and maternal vascular systems (Fig. 8.5a; Adamson et al. 2002). The capillary units near term show no linkages (Fig. 8.5a). In vivo the capillary units are probably linked together by the outermost layer I of their enveloping tro-phoblast. The corrosion casts of the maternal blood space (Fig. 8.5d, e) show that it is labyrinthine. It is bounded only by the trophoblast layer I. and is clearly differ-ent architecturally from the individual fetal capillary units (Fig. 8.5a).

The main maternal arterial channels run independently through the whole thick-ness of the placenta directly to the subchorial lake at the base of the fetal placenta, from where the maternal blood flows back through the labyrinthine network formed by the fetal capillary growth (Fig. 8.5).

Growth of a haemotrichorial placenta like that of the rat or mouse presumably is based upon small groups or individual mitotically active cells within the main attenu-ated layers (I, II and III) which separate fetal and maternal blood. Layers II and III are in fact syncytial from the earliest stage and, as no nuclear division has yet been reported in such syncytia, some system for growth and extension of the layers is nec-essary (Davies and Glasser 1968; Metz 1980). However the initial rapid proliferation of the layers probably only lasts until ~13 dpc and the subsequent fourfold to fivefold expansion in the area of the trophoblast depends upon thinning and expansion of the cytoplasm of the layers (Cross 2007 personal communication). Layers II and III are very closely apposed and connected by numerous gap junctions (Metz et al. 1976) one function of which is probably to facilitate maternofetal glucose transport (Shin et al. 1997). The combined layers form a complete barrier to tracers introduced into mater-nal (Metz et al. 1978) or fetal (Aoki et al. 1978) circulations. The outermost layer (I) is cellular and fenestrated, and forms no barrier to tracers added to the maternal circu-lation. Enders (1965) has suggested that this layer I functions by slowing the maternal

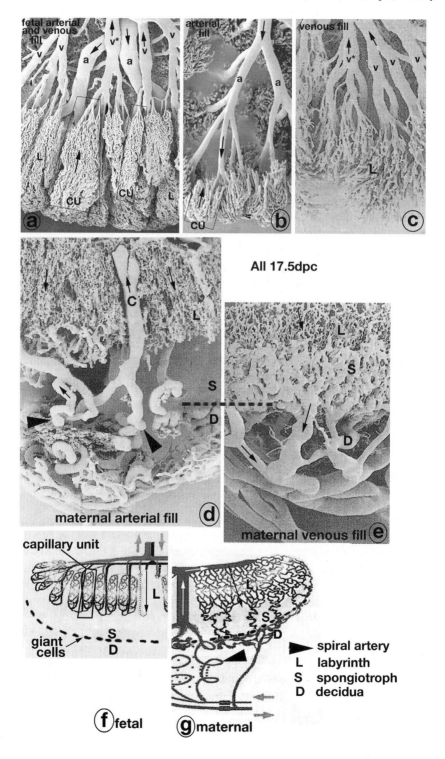

blood flow and forming local regions of blood stasis behind the fenestrations which facilitate fetomaternal transport. There may be special difficulties in fetal uptake from maternal blood which are alleviated by such a slowing of the flow. Clearly a balance is needed between rapid flow to maintain solute concentrations and efficient capture of the substrates by the membrane transporters in the plasmalemma.

Morphometric studies in the mouse (Coan et al. 2004) show an increase in the volume of the labyrinth up to 18.5 dpc, with the fetal capillaries increasing in length and surface area but decreasing in diameter. The volume and surface area of the maternal blood space also increase rapidly between 14.5 and 16.5 and then stay fairly constant. However the spongiotrophoblast zone shows a 15% decrease in volume at 18.5 dpc from a peak at 16.5 dpc, due to migration of the glycogen cells (see below). The maternofetal 'diffusion barrier' decreases from 12.5 to 5 μm between 11 and 18.5 dpc increasing the theoretical diffusing capacity.

The ultrastructure of the maternofetal barrier changes little during the thinning of the layers during pregnancy except for a significant increase in the size and DNA content restricted to the nuclei of the trophoblast layer I cells (Coan et al. 2005). These cells have recently been shown to be one of the four classes of polyploid trophoblast giant cells characterised by their different location and functions in the mouse placenta (Whitely et al. 2002; Simmons et al. 2007). Parietal giant cells (PTGC) are the primary and secondary giant cells discussed above, spiral artery giants (SpaTGC) are the cells which invade the spiral arteries from an early stage, maternal arterial canal giants (CTGC) are the cells lining the conduits/channels through the labyrinth and the cells of layer I (STGC) are the final category. Each class produces a different range of Prolactin or Placental lactogen isoforms or paralogs (Hemberger et al. 2003; Ain et al. 2003) and placental specific cathepsins (Simmons et al. 2007) at different stages of pregnancy. Rat and mouse Parietal giant cells also contain 3βhydroxysteroiddehydrogenase activity indicating that they produce steroids (Deane et al. 1962). Whether other rodents have the same variety of giant cells remains to be investigated. Certainly Hamster primary and secondary parietal placental giant cells at 12 dpc (term 16 dpc) have been shown to contain

Fig. 8.5 Hemochorial placentation. Corrosion casts of mouse placental vasculature at 17.5 dpc (term 18.5 dpc). Blood flow directions are indicated by the arrows. (**a**) Complete filling of arterial (a) and venous (v) vessels from the umbilical artery. The capillary units (CU) drain into the veins (v). The capillary units link artery to vein but this is not obvious on this micrograph because the capillary meshes in the labyrinth zone (L) hide the course of the arteries. The arterial course can be clearly seen on (**b**), where the vasculature was filled from the umbilical artery only to the start of the capillary units (CU), demonstrating that the fetal arteries deliver blood directly the tips of the capillary units in the labyrinth. From this point the blood would flow back down the capillary meshwork and link to a vein as in (**a**). (**c**) Fetal venous vasculature filled from the umbilical vein only as far as the bases of the incompletely filled capillary units (*asterisks*). The pattern of venous outflow (e.g. v*) is very similar to that shown at v* in (**a**). (**d**) Maternal spiral arteries (*arrowhead*) filled from a uterine artery and supplying a placental arterial conduit (C). This delivers blood to the base of the fetal placenta and from there back down through the initial labyrinthine drainage channels (L), (**e**) Maternal venous spongiotrophoblast drainage channels (S) and basal labyrinth (L) filled from the uterine vein. (**f,g**) Schematic reconstruction of the fetal and maternal blood flows. All modified from Adamson et al. (2002)

Placental lactogen II and relaxin, as did large cells in the wall of their mesometrial arteries (Renegar et al. 1990).

A further specific population of non-giant but migratory trophoblast cells have been identified in the mouse and rat – the Glycogen Cells (GC). These are first recognisable as small islands of small cells in the spongiotrophoblast layer with appreciable amounts of glycogen (Vercruysse et al. 2006). as well as other specific markers (protocadherin, Bouillot et al. 2006; connexin 31, Coan et al. 2006).They increase in number 80 fold between 12.5 and 16.5 dpc and migrate interstitially in increasing numbers from 11 dpc into the decidua as deep as the myometrium. They form extensive cuffs around, but not in, the decidual spiral arteries (Ain et al. 2003) after these arteries have been remodelled and widened by the spiral artery endovascular giant cells and uNK cells. As they migrate, the decidual uterine NK cells coincidentally (?) are decreasing in numbers and in uNK KO mice the GC cell migration is considerably greater (Ain et al. 2003). The GCs vacuolate and lose or release their glycogen between 17 and 18 dpc and this glycogen is suggested to be an important source of nutrient for the fetus (Bouillot et al. 2006) presumably via the spiral arterial inflow. Whether the GCs play an essential role in pregnancy has yet to be established.

The Parietal secondary TGC become very large with a characteristically enormous, deeply divided nucleus, of variable but usually very high DNA content (up to 1,024 N) whereas the other TGC all show the same moderate polyploidy (Simmons et al. 2007). As the embryo grows the giant cells become stretched, and some of the secondaries are replaced by cell division in the underlying spongiotropholast tissue (Ilgren 1983; Kaufmann 1983). These secondary giant cells are rarely as large as the primaries and grade into the diploid cells of the spongiotrophoblast layer which form a meshwork through which the maternal blood flows. Unlike the adjacent labyrinthine layer there is no fetal vascularization and the trophospongium has a fine structure very different from the giant cells, but both are of the layer I population.

Layers II and III are dependent on the fetus for continued growth and maintenance, since fetectomy at 12 dpc results in their rapid degeneration together with the fetal villi. However, layer I continues to growth after fetectomy as a residual meshwork in which the maternal blood circulates (Davies and Glasser 1968).

8.3   Haemodichorial: Rabbit, Oryctolagus cuniculus

Oestrous cycle: none
Ovulation: on coital stimulation
Uterus: duplex
Litter: 8–12 gestation: 30–32 days
Implantation: superficial, antimesometrial
Amniogenesis: folding

Yolk sac: (1) Large bilaminar non-vascular yolk sac persists until 16 dpc, then finally ruptures. Consists of giant trophectoderm cells, a basement membrane (much thinner than the equivalent Reichert's membrane in the rodents) and endoderm.

(2) Inverted yolk sac placenta, vascularized villus folds adjacent to embryo. Active in uptake of macromolecules, especially immunoglobulins, during pregnancy.

Chorioallantois: definitive placenta vascularized by mesoderm of allantoic sac
Shape: discoid, labyrinthine internal structure
Decidual development: considerable
Interhaemal membrane: haemodichorial

8.3.1 Fetal Membranes and Placental Development in Rabbit

The early conceptus development is very different from the rodents, but differentiation of the yolk sac and chorioallantoic placenta is essentially similar (Amoroso 1952).

There is considerable blastocyst swelling prior to central implantation between 7 and 8 dpc (Fig. 8.6a)

Even spacing of blastocysts is achieved by the usual uterine muscle contractions refined eventually by interaction with the expanded 4–5 mm diameter blastocysts. Sea urchin eggs introduced into the pseudopregnant uterus are also evenly distributed although they are much smaller than the rabbit blastocyst (Boving 1971).

Changes in the uterine epithelial glycocalyx at implantation are detailed in Anderson and Hoffman (1984) and Anderson et al. (1986).The EGF family and its receptors are expressed in trophoblast and uterine epithelia and are closely involved in the implantation process (Klonisch et al. 2001). The zona pellucida has broken down and been replaced by an acellular capsule around the expanded blastocyst as in the horse (see Sect. 5.3.1). The trophectoderm forms knobs of syncytium which penetrate through the capsule and contact the uterine epithelium. The first cellular changes at implantation (Fig. 2.2) are fusion between the knobs and the cellular uterine epithelium, initially at the antimesometrial side and subsequently over the mesometrial side (where the chorioallantoic placenta eventually forms) (Larsen 1961; Enders and Schlafke 1971a). The uterine epithelium then fuses into syncytial plaques, each with a fairly constant number of nuclei and characteristic ultrastructure. This change also occurs in pseudopregnancy and can be induced with the correct hormonal regimen with no conceptus present (Davies and Hoffman 1975; Davies and Davenport 1979).

Over the antimesometrial half of the uterus the fetomaternal hybrid syncytiotrophoblast displaces the uterine epithelial syncytial plaques which degenerate into symplasmic masses which are eventually sloughed and removed by phagocytosis. In the absence of a blastocyst or physical trauma the syncytial plaques disappear in a much more controlled fashion. Reversion of individual multinucleate cytoplasmic areas to single uninucleate cells has been claimed (Busch et al. 1986). However, restoration of the uninucleate epithelium seems more likely to be by proliferation of residual uninucleates as in ruminant intercotyledonary areas.

The number of fetal nuclei incorporated into the maternal syncytiotrophoblastic knobs is not known, nor is their fate. Larsen (1961) says they 'dissolve' but offers no pictorial evidence for this. The (feto?)maternal syncytiotrophoblastic knobs which

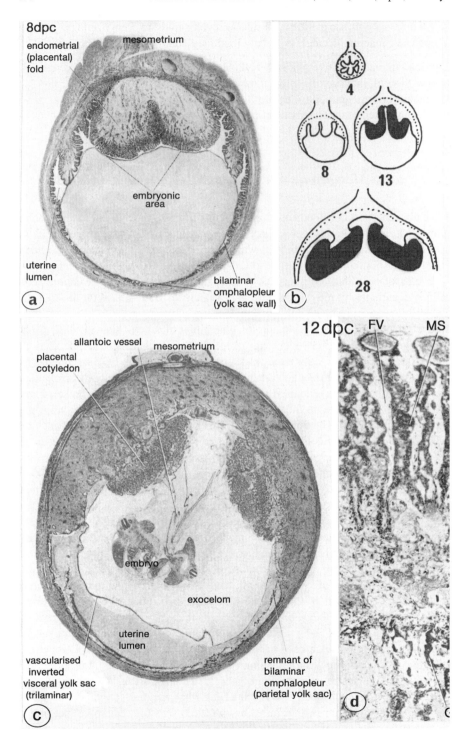

8dpc

endometrial (placental) fold

mesometrium

embryonic area

uterine lumen

bilaminar omphalopleur (yolk sac wall)

(a)

(b)

4

8

13

28

12dpc   FV   MS

allantoic vessel   mesometrium

placental cotyledon

embryo

exocelom

uterine lumen

vascularised inverted visceral yolk sac (trilaminar)

remnant of bilaminar omphalopleur (parietal yolk sac)

(c)

(d)

penetrate the antimesometrial uterine epithelium basement membrane seem to be directly involved (Blackburn et al. 1989) in the production of the uninucleate giant cells which form all round the lateral and antimesometrial aspects of the conceptus (Amoroso 1952). Yolk sac development, with an antimesometrial bilaminar segment which ruptures at 16 dpc, leaving a fully inverted yolk sac placenta, is essentially similar to the rodents.

At the mesometrial pole once the amniotic folds have fused, the trophectoderm over the embryo invades and displaces the uterine epithelial syncytial plaques over the two main endometrial folds (Fig. 8.6a) and subsequently penetrates the maternal endothelium as it did over the antimesometrial region (Hoffman et al. 1990a,b). The invading fetal tissue does not form any structure equivalent to the ectoplacental cone in rodents but presents a broad syncytial front toward the folds of decidualizing maternal tissue, backed by and presumably formed from the cellular cytotrophoblast (Fig. 8.6a). Thus, as soon as the fetal processes reach the maternal blood spaces they have the haemodichorial structure; maternal blood, fetal layer I syncytial, layer II cellular, fetal endothelium which characterizes the definitive placenta (Fig. 8.7; Enders 1965; Enders and Schlafke 1971a). The allantoic vesicle has by now (9 dpc) reached the mesometrial trophectoderm and it then actively vascularizes the extending front of fetal tissue. As in the rodents the larger arteries are invaded by trophectoderm (Pijnenborg et al. 1981) but persist as channels running through the full depth of the forming labyrinthine haemodichorial placenta to supply maternal blood to the subchorial lake at the base of the fetal placenta (Fig. 8.1) (Hafez and Tsutsumi 1966; Carter et al. 1971). From here it flows in the newly formed fetal tissue channels and drains via maternal veins. Considering the final dimensions, the placenta must grow by extension of the fetal tissue just as much as by erosion of the endometrium (Fig. 8.6).

The definitive placenta of the rabbit has fewer fetal giant cells (Fig. 8.8) between the mesometrial decidua and the syncytial spongiotrophoblast than the rodents, and there is no evidence for any placental lactogen production. However granules in the syncytium have been shown to contain the hormone relaxin (Lee et al. 1995). The haemodichorial interhaemal membrane persists but thins gradually to term (Fig. 8.7, Samuel et al. 1975).

Glucose transporter 1 is present in trophoblast fractions (Kevorkova et al. 2007) but no localization has been shown as yet. Layer II is normally much thinner than I, but both vary considerably (Fig. 8.7) and can be equally thin (Fig. 8.7c).

Fig. 8.6 Hemochorial placenta of the rabbit. (**a**) The blastocyst swells considerably before implantation and implants mesometrially on the two largest decidualised endometrial folds. (**b**) The two cotyledons of the chorioallantoic placenta increase considerably in extent during pregnancy (all at same magnification). (**c**) Early development of the two placental cotyledons; the histological structure of this 12 dpc placenta (term 30–32 day/s) is shown in (**d**). The fetal villi (F) and the maternal blood sinuses they enclose (W) increase in length continuously so that invasion of the maternal endometrium only accounts for a small proportion of total final depth of the placenta. Note the symplasmic remnant (Y) of a gland in the junctional zone. (**a**) 8 dpc, ×7. (**b**) traced from Hafez and Tsutsumi (1966), the numbers indicate dpc, term ~31 dpc. (**c**) 12 dpc, ×6.5 (**d**) 12 dpc, ×60

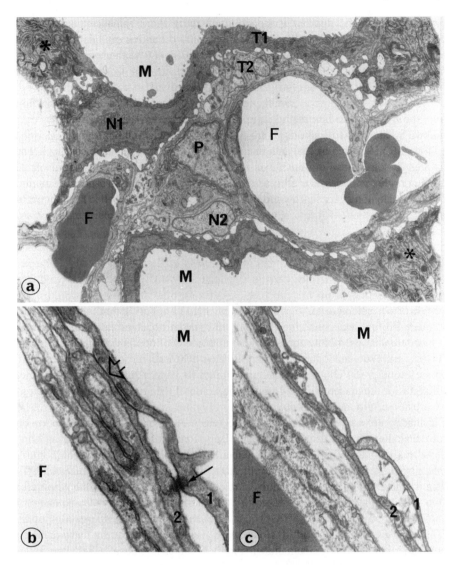

Fig. 8.7 Hemodichorial rabbit placenta. (**a**) Definitive labyrinth structure. The maternal blood space (M) is bounded by two layers of trophectoderm (Tl, T2) each of which show nuclei (N1, N2), T1 is a syncytium, T2 unicellular. T1 has far more rough endoplasmic reticulum (*asterisks*) than T2 which is considerably thinner. Fetal blood capillaries, F; fetal pericyte, P. (**b**) There are frequent gap (*open arrow*) and small desmosomal (*small arrow*) junctions between layers 1 and 2. (**c**) Although layer 2 is usually thinner both layers frequently reduce to an equivalent exiguous extent. (**a**) 22 dpc, ×4,000, (**b**) 22 dpc; ×35,000. (**c**) 22 dpc; ×15,000

They are frequently connected by small desmosomes and gap junctions (Fig. 8.7b). The structure of the spongiotrophoblast is syncytial and continuous with layer I of the labyrinth but with quite different ultrastructure to that in the adjacent giant cells (Fig. 8.8).

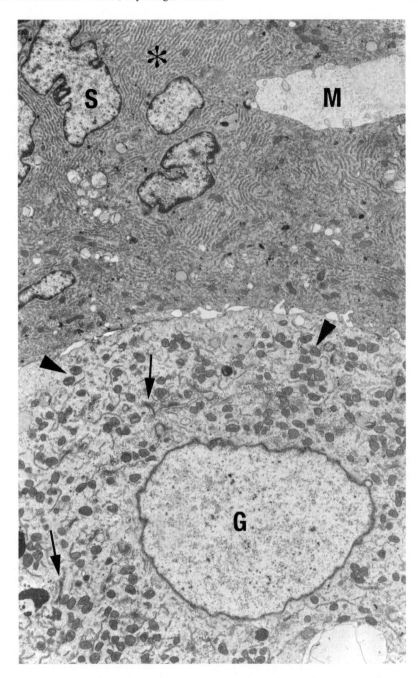

Fig. 8.8 Hemodichorial rabbit placenta. The fine structure of the labyrinthine layer T1 on Fig. 81 is similar to that in the syncytial spongiotrophoblast (S) shown here with massive arrays of endoplasmic reticulum (*asterisk*). The giant cells (G) have a very different ultrastructure with larger and more numerous mitochondria (*arrowheads*) many golgi bodies (*arrows*) and scattered individual strands of endoplasmic reticulum. *M* maternal blood space

Since the maternal blood flows back from the fetal subchorial lake and the fetal from the opposite maternal side of the placenta, the flows are grossly countercurrent.

Just before parturition (28 dpc) a discrete thin lamina of uninucleate decidual cells forms on the myometrial side of the decidua basalis (Mossman 1987). This forms the zone of separation of the placental stalk, which is considerably narrowed at parturition. These two processes, together with contraction of the myometrium, reduce the placental scar to minimum dimensions (Amoroso 1952). The myometrial contractions are considerably enhanced at this time by the increase in frequency of gap junctions between individual smooth muscle cells (Cole and Garfield 1989).

8.4   Haemomonochorial: Human, Homo sapiens

Menstrual cycle: 28 days
Ovulation: spontaneous
Uterus: simplex
Litter: 1–3
Implantation: interstitial, antimesometrial
Gestation: 270 days
Amniogenesis: cavitation

Yolk sac: The endoderm develops in an unusual manner and initially forms a latticework in the lumen of the blastocyst, only secondarily forming a sac (Luckett 1978). The secondary yolk sac never makes contact with the chorion, but may be involved in uptake from the exocoelomic fluid in the first few weeks of pregnancy (see below). It degenerates and disappears by 12 weeks (Fig. 1.2).

Chorioallantois: this forms the definitive placenta
Shape: discoid
Decidual development: considerable
Interhaemal membrane: haemomonochorial

8.4.1   Fetal Membranes and Placental Development in Human

The morula enters the uterus on 4–5 dpc, transforms to a blastocyst, loses its zona pellucida and normally implants in the midline of the upper part of the anterior or posterior wall on 6 dpc. It is not clear how it is so positioned, but it can remove the MUC1 glycoprotein locally (Meseguer et al. 2001), and other implantation-blocking glycoproteins are downregulated generally (Gipson et al. 2008) at this receptive stage (see Chap. 2.1.2). Uterine closure ensures no significant further movement. Implantation occurs prior to any blastocyst swelling, the cellular trophectoderm produces a syncytial cap which rapidly intrudes through the uterine epithelium, and the whole blastocyst passes into the endometrium. The details of this process are not fully understood (Fazleabas et al. 2004), but the uterine epithelium is re-established over the conceptus around 11 dpc. Once the conceptus is established in the endometrium (Figs. 8.9a, b) active proliferation of the cytotrophoblast provides the basis for rapid formation of the syncytiotrophoblast (Potgens et al. 2002).

Nuclear division has never been reported in the human syncytiotrophoblast, but autoradiographic and morphological studies have demonstrated that the cytotrophoblast is the precursor of the syncytiotrophoblast throughout pregnancy (Tedde and Tedde-Piras 1978; Kurman et al. 1984) and that all the nuclei of the syncytiotrophoblast are diploid (Galton 1962). The control of cytotrophoblast proliferation is uncertain, but division is stimulated strongly by hypoxia both in vivo and in vitro (Fox 1964; Esterman et al. 1997). Epidermal Growth Factor is also a powerful mitogen when applied to very early placental explants in vitro (Maruo et al. 1992). The mechanisms underlying trophoblast fusion are not fully understood either, but the formation of gap junctions (Cronier et al. 2003), activation of the apoptotic cascade (Black et al. 2004), and expression of endogenous retroviral proteins (Frendo et al. 2003), in particular the envelope protein of HERV-W or syncytin, have all been implicated. The syncytiotrophoblast thickens all around the conceptus, and between 9 and 11 dpc lacunae fringed with microvilli develop within its bulk (Fig. 8.9). These are the forerunners of the intervillous space. The syncytiotrophoblast is actively invasive and soon breaches maternal capillaries in the superficial endometrium, allowing maternal blood into the lacunae (Fig. 8.9c) (Boyd and Hamilton 1970; Enders 1989).

By 12 dpc columns of cytotrophoblast cells proliferate into the syncytial trabeculae separating the lacunae to initiate fetal villous development over the whole conceptus (Fig. 8.9d). When they reach the endometrium, beginning at 14 dpc, the cytotrophoblast cells spread laterally and merge with neighbouring columns to establish the cytotrophoblastic shell at the materno-fetal interface (Fig. 8.9d). The allantoic vesicle is rudimentary in the human but there is vigorous growth of allantoic mesoderm, which penetrates between the primary yolk sac endoderm lining the blastocystic cavity and the trophoblast outer layer (Fig. 1.2). Mesodermal cells derived from this layer also penetrate the cytotrophoblastic columns. With subsequent vascularisation of the mesodermal cores the rudiments of the fetal villi are established. Repeated branching throughout pregnancy leads to elaboration of the villous trees, projecting into the intervillous space.

The cytotrophoblastic shell initially represents the outer limits of the conceptus, but soon cells migrate from its outer surface into the endometrium. As they play no direct role in the development of the villous trees these are referred to as extravillous cytotrophoblast cells. They can be further sub-divided into interstitial cells that migrate through the endometrial stroma, and endovascular cells that migrate down the lumen of the maternal spiral arteries. Both populations are non-proliferative, probably because they become aneuploid (Weier et al. 2005). The cells are rich in glycogen, and express an unusual combination of HLA-C, HLA-G and HLA-E molecules (Moffett-King 2002; Chap. 9.1). In normal pregnancies the interstitial cellular invasion extends into the myometrium and disrupts the muscle sheath of the myometrial spiral arteries before the endovascular EVT arrives there to erode the

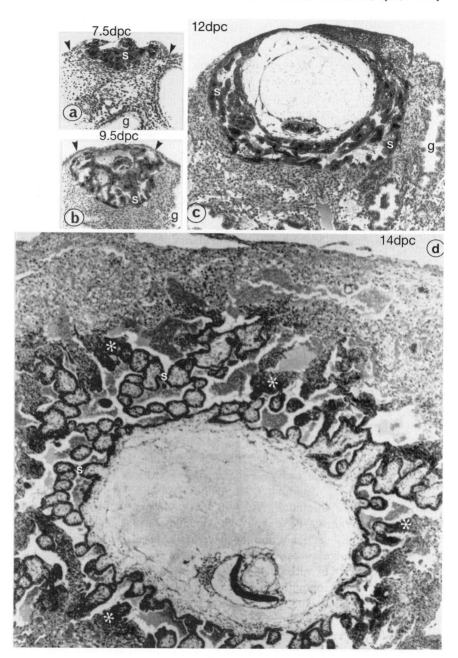

Fig. 8.9 Hemochorial placenta in man. Early stages of development, all at the same magnification. (**a**) The blastocyst passes through the uterine epithelium (**a, b**, *arrowheads*) and (**b**) establishes itself in the endometrium. (**b, c**) The syncytiotrophoblast (S) forms the initial boundary around the conceptus but (**d**) with the formation of the chorionic villi all round the periphery a cellular cytotrophoblastic shell (*asterisks*) develops around the conceptus and facilitates further villus growth (**a**) 7 ¹/₂ dpc, ×55. (**b**) 9 ¹/₂ dpc, ×55. (**c**) 12 dpc, ×55. (**d**) 14 dpc, ×55, from Boyd and Hamilton (1970)

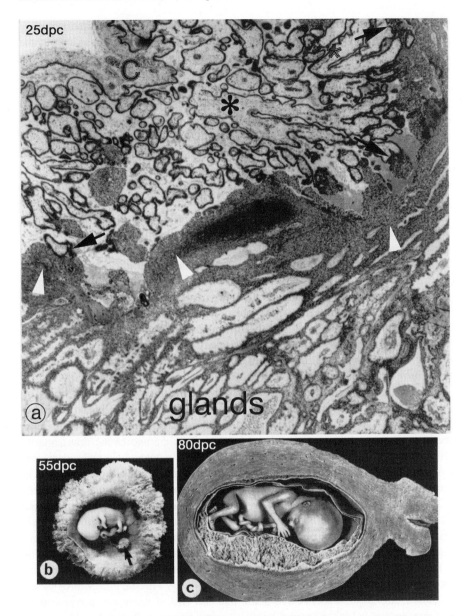

Fig. 8.10 Hemochorial placenta in man. (**a**) The chorionic villi increase in length and degree of branching (*asterisks*) as pregnancy progresses. The villi insert into the cytotrophoblastic shell (*arrows*) above a zone of decidual cells (*white arrowheads*). The endometrial glands (bottom third of figure) are eroded and distorted by the growth of the placental disc but they still appear to be actively secreting material. *C* chorionic plate, (**b, c**) Halved conceptuses (**c** still in its uterus) illustrate the reduction in trophectodermal villi from a uniform halo to a localised disc between 55 dpc and 80 dpc. Note the vestigial yolk sac (*arrow*) in (**b**). (**a**) 25 dpc, ×10. (**b**) 55 dpc, ×0.5. (**c**) 80 dpc, ×0.5. From Boyd and Hamilton (1970)

endothelium (Pijnenborg et al. 1981, 1983). The interstitial EVT eventually fuse to form multinucleate giant cells in the deep endometrium (Al-Lamki et al. 1999) and the inner myometrium.

Interactions with maternal immune cells within the decidua, in particular the natural killer cells, may play a role in limiting invasion (Moffett-King 2002; Chap. 9.1).

Together, the extravillous trophoblast cells are associated with physiological conversion of the arteries, leading to loss of the smooth muscle cells within their walls and their replacement with fibrinoid material (Pijnenborg et al. 2006). The molecular mechanisms underlying this process are still uncertain, but endocrine priming appears to be important (Brosens et al. 2002). There is also increasing evidence that interactions between HLA-C antigens expressed by the extravillous trophoblast cells and KIR receptors on the uterine natural killer cells plays a role through the release of cytokines and growth factors (see Chap. 9.1 and Moffett-King and Loke 2006; Hanna et al. 2006). The end result of conversion is that the spiral arteries lose their vasoreactivity, ensuring a constant blood supply to the placenta. In addition, their terminal segments dilate (Harris and Ramsey 1966), which will reduce both the rate and the pressure with which the maternal blood enters the intervillous space to levels optimal for placental exchange (Karimu and Burton 1994; Moll 2003). Conversion of the arteries is not an all or nothing phenomenon, but failure to widen suficiently can result in too high a blood pressure inflow. This potentially can cause significant mechanical and shear force damage to the villi and this is associated with fetal growth restriction and other hypertensive complications, such as preeclampsia (Kim et al. 2003; Ness and Sibai 2006).

The presence of maternal erythrocytes within the syncytiotrophoblastic lacunae around 11 dpc was in the past taken as evidence of a maternal intraplacental circulation. However, it is now recognized that free communication between the spiral arteries and the intervillous space is not established until towards the end of the first trimester (Burton et al. 1999). Prior to this time the volume of the endovascular trophoblast invasion into the spiral arteries is sufficient to block the arterial openings into the intervillous space. In contrast, connections between the developing intervillous space and the maternal venous system are established soon after implantation is complete, allowing the possibility of some maternal venous ebb and flow within the placenta. At this early stage, the conceptus is supported by histiotrophic uptake by the villous syncytiotrophoblast of secretions from the endometrial glands (25 dpc Fig. 8.10a, 32 dpc; Fig. 8.11a) lying beneath the implantation site and opening through the basal plate into the intervillous space (Burton et al. 2002). By 42 dpc (Fig. 8.11b) the glands have mostly collapsed, although remnants of glands that are still synthetically active can be observed embedded in the decidua basalis at term (Nelson et al. 1986).

Growth factors and nutrients contained within the gland secretions potentially regulate placental cell proliferation and differentiation (Burton et al. 2007), but cannot reach the embryo quickly since the villous circulation is not functional during the first trimester. However growth factors and nutrients can be taken up and passed across the trophoblast of the chorionic plate and accumulate in the exocoelomic fluid (Jauniaux and Gulbis 2000). From there they may be transported to the fetus through uptake by the outer mesodermal layer of the yolk sac which is the first extra-embryonic membrane to be vascularised (Fig. 1.2) and fetal blood circulation starts at about 21 dpc. This mesoderm layer possesses numerous micro-villi, coated pits and other morphological features of an absorptive epithelium (Jones 1997), and also expresses specific transporter proteins (Jauniaux et al. 2004). The mesodermal cells transport nutrients and growth factors into the vitelline veins which drain directly to the fetal liver. In addition, the endodermal epithelium lining the yolk sac cavity synthesizes many proteins in common with the liver, such as α-fetoprotein (Gitlin and Perricelli 1970). Once the fetal chorioallantoic villus circulation is fully established at about 10 weeks of pregnancy the yolk sac rapidly regresses.

As soon as the cytotrophoblast shell has formed around the conceptus (21 dpc) fibrinoid material develops between the decidua and the shell, and the shell and the intervillous space forming the 'striae' (layers) (Boyd and Hamilton 1970: Huppertz et al. 1996, 1998), 'Rohr's stria' forms the fetal surface of the basal plate (which consists of the cytotrophoblast shell) facing the intervillous space and interrupted by the anchoring villi. 'Nitabuch's stria' lies on the decidual side of the basal plate where maternal and fetal cytotrophoblast cells come into contact. The former is composed of fibrin-type fibrinoid, and is most likely derived principally from maternal blood fibrinogen. By contrast, 'Nitabuch's stria contains more components of extracellular matrix origin, and is thus termed matrix-type fibrinoid. It is thought to be derived from both decidual cells and the extravillous trophoblast cells, which it frequently surrounds (Kisalus et al. 1987; Huppertz et al. 1996, 1998). Various functions have been attributed to the fibrinoid deposits, with many authors favouring a role as a store of immobilized molecules with tissue remodelling or immunological potential (Huppertz et al. 1998).

Initially, villi are formed over the entire surface of the chorionic sac, generating the chorion frondosum (Figs. 8.9d and 8.10b). By approximately 6 week/s of gestation the villi over the superficial pole begin to regress, while those over the deep pole elongate and branch (Figs. 8.11b and 8.10c). It has recently been proposed that this pattern reflects the onset of the maternal arterial circulation of the placenta (Jauniaux et al. 2003). When onset begins towards the end of the first trimester it starts at the periphery of the implantation site, where extravillous trophoblast invasion, and hence plugging of the spiral arteries, is least. Onset is associated with a three-fold rise in oxygen concentration, and villi in the peripheral region display high levels of oxidative stress, a potent stimulator of apoptosis. Gradually, these villi become avascular and atrophic, and their syncytiotrophoblastic covering is lost. Finally, the chorionic and basal plates fuse over the superficial pole of the chorionic sac, trapping any villous remnants in a multilayered cellular membrane, the smooth chorion or chorion laeve. The remaining, definitive villi covering the embryonic pole continue to proliferate and form the discoid placenta (Fig. 8.11). Regression of early villi to form the definitive placenta is a phenomenon that appears to be unique to those species displaying interstital implantation, i.e. the human and great apes. Abnormalities in the process caused by local perturbations in the onset of maternal blood flow may cause excessive villous regression and hence abnormal placental shapes, or failure of regression and the formation of accessory ("succenturiate") lobes.

At approximately the same time the amniotic cavity enlarges to the extent that the amnion is brought into contact with the inner surface of the chorionic sac,

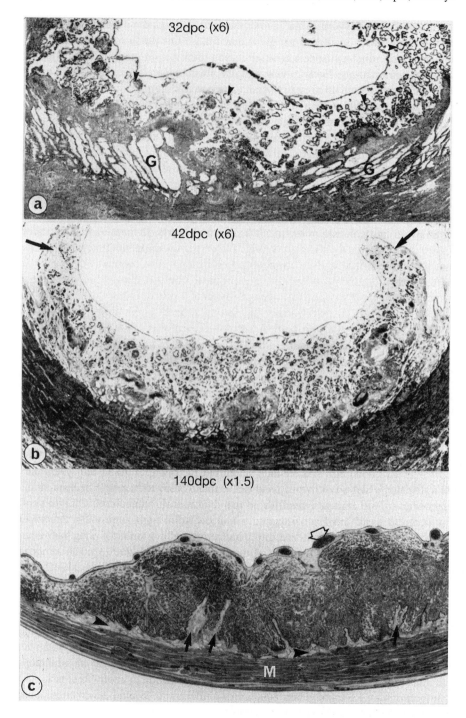

obliterating the exocoelomic cavity by around 13 week/s and forming the chorio-amniotic membrane (Fig. 8.12).

With continual enlargement of the conceptus the endometrium overlying the implantation site, the decidua capsularis, bulges increasingly into the uterine lumen. During the fifth month of gestation it comes into contact, and fuses with, that of the opposite wall, the decidua parietalis, thus obliterating the uterine lumen (see Fig. 8.12). From then on, the smooth chorion makes extensive contact with the decidua parietalis, and may function as a paraplacental area for exchange. However, as the smooth chorion is avascular, any exchange between the fetus and the decidua must occur through the amniotic fluid.

As with the other haemochorial placentas discussed above, it is clear that most of the villous elongation is due to expansion of the fetal tissue after the initial erosion of the maternal tissue has occurred, since a considerable thinning of endometrial substance must also be caused by the stretching needed to accommodate the increase in conceptus size. Ultrasound measurements indicate that an endometrial thickness of approximately 8 mm is required for successful implantation (Basir et al. 2002), but this has reduced to approximately 1 mm by the end of the first trimester (Hempstock et al. 2004). Mossman (1987) has evidence that there is very little increase in thickness of the cytotrophoblast shell or the decidua from as early as 25 dpc to term, but the villi increase in length eightfold in that time.

Differences in the extent of fetal erosion are thought to account for the shallow ridges (septae) of maternal tissue that are covered with cytotrophoblast shell (Figs. 8.11c and 8.12). These delineate the lobes of the placenta, which incompletely subdivide the intervillous space into 10–40 compartments. Each compartment may contain one or more aggregates of fetal villi, termed lobules (Figs. 8.11c and 8.13).

A lobule comprises a stem villus arising from the deep surface of the chorionic plate that then branches repeatedly to form a hollow cluster centered over the opening of a spiral artery. Angiographic studies indicate that maternal blood spurts as a 'puff' into each hollow lobule from a spiral arterial channel (Fig. 8.13c) and drains by venous exits at the sides and bases of the septa (Fig. 8.12) (Ramsey and Donner 1980). Regional differences in antioxidant enzyme activity indicate the presence of an oxygen gradient across each lobule (Hempstock et al. 2003), and this

---

Fig. 8.11 Hemochorial placenta in man. Development of the definitive placental disc. (a) The initially broad villi (*arrowheads*) are arranged evenly around the conceptus at 32 dpc and the basal uterine glands (G) are still extensive. At 42 dpc, (b) the villi are thinner and more frequent in the placental disc area, and have virtually crowded out the uterine glands basally, but are regressing outside the placental disc (*arrows*). (c) By 140 dpc. the villi are much more extensive and closely packed. (This micrograph is only one quarter of the magnification of (a) and (b)). Septa (*arrows*) are now obvious. They are continuous with the thin decidual layer (*arrowheads*) which lies above the myometrium (M). The septa divide the placental disc into lobules or "cotyledons" (see Figs. 8.15a, b). Open arrow, fetal blood vessel in chorionic plate. (a) 32 dpc, ×6. (b) 42 dpc, ×6. (c) 140 dpc, ×1.5, from Boyd and Hamilton (1970)

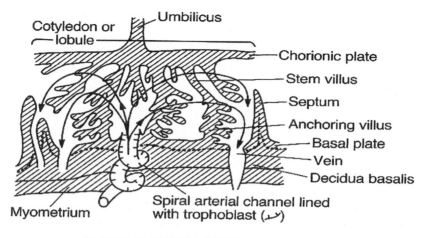

BLOOD FLOW IN LOBULE

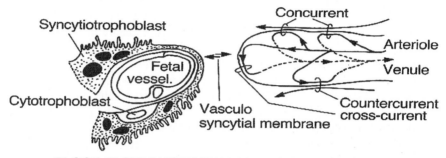

BLOOD FLOW IN TERMINAL VILLUS STRUCTURE

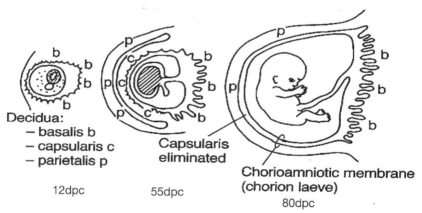

DECIDUAL DEVELOPMENT

Fig. 8.12 Blood flow in the definitive human placenta and development of the decidual tissues

may account for changes in villous maturity and morphology between the centre and the periphery (Wigglesworth 1969; Schuhmann 1982; Critchley and Burton 1987). Each lobule may therefore be considered to serve as individual materno-fetal exchange units equivalent to the cotyledons of a ruminant placenta (compare Figs. 8.13 and 6.3). Given the three-dimensional branching pattern of the villous trees, it is likely that the relative directions of the maternal and fetal blood flows will vary from point to point around the villous surface. The overall result is multivillous (network) flow (Fig. 8.12) (Dantzer et al. 1988).

An individual villus consists initially of a solid column of cytotrophoblast cells, but this is invaded by the allantoic mesodermal vasculature by 18 dpc. The young stem villus has a cytotrophoblastic tip at its insertion into the cytotrophoblastic shell (Fig. 8.10a). Behind this, where the villus is bathed in maternal blood, the cytotrophoblast is covered by a continuous layer of syncytiotrophoblast.

The villus elongates and branches, becoming tree-like, but usually retains at least one connection via an anchoring branch to the basal plate as well as its stem inserted into the chorionic plate (Figs. 8.10a and 8.13d) (Boyd and Hamilton 1970; Kaufmann 1981b; Burton 1987, 1990; Mossman 1987; Castellucci et al. 1990).

There are three major levels of villus organization, stem, intermediate and terminal. Stem villi are the supporting and distributing framework, containing arterial and arteriolar vessels derived from the umbilical arteries. Morphometrically, as one moves down the villous tree there is a progressive loss of connective tissue, an increase in the volume occupied in the villous core by fetal blood vessels, and a decrease in thickness of the syncytiotrophoblast over the closely adjacent fetal vessels (Kaufmann et al. 1985) (Figs. 8.14 and 8.15).

This reaches an extreme in the so-called 'vasculosyncytial' membrane in the terminal villi where the distance from the maternal to fetal blood is reduced to less than 3 μm over 40% of the total area of the terminal villus (Figs. 8.12, 8.14 and 8.15; Kaufmann 1985; Mayhew et al. 1986) The underlying fetal capillary is often sinusoidally dilated (Fig. 8.14), slowing blood flow so that conditions are optimized for diffusional exchange. From an early stage the cytotrophoblast layer becomes discontinuous, although the ratio of cytotrophoblastic to syncytial nuclei remains constant at 1:9 throughout gestation (Simpson et al. 1992). The definitive placenta is therefore considered haemomonochorial, with only the syncytiotrophoblast forming a significant boundary layer (Fig. 8.15). For a review of human placental ultrastructure see Jones and Fox (1991).

Formation of terminal villi occurs principally from 20 weeks of gestation onwards (Jackson et al. 1992), providing an exchange area of $12–14 \, m^2$ at term (Burton and Jauniaux 1995). It is thought to be driven by angiogenesis within the villous core (Kaufmann et al. 2004), although the transcription factor Gcm1 (Rossant and Cross 2001) and signaling pathways involved in branching morphogenesis in other systems, such as the Fibroblast Growth Factor, Hepatocyte Growth Factor and wingless-related (Wnt) pathways have also been recently implicated.

Since the fetal trophoblast is bathed in maternal blood there is no need for any specialized haemophagous zone, and iron is probably transported from the maternal blood using the transferrin binding protein (King 1976). There is no evidence for

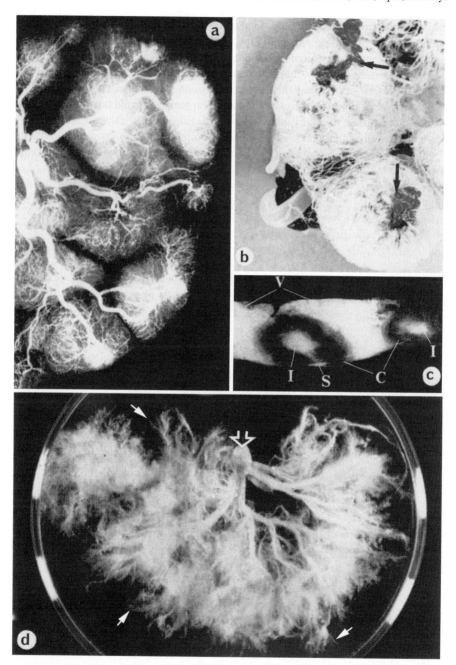

Fig. 8.13 Hemochorial placenta of man. Vascular relationships. (**a**) Injection of the fetal circulation of a term placenta viewed from the fetal side. This illustrates the supply to individual lobules or cotyledons. (**b**) Latex casts of the fetal circulation (*white*) and of the maternal spiral arteries (*black*) of a term placenta viewed from the maternal side. One maternal spiral artery (*arrows*)

significant phagocytosis of red blood cells by the fetal syncytiotrophoblast. Human chorionic gonadotropin (hCG) and placental lactogen hormones are synthesized and stored in different granules in the syncytiotrophoblast (Fig. 8.16) (Johnson and Wooding 1988; Morrish et al. 1988; Billingsley and Wooding 1990), which also shows immunoreactivity for α,β and γ interferons (Bulmer et al. 1990), corticosteroid-releasing hormone (Riley et al. 1991), EGF (Hofmann et al. 1991) and IGF-1 and -2 and their binding proteins (Zhou and Bondy 1992; Hill et al. 1993).

The syncytium contains a wide variety of transporters for nutrient uptake, see Chap. 2.

Unlike most animals, there is no fall in progesterone levels prior to parturition in man, nor is there any change in relaxin levels, which anyway are much lower than they were during the first 3 month/s of pregnancy (see Sherwood 1988, for an excellent review). Labour in man is also unusual in that it evolves gradually over a period of many days without an abrupt onset as in most species studied. Oxytocin is said to play no significant role until the second (expulsive) stage of labour, when it ensures retraction of the uterus and prevents excessive blood loss (Casey et al. 1983). The trigger for human parturition remains controversial but may involve oestrogen precursor production from the fetal adrenal and prostaglandin secretion by amnion and the parietal chorion (Casey et al. 1983; Gibb and Challis 2002).

8.5 Haemomonochorial: Other Anthropoids

8.5.1 Apes and Old World Monkeys

As far as they have been investigated the implantation, development and definitive placentation of man and the great apes (chimpanzees, gorillas, orangutans) are essentially similar (Enders 2008). The Old World monkeys (catarrhines) have a very different implantation and development, although the definitive structure of the placenta is almost indistinguishable at macroscopic and ultrastructural levels (Amoroso 1952; Ramsey 1982; King 1986).

◀───

Fig. 8.13 (continued) enters the centre of each lobule or cotyledon. (**c**) Section of a term placenta injected via the spiral arteries (S) with X-ray opaque medium. The injection fills a central area (I) in the two lobules or cotyledons (C) but is excluded from the zone immediately surrounding the injection by the densely packed villi. The presence of opaque medium in the intervillous space (V) shows that it was able to pass across the villi but not to leak back. From Wigglesworth (1969). (**d**) Hemochorial placenta of man. Stem villus dissected after trypsin digestion from a term placenta, the chorionic plate insertion was at the open arrow. Two or three of such villi would form one lobule or cotyledon. The finest "twigs" (*arrows*) correspond to the terminal villi shown on Fig. 8.14. From Ramsey and Donner (1980)

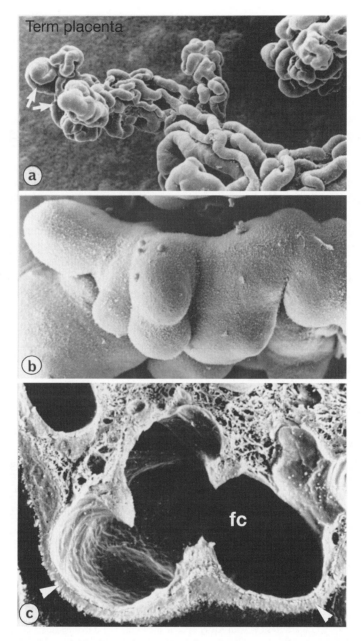

Fig. 8.14 Hemochorial placenta of man. (**a**) Scanning electron micrograph of a corrosion cast of the blood capillary meshwork in the terminal villi, note the sinusoidal swellings at the ends of the villi (*arrows*). (**b**) Scanning electron micrograph of the surface of an intact perfused terminal villus emphasizing that the villus consists almost entirely of blood capillary volume. (**c**) Scanning electron micrograph of a cross fracture through a specimen similar to that in (**b**). Note the considerable size of the fetal capillary (fc) and the thin investment of syncytiotrophoblast (*white arrowheads*). (**a**) Term placenta; ×200 (**b**) Term placenta; ×400. (**c**) Term placenta; ×1,400, from Burton (1987)

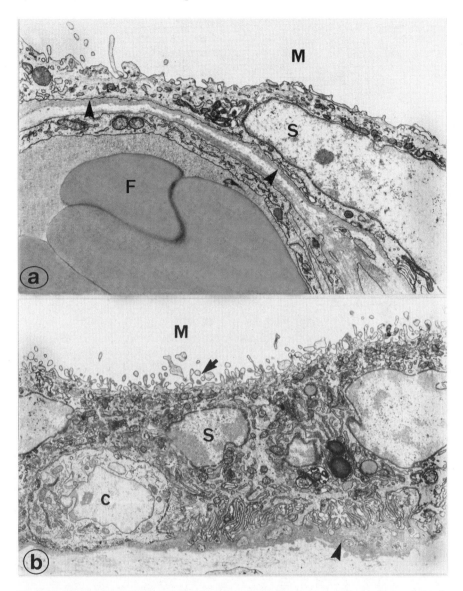

Fig. 8.15 Hemochorial placenta of man. Transmission electron micrographs showing variation in local structure. (**a**) Over the fetal blood vessel (F) the syncytiotrophoblast layer (S) is extremely thin as is the basement membrane (*arrowheads*). (**b**) Away from the capillaries the fetal syncytium is thicker, and in places the basal plasmalemma shows a complex folding (*white asterisk*) increasing the membrane area as does the microvillar elaboration (*arrow*) of the apical membrane. The basement membrane under the basal folds is considerably thickened (*arrowhead*) compared to that under the cytotrophoblast cell (C). *M* maternal blood space. (**a**) Term placenta, ×9,000. (**b**) Term placenta; ×7,500

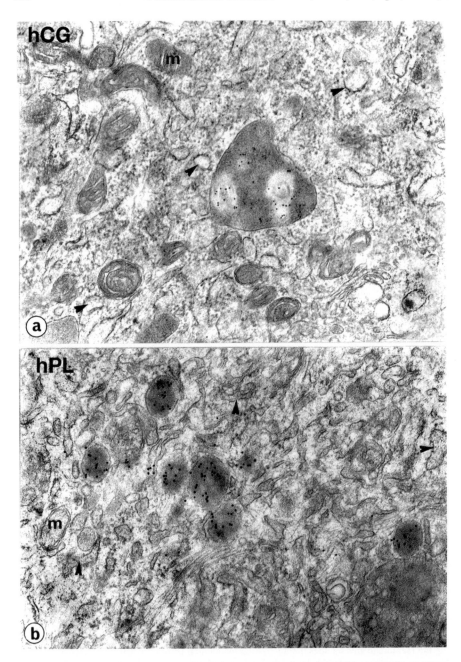

Fig. 8.16 Hemochorial placenta of man. Micrographs of syncytiotrophoblast showing localisation of hormone storage with immunogold techniques (**a**) hCG is found in large irregular shaped membrane bounded granules, (**b**) hPL in smaller more uniform sized granules. Material fixed in glutaraldehyde and osmium and araldite embedded. *m* mitochondria; *arrowheads* rough endoplasmic reticulum. (**a**) First trimester placenta; ×28,000. (**b**) Term placenta; ×39,000

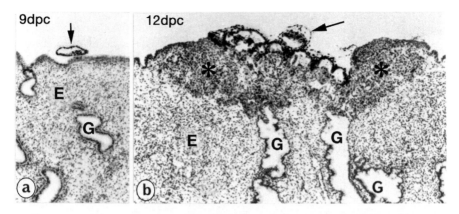

Fig. 8.17 Hemochorial placental development in Rhesus monkeys. The micrographs on Figs. 8.17 and 8.18 are all at the same ×50 magnification. The major differences with human are the superficial implantation (8.17a, b; *arrow*, conceptus) and development of epithelial plaque (*asterisks* in 8.17b and 8.18a) by local proliferation of the uterine epithelium. There is little evidence for decidualisation in the endometrium (E) in which the glands (G) appear active. (**a**) 9 dpc, ×50 (**b**) 12 dpc, ×50

The best-studied Old World monkey is the macaque or rhesus, *Macacus rhesus*. Many details of the reproduction are identical to those for Human (Chap. 8.4) but the gestation is only 160 day/s, the placenta is bidiscoid, implantation is superficial and there is little stromal decidual development but a considerable proliferation of uterine epithelial plaque (Figs. 8.17 and 8.18 and see chap. 1.7.4.7) (Enders et al. 1985, 2001). Plaque only persists for the first month of pregnancy and may be one reason for the limited invasion of the endometrium. It can also be produced by scratch trauma as a prelude to deciduoma formation in the macacque (Ghosh et al. 2004) or by the infusion of hCG in the baboon (Jones and Fazleabas 2001).

The invasive syncytiotrophoblast of the adherent blastocyst needs only to penetrate into the superficial, greatly hypertrophied endometrial blood vessels to establish a good blood supply (Fig. 8.18a) (Enders 1989; Enders and King 1991; Enders and Welsh 1993; Enders et al. 1996). However, as in man, the cytotrophoblast proliferation blocks the invaded vessels initially and maternal blood flow through the spiral arteries into the intervillous space surrounding the placental villi only becomes significant between 25 and 28 dpc. This is also true for the baboon (Enders and King 1991; Simpson et al. 1997; Enders 2007). Once this is accomplished extensive exclusively fetal villus growth can produce the definitive placenta (Fig. 8.18e) with a villus structure equivalent to man (compare Figs. 8.10a and 8.18e).

There is no need for any deeper invasion of the endometrium since the maternal blood supply is always at the base of the placenta, as in the human. Fetal cytotrophoblast does invade the spiral arteries replacing the endothelium as far back as the myometrium but there is less erosion of the arterial elastic sheath than in man. This is probably because the rhesus cytotrophoblastic shell, although it does synthesize connective tissue components (Blankenship et al. 1992), is not at all invasive so

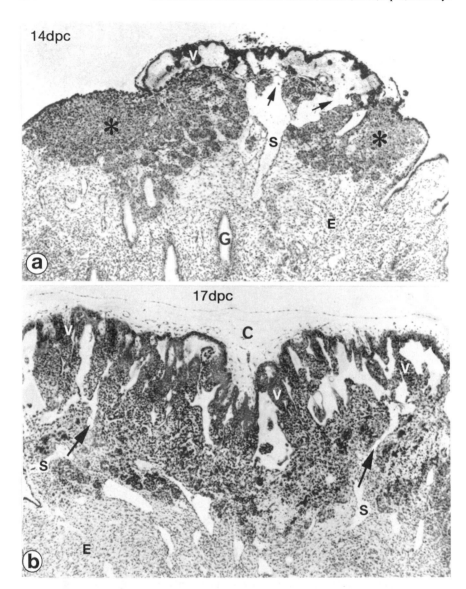

Fig. 8.18 Hemochorial placental development in Rhesus monkeys. (**a, b**) At 14 dpc and 17 dpc the superficial endometrial blood sinusoids (S) connect to the spaces within the syncytiotrophoblast (at the *arrows*). Once the fetal cytotrophoblast and mesoderm have formed the initial broad villi (V), their subsequent growth is by lengthening and subdivision (**b–e**). There is little evidence for any significant further invasion of the endometrium (E). The cytotrophoblastic shell (X) which proliferates from the tips of the villi from ~20 dpc (Fig. 8.18c) is particularly well demonstrated. *C* chorionic plate, *G,06* endometrial gland, *V* placental villus. (**a**) 14 dpc; ×50. (**b**) 17 dpc; ×50. (**c**) 20 dpc, ×50. (**d**) 29 dpc, ×50. (**e**) 50 dpc, ×50. Figs. 8.17 and 8.18 from Wislocki and Streeter (1938)

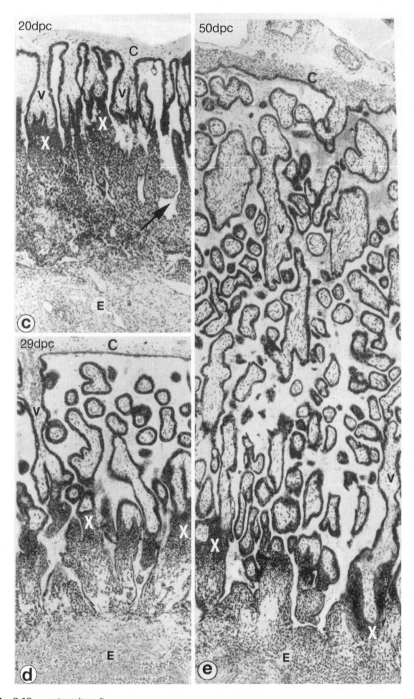

Fig. 8.18 c–e (continued)

there are very few if any cytotrophoblast-derived cells in the endometrial connective tissue to attack and erode the artery wall to the extent seen in man (Ramsey et al. 1976). However, as in man, there is a 'dramatic' influx of macrophages and uNKs into the endometrium. In macaque this occurs at the time (15 dpc) when the cytotrophoblast is starting to invade (Blankenship and Enders 2003; Slukvin et al. 2004), and macrophages may serve the role played by LGLs in man. There seems to be no simple relationship between epithelial plaque or decidua formation and the depth of implantation since the baboon has superficial implantation and little epithelial plaque or decidual development; the macaque has superficial implantation, extensive plaque and limited decidual development; and man has interstitial development, no epithelial plaque and massive decidualization (Ramsey et al. 1976). All primates are said to have 'extensive decidua' by mid-pregnancy (Enders 1991).

Rhesus and the African green monkey Cercopithecus (Owiti et al. 1989) are good models for studying the functions of the definitive human placenta since the structures are closely similar. However, the Old World monkeys are poor models for events at implantation and in early pregnancy in humans because interstitial or superficial implantation with decidua or plaque development are such very different processes (Enders 1989).

8.5.2   New World Monkeys

There are two families: Cebidae, including howler, squirrel and spider monkeys; and Callithricidae, the marmosets. There are only isolated studies of the cebids and all prior to the advent of the electron microscope (Wislocki 1929, 1930; Hill 1932). Recently studies have concentrated on the marmoset (*Callithrix jacchus*) since it is relatively easy to maintain in breeding colonies (Moore et al. 1985; Merker et al. 1987; Smith et al. 1987).

There is no overt menstruation in New World monkeys which correlates with an absence of the typical spiral form of artery in the endometrium (Ramsey 1982). Cyclic proliferation and degeneration of the spiral arteries is thought to be the basis of the menstruation characteristic of other anthropoids (Ramsey 1982).

Implantation is superficial, usually with considerable epithelial plaque formation (Hill 1932) although there is very little in marmosets (Moore et al. 1985). The blastocyst trophoblast expresses Thioredoxin, an antioxidant essential for mouse implantation (Lopata et al. 2001).

After attachment (11–12 dpc) the blastocyst cytotrophoblast produces syncytium which penetrates the uterine epithelium (Enders and Lopata 1999). Once the stroma is reached the cytotrophoblast proliferates to produce a thick pad of syncytiotrophoblast which surrounds but does not penetrate the superficial maternal capillaries (23 dpc) (CA. Smith et al. 1987). By this time the epithelial plaque is regressing. The syncytiotrophoblast then develops lacunae within its bulk, but in the marmoset no maternal blood is observed within these spaces until 80 dpc, halfway through pregnancy, when the syncytiotrophoblast finally erodes the maternal endothelium. At this stage the fetal mesoderm pushes into the trophoblast layers, forming vascularized trabeculae within the meshwork of syncytium (Smith and Moore 1990). The syncytio-trophoblast not only appears very similar ultrastructurally to that of other primates (Smith and Moore 1990; Enders and Lopata 1999) but also has equivalent C-type retro-virus particles budding from the syncytiotrophoblast basal plasmalemma opposite the cytotrophoblast cells. The cytotrophoblast layer becomes discontinuous very early, and these fetal trabeculae are bounded largely by syncytiotrophoblast. Further development of the placenta is controlled by the interplay of VEGF and other angiopoietins produced by the trophoblast and endothelial cells (Wulff et al. 2002) and results in the continual extension and subdivision of the trabeculae to form a complex meshwork bounded by syncytium. The trabeculae become more individual and more slender during later pregnancy (Fig. 8.19) (Wislocki 1929, 1930; Amoroso 1952; Smith and Moore 1990).

In the spider monkey, Ateles, the thickness of the placental pad does not increase significantly as the fetus grows from 50 to 180 mm (term) crown-rump length (Wislocki 1930). The process is not villus elongation but one of the continual sub-division of a coarse trabecular meshwork to a finer one with a considerably greater surface area for fetomaternal exchange. The New World monkeys thus achieve a structural basis for a significant blood circulation much more slowly than the Old World monkeys. Possibly in compensation for this, the New World monkeys show a massive uterine glandular activity throughout pregnancy (Wislocki 1929), whereas in the Old World monkeys, apes and man this is restricted to the first third of pregnancy. However the glucose transporter GT1 is highly expressed on apical and basal trophoblast membranes of the syncytium and fetal endothelium in mar-mosets and man (Hahn et al. 1995; Illsley 2000) and the gestation length and pla-cental weight for equivalent sized neonates is similar in the two groups; there is no evidence that one is more 'efficient' than the other, although the size and degree of fusion of the trabeculae or villi vary considerably (Fig. 8.19).

The efficiency of the fetomaternal exchange in New World monkeys may also be influenced by the persistence, uniquely in anthropoids, of one or two wide maternal arterial channels per lobule which run right through the entire thickness of the placental disc and terminate just under the chorionic plate (Fig. 8.1). Presumably each channel is formed by modification of an original endometrial blood vessel which was surrounded by the initial syncytiotrophoblast proliferation. These chan-nels have occasional openings distributed apparently randomly along their walls and at their tips which are present from 45 dpc to term. The walls of the channels consist of a continuous sheet of cuboidal epithelial cells which could be of either fetal or maternal origin, certainly they do not express the GT1 transporter found on the trophoblast (Hahn et al. 1995) These cells secrete a thick basement membrane like material around the vessels, which persist to term (Wislocki 1930; Hill 1932; Gruenwald 1973; Merker et al. 1987; Smith and Moore 1990).

These peculiar perforated maternal blood channels may indicate one possible route from the rodent-style narrow-meshed labyrinthine haemochorial placenta with a few maternal arterial channels supplying a subchorial lake directly to the widely spaced villous haemochorial placenta of man and the Old World monkeys with maternal blood supplied only from the basal plate (Fig. 8.1).

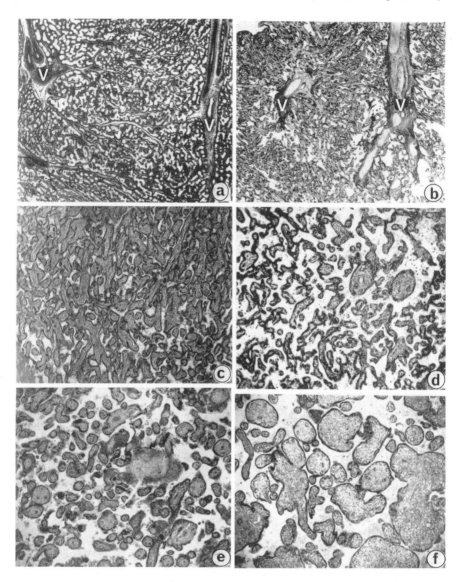

Fig. 8.19 Hemochorial placentation. Comparison of the size of the fetal trabeculae or villi in "mature placentae at approximately equivalent stages". Description and plate from Wislocki (1929). Villi are individual; trabeculae are villi with interconnected cores forming a meshwork (or labyrinth). (**a**) *Cynocephalus* (Galaeopithecus), a Dermopteran probably most closely related to the Insectivores. The placenta consists of narrow interconnected trabeculae very similar to that of the new world monkey *Ateles*, shown in (**b**) at identical magnification. V, major fetal villi carrying blood directly to the maternal side. (**c–f**) All at the same magnification to show the range of placental architecture. (**c**) The new world monkey *Alouatta* has a placenta of narrow trabeculae (**d**) The old world *Colobus* monkey shows a mixture of villi and narrow trabeculae. (**e**) The old world *Gibbon* has larger villi with frequent syncytial, but not core, connections. (**f**) The anthropoid gorilla shows a completely free villous structure, as in Man. (**a**) *Cyncocephalus*, 84 mm crown rump length; ×14 (**b**)*Ateles*, 187 mm CR; ×14 (**c**) *Alouatta*, 130 mm CR; ×42 (**d**) *Colobus*, 143 mm CR; ×42 (**e**) *Hylobates*, 102 mm CR; ×42 (**f**) *Gorilla*, 89 mm CR; ×42, from Wislocki (1929)

Yet again, as many times before in this review, the variety of structures evolved by the different groups far outstrips our ingenuity in justifying them in terms of placental function.

8.6   Haemochorial Placentation: Variety of Interhaemal Layers

All placentas so far examined seem to possess at least one persistent continuous trophectodermal layer as well as fetal endothelium to separate fetal and maternal circulations. The most efficient arrangement for maternofetal transport is for the maternal blood to circulate against the thinnest trophectodermal barrier compatible with immunological and other constraints. The most complete barrier is a syncytial sheet since there are no paracellular, non-selective routes possible: all molecules reaching the fetus can be screened for suitability by the trophectoderm. A haemomonosyncytiochorial structure of this sort is found in many different orders (Sect. 8.2), but despite its apparent advantages there exist also a remarkable variety of haemochorial trophectodermal organizations with one, two or three layers, each of which may be syncytial or cellular. There is no evidence that nuclei in any vertebrate syncytium can divide. Evidence from human (Simpson et al. 1992) indicates that the ratio of cytotrophoblast to syncytiotrophoblast nuclei is constant throughout gestation. Cytotrophoblast division followed by fusion into the trophoblast syncytium is the normal means of increasing its bulk and therefore its potential area. Since a syncytium has considerable plasticity the cytotrophoblast inputs can be scattered, and no continuous cellular layer is necessary. However, some such input is essential if the fetomaternal exchange surface is to enlarge continuously to the enormous extent reported in the second half of pregnancy (Baur 1977) to support the fetal weight increase. In man the syncytiotrophoblast is produced initially from a continuous cytotrophoblast layer. This soon becomes discontinuous and the syncytium can differentiate locally to form the very thin vasculosyncytial membrane over fetal capillaries for example (Figs. 8.12 and 8.15).

Cytotrophoblast cells persist and at term form an attenuated meshwork connected by thin processes covering about 40% of the basement membrane (Jones et al. 2008).

They are still capable of cell division even after the placenta is shed at term.

The characteristic mono-, di-, or trihaemochorial structure is established soon after implantation and, although the layers become thinner and the fetal capillaries indent the layers more deeply as pregnancy proceeds, there is no evidence for any structural simplification (tri- to di- to monochorial or the reverse) during development of any definitive placenta. At present little is known about the reasons for the many different patterns of layering found in haemochorial placentas. The more layers there are the more complete the spatial separation of functions can be. In a haemomonochorial placenta like that of man, numerous different proteins including hormones are synthesized and secreted independently (see Fig. 8.16) but all use the same subcellular machinery (Pattillo et al. 1983; Morrish et al. 1987). This must require a very complex hierarchy of controls which could be considerably simplified if several different types of cells are available, and there is good evidence for very different ultrastructure in the different layers of the same placenta (King and Hastings 1977; Simmons et al. 2007). Three layers obviously form a more complete barrier to invasion by maternal cells than one, and it is notable that haemomonochorial placentas always have a substantial fetal endothelial cell layer, whereas haemotrichorial placentas frequently show fenestrated fetal capillaries. The reason for the variety of haemochorial structures seems likely to lie in the multiplicity of functions served by the trophectoderm. With our present limited information we can only record the range and wonder.

8.6.1 Haemomonochorial

8.6.1.1 Cellular Trophoblast

Myomorph rodents: Jaculus (desert jerboa) and Zapus (jumping mouse) At implantation the germinative cytotrophoblast (layer III) produces a layer of syncytiotrophoblast (layer II), but both are very soon overgrown and disrupted by giant cells (layer I). These giant cells form the channels of the labyrinth in which the maternal blood circulates (Figs. 8.4 and 8.20).

They are derived from cytotrophoblast cells at the edge of the placental disc (King and Mossman 1974; Mossman 1987).

Hyracoidea: Procavia (rock hyrax) This order is thought to be related to the elephants on skeletal evidence, and both are classified in the superorder Afrotheria. The placental structures and their development are very different although the Elephantidae do have a monocellular trophoblast but an endotheliochorial structure (Wooding et al. 2005a,b). The hyrax cytotrophoblast produces a thick layer or 'pad' of cells replacing the upper layers of the endometrium in the implantation chamber (Sturgess 1948). By the time the fetus has reached a crown-rump (CR) length of 12 mm (midgestation; Wislocki and Westhuysen 1940) this cellular layer is almost as thick as is the definitive placenta near term (50 mm CR). The cellular layer of trophoblast transforms into a trabecular meshwork forming channels with cellular walls (Oduor-Okelo et al. 1983) through which the maternal blood drains back to the uterine veins (Fig. 8.20a). The blood is delivered to the fetal side of the placenta by persistent maternal arterial channels which pass right through the fetal cell layer. The fetal allantoic arterioles invade the cytotrophoblastic trabeculae as they develop, delivering fetal blood to the maternal side of the placenta, from whence it drains back through a capillary meshwork establishing a grossly countercurrent system of fetal and maternal blood flows. This developmental pattern is very similar to that described (Chap. 8.5.2) for the New World monkeys but they are haemomonosyncytiochorial.

Fig. 8.20 Hemochorial placentation. Monochorial cellular trophectoderm in (**a**) Hyrax and (**b**) the rodent Zapus. The interhemal membrane is clearly cellular in both with apical tight junctions (*arrows*) obvious in these placentas. *FC* fetal capillary, *MBS* maternal blood space, *T* trophectoderm. (**a**) *Hyrax,* mid pregnancy; ×6,000. (**b**) *Zapus,* late pregnancy; ×8,000, from King and Mossman (1974)

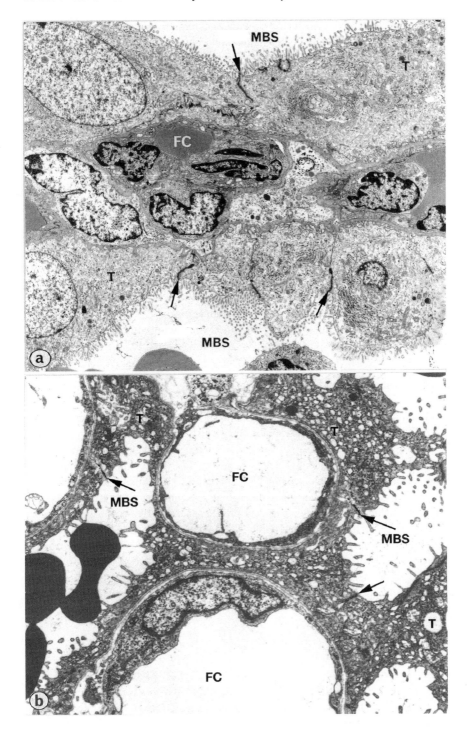

Tenrecs: Newly classified in the superorder Afrotheria (Carter et al. 2004a), the Malagasy Tenrecs have a labyrinthine hemomonocellularchorial placenta that develops from an avascular cytotrophoblast pad which replaces the upper layers of the endometrium (Enders et al. 2007). This is very similar to the hyrax but the methods by which these 'pads' transform into the definitive placenta (Carter et al. 2004b, 2005) are not clear.

Bats: Most bats are haemochorial but two orders, Vespertilionidae and Molossidea, have genera (Miniopteris, Kimura and Uchida 1984; Tadarida, Stephens 1962, 1969) which initially support early embryonic development with a diffuse chorioallantoic endotheliodichorial placenta. However this becomes non functional by mid pregnancy as it is replaced by growth of a separate discoid chorioallantoic hemochorial definitive placenta. This is based on invasive cytotrophoblast enveloping a local proliferation ('tuft') of maternal decidual blood vessels. The cytotrophoblast removes and replaces their endothelium and by rapid growth expands and elongates the original 'tuft' of maternal blood channels. Simultaneously solid cytotrophoblast rods grow from the blood channel walls into the connective tissue surrounding them. The rods fuse with others and canalize, forming capillary meshworks between the arterial inflow and venous outflow of the original 'tuft', now enormously extended. Simultaneously fetal blood vessels infiltrate the growing cytotrophoblast meshwork and vastly increase the fetomaternal exchange potential.

These processes produce a labyrinth lined only with cellular cytotrophoblast (Fig. 1.8d) (Stephens and Cabral 1971, 1972; Badwaik and Rasweiler 2000).

8.6.1.2 Syncytiotrophoblast

A wide variety of orders have members with haemomonosyncytiochorial membranes in their definitive placentas: all anthropoids (Fig. 8.15); one carnivore, hyena (Fig. 1.8a) (Oduor-Okelo and Neaves 1982); two insectivores, cane rat and elephant shrew (Oduor-Okelo 1984a,b); several rodents such as marmots (Fig. 1.8b), chipmunk (Fig. 1.8c), squirrels (Mossman 1987), guinea pig (Figs. 1.6 and 1.7) (Kaufmann 1981b) chinchilla (King and Tibbitts 1976)and capybaras and agoutis (Miglino et al. 2004). This group also includes Armadillos (Dasypodidae; Enders 1982) anteaters (Myrmecophagidae; Becher 1931; Walls 1938); and several bats (Bhiwgade 1990; Badwaik and Rasweiler 2000).

Although nominally haemomonochorial in practice there are probably always cytotrophoblast cells scattered over the fetal side of the syncytiotrophoblast (Fig. 8.15). These are necessary to support the vast increase in area by continual division and fusion into the syncytium, because there is no evidence that nuclei in vertebrate syncytia can divide.

8.6.2   Haemodichorial

There are two types. In the first, the cytotrophoblast which generates the syncytiotrophoblast persists as a continuous layer. The circulations are separated by fetal endothelium, cytotrophoblast and syncytiotrophoblast

This is characteristic of several orders of bats for example, Thyropteridae (Wimsatt and Enders 1980) (Fig. 8.21a), Vespertilionidae (Enders 1982) and Desmodontidae (Bjorkman and Wimsatt 1968).

All these bats show a characteristic intrasyncytial lamina close to the syncytiotrophoblast apex (Fig. 8.21).

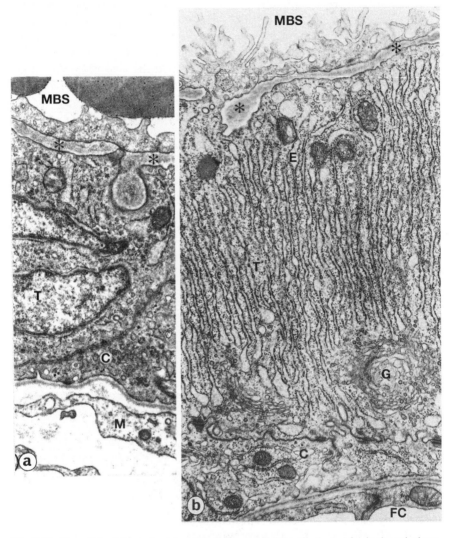

Fig. 8.21 Hemodichorial bat placentae. The intra syncytial lamella (*asterisks*) in the apical syncytiotrophoblast (T) is much more uniform in thickness, texture and separation from the maternal blood (MBS) than those shown in Fig. 1.8. Compare the structure to that of the endotheliochorial placenta (cat) in Fig. 7.8a which is said to be characteristic of a few genera of bats. *C* cytotrophoblast, *FC* fetal capillary, *G* golgi body, *N* nucleus, *M* mesoderm cell (**a**) *Thyroptera* near term placenta; ×17,500, from Wimsatt and Enders (1980). (**b**) *Myotis* near term placenta; ×22,000, from Enders (1982)

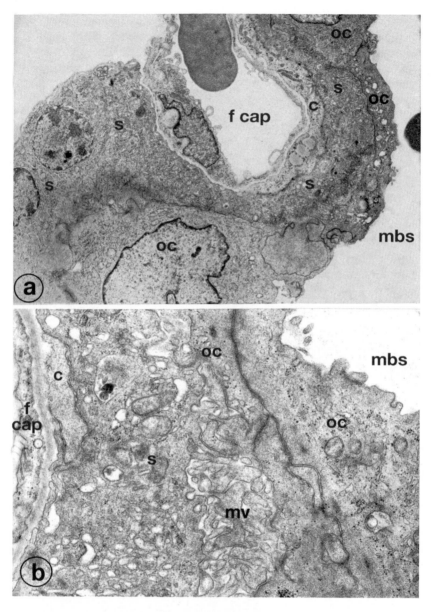

Fig. 8.22 Hemodichorial (??) placenta in Beaver, The maternal blood space is bounded by two layers, an outer cellular later (OC) and an inner syncytiotrophoblast layer (S) which includes an occasional, basal cytotrophoblast cell (C). It is not clear whether the outer cellular layer originates from the maternal endothelium or is trophectodermal. The cellular layer is flatly apposed to the syncytium or separated by a loose microvillar meshwork (mv). Since there is no trace of any basement or 'interstitial' membrane between the two layers, a hemodichorial nature seems more likely than endotheliochorial, but there is only one EM report so far from which these micrographs were taken (Fischer 1971). Late pregnant, (**a**) ×1,500; (**b**) ×11,400

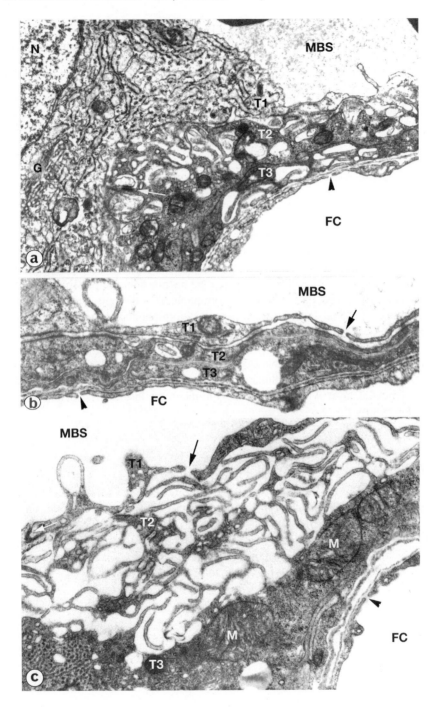

Fig. 8.23 Hemotrichorial placentation in Myomorph rodents. The triple layering is remarkably consistent in Muridae (rats, mice) and these examples of Cricetidae. Bounding the maternal blood space (MBS) layer T1 is usually cellular but very irregular with thin discontinuous areas (*arrows* on **b** and **c**) next to regions containing the nucleus (N on a) and abundant cytoplasm with a lot of rough endoplasmic reticulum (E) and golgi bodies (G). Layer T2 is syncytial, connected to Tl via the occasional desmosome (**a**, *white arrow*) and is usually the widest, frequently showing considerable elaboration of the apical plasmelemma (**a, c**) T3 is thinnest and most uniform, linked to T2 via desmosomes and gap junctions and occasionally having proliferations of the basal plasmalemma (a) adjacent. to the fenestrated (*upward arrow heads*) fetal capillaries (FC). *M* mitochondrion. White arrow in (**c**), cell boundary junction on T1. (**a**) *Peromyscus*, near term placenta; ×17,500. (**b**) *Lemmus*, near term placenta; ×22,600. (**c**) *Cleithryonomys*, near term placenta; ×21,300. All from King and Hastings (1977)

The rabbit has a similar dichorial arrangement but the cellular layer is very different in structure from the rather uniform columnar cellularity of the bats (Fig 8.7; Enders 1965; Samuel et al. 1975). It may be that both rabbit layers are independently generated by cellular inserts on the haemotrichorial pattern.

In the second type, the generative cytotrophoblast is discontinuous but there is an additional cellular layer between the syncytiotrophoblast it generates and the maternal blood. This pattern has been described only in the beaver.

There is a considerable initial cytotrophoblast proliferation at the placental site (Fig. 8.22). The syncytiotrophoblast-covered fetal villi grow into this layer, producing a definitive placenta with a cellular outer layer bathed in maternal blood underlain by a syncytial layer with occasional cytotrophoblast inserts. The details of the development of the two layers are not clear (Fischer 1971).

8.6.3   Haemotrichorial

This is found only in myomorph rodents with just one pattern of layering in the cricetids (Fig. 8.23) (gerbil) and murids (rats, mice, hamsters) (Enders 1965; King and Hastings 1977).

The circulations are separated by fetal endothelium, two syncytial layers and an outer fenestrated cellular layer facing the maternal blood (Fig. 8.4).

Chapter 9
Placental Immunology, Viviparity, Evolution

9.1 Immunology

9.1.1 Introduction

The immunological defences of vertebrates are based on two systems, the innate [non specific] and the adaptive [specific]. The innate uses largely monocytes, macrophages, granulocytes, NK [natural killer] and mast cells, with complement and acute phase proteins as humoral effectors. This is integrated with the adaptive system which employs the MHC controlled T and B lymphocytes and humoral antibodies. The innate is more primitive but both had evolved at least 220 million years ago [mya]. Vertebrate matrotrophic viviparity, based on internal fertilisation, retention of an allogeneic conceptus, and the development of a placenta in the uterus evolved 150–100 mya and was therefore always constrained by immunological considerations (Sacks et al. 1999; Parham 2004). These include the need for local tolerance in the uterus but maintenance of a normal level of systemic immunological surveillance.

Initially the brief toleration of the spermatozoa and seminal fluid components is based on the uterine epithelium providing a significant cellular barrier together with the local immunomodulation by the progesterone secretion from the corpus luteum developed from the ovulated ovarian follicle. After fertilisation the laying down of a shell around the zygote will provide a solely maternally derived immunological camouflage. Retention of the shelled yolky egg for ovoviviparity could probably be ignored immunologically, as in the monotremes.

However in order to dispense with the eggshell and yolk, the outer cellular layer of the conceptus [the eutherian trophoblast] needs to be apposed closely to the uterine epithelium to facilitate nutrient transfer. Such simple epithelial apposition is seen in most brief marsupial gestations, with insufficient time to develop any significant immune recognition.

Longer pregnancy always depends on the maintenance of progesterone secretion with its vital endocrine and immunological effects on the growth and maintenance of the placenta (Arck et al. 2007).

P. Wooding, G. Burton, *Comparative Placentation*,
© Springer-Verlag Berlin Heidelberg 2008

Eutherian placentas range from epitheliochorial through endotheliochorial to hemochorial and are defined by the depth of trophoblast invasion into the uterine tissues. The trophoblast which interacts with the uterine epithelium or stroma is always the outermost layer of the conceptus. The immune reaction will depend on the antigens presented by the trophoblast surface (not the fetus) and their recognition as allogeneic by the innate and adaptive systems.

In the few examples investigated so far (pigs, horses and ruminants [epitheliochorial], rodents and humans [hemochorial]) the trophoblast of the bulk of the placental surface does not express MHC I or II antigens (see references in Bainbridge et al. 2001; Moffett-King and Loke 2006) and therefore would not trigger an immune response directly. However in most cases, there is a smaller subpopulation of trophoblast cells which do express MHC, although usually of a non classical type often with restricted polymorphism.

9.1.2   Human

Most detail is known for the human, where the syncytiotrophoblast covering the blastocyst mediates invasion through the uterine epithelium into the decidualised endometrial stroma. The villous placenta then develops within the maternal tissue. This villous placenta has an MHC negative syncytiotrophoblast covering and the villi insert into a cellular layer of extravillous trophoblast [EVT] apposed to the decidualised endometrium. EVT cells express three MHC determinants, the classical HLA- C and the non-classical E and G which show little polymorphism. The cells do not express the classical HLA-A and HLA-B proteins which are the dominant T cell ligands mediating an immunological rejection response to infections and allografts. Cells from the EVT invade the decidualised endometrium and recent work (Moffett-King and Loke 2006) suggests that the HLA-C ligand reacts primarily with receptors on a uterine specific population of stromal NK cells [uNK] of the innate immunological system. Progesterone plays a role in the uterine homing and proliferation of these uNK cells in large numbers, up to 70% of the total leucocytes of the decidua, when the uterus decidualises post oestrus. If the balance between the EVT ligand and uNK receptor variants is optimal, this interaction produces cytokines and growth factors which could direct and regulate the EVT invasion and elimination of the muscle sheaths of the narrow spiral arteries supplying the uterus, producing wide flaccid conduits within the uterine decidua. This process is essential to increase the blood flow suitably for optimal placental perfusion and nutrient supply for the remainder of pregnancy, and the uNK cells decrease and disappear in the second half of pregnancy. The EVT – uNK reaction is thought to be finely balanced between too great an invasion by the EVT causing hemorrhage and too little, resulting in insufficient blood flow to the placenta. Coincident with these structural changes are the equally important immunomodulatory effects of progesterone and other reactions mediated by HLA-C, -E and -G with a wide variety of cells in the stroma, including decidualised fibroblasts, macrophages, dendritic and myelomonocytic types (see Chap. 1.7). These complex interactions have the potential either to bias the few T and B cells of the adaptive system present in the decidua toward Th2 and regulatory T cells providing tolerance throughout pregnancy, or to promote T cell immunorejection (Dietl et al. 2006; Blois et al. 2007; Hunt et al. 2007). In hemochorial placentation the maternofetal immunomodulation is multifactorial, always aiming toward a balance between the wide variety of cytokines, growth factors and hormones potentially available from decidua and trophoblast. The first requirement is to optimise the EVT remodelling of the maternal arteries. Subsequently, throughout pregnancy, it remains necessary to ensure sufficient immunomodulation for optimal growth and development of the uterus and the virtually wholly fetal placental tissue in a potentially hostile environment. This immunomodulation is largely local. There are a few systemic reactions, probably induced by exosomes and microparticles released by the trophoblast. The antibodies they express bias T cells of the circulating adaptive system toward Th2 and regulatory T cells rather than cytotoxic T cells. This however does not compromise the ability of the mother to resist infection (Sacks et al. 1999; Redman and Sargent 2007).

Other primates show much less EVT invasion, with a much more superficial implantation (Carter 2007) but the spiral artery modification by trophoblast, although differing in degree, is common to all hemochorial placentas investigated so far.

9.1.3   Rodents

In rodents the subpopulation of the spongiotrophoblast adjacent to the decidua expresses an as yet ill defined MHC type(s). uNK cells are present in the decidua in considerable numbers and in mouse play a much more direct role in remodelling arteries but in the rat and guinea pig the invasive trophoblast is the main effector and penetrates further into the arteries than in the mouse (Ain et al. 2003). However uNK gene Knock Out (KO) mice with no uNK cells become pregnant, no arterial remodelling occurs, but they still produce viable, though fewer and smaller neonates from smaller placentas (Croy et al. 2002). This emphasizes the complexity of the system for numerous genomic and KO studies in mice indicate that many other genes are involved in maintaining the correct balance between invasion and rejection but the relative importance of the MHC expression and the exact roles of the innate and adaptive systems are still unclear (Moffett-King and Loke 2006). The complexity may be necessary to ensure a viable compromise between the paternal genetic drive to maximise and prioritise fetal growth and the maternal to limit the fetal demand to a level the mothers' metabolism can sustain without damage. As with the human, the placenta is virtually 100% fetal tissue.

9.1.4   Pig

In epitheliochorial placentation there is a similar lack of MHC expression on the trophoblast surface as in the hemochorial type, and the microvillar interdigitation of the apical surfaces of the uterine epithelium and trophoblast in the pig, for example, produces no decidualisation in the uterine stroma. However there is a conceptus dependent increase in the stromal lymphocyte population in the 10–22 dpc implantation period, but with no recognisable subpopulation morphologically equivalent to the human or rodent uNK phenotype (Dimova et al. 2007). One or more of the subpopulations isolated from the endometrium at this time can produce more angiogenic factors [VEGF, HIF] than the trophoblast from the same conceptus (Croy et al. 2006) and together they may be as important as human or rodent uNKs in maternal vasculature modification – in this case initiating proliferation of the capillary bed, which starts at 15 dpc (Dantzer and Leiser 1994).

The massive increases in blood flows and placental exchange surfaces required during pregnancy is accommodated by mutual growth of VEGF expressing trophoblast and uterine epithelium and their separate underlying capillary networks (Winther et al. 1999a,b). The placental maternofetal interface is always 50% maternal.

Again a balance is achieved between the maternal immune defences, pregnancy hormones and the growth of the placenta and conceptus in the pig, relying on the absence of any trophoblast MHC expression (Ramsoondar et al. 1999).

In contrast, the horse epitheliochorial and ruminant synepitheliochorial placentas show a small subpopulation of MHC positive invasive trophoblast cells among the passive MHC negative majority.

9.1.5   Horse

In the horse, at 25–35 days of pregnancy prior to implantation, the blastocyst develops a localised narrow equatorial girdle of trophoblast cells, These express MHC I, become binucleate, synthesize equine chorionic gonadotropin (eCG), locally remove the uterine epithelium to which they are apposed and invade the uterine stroma to form endometrial cups (see Chap. 5.3.1). Once in place they switch off expression of MHC but are surrounded by maternal T lymphocytes and leucocytes. Their secretion of large amounts of eCG into the maternal circulation for the next 70 days is necessary to initiate new corpora lutea and maintain a hormone balance suitable for implantation and placental growth between the MHC negative trophoblast and the uterine epithelium. As soon as the cup cells secretion of eCG decreases the encircling lymphocytes and leucocytes infiltrate and kill the cup cells. At this stage, 120–130 dpc, the placenta (350 dpc at term) is presumably sufficiently well established to support its own considerable further expansion of the maternofetal interface. All equids require cup formation to ensure a successful normal pregnancy. Also normal is the generation of antipaternal antibodies and down regulation of the cytotoxic lymphocyte activity in the maternal circulation. Presumably all these processes are needed to achieve a suitable balance between rejection and local tolerance as in the pig.

9.1.6 Ruminants

Ruminants also have a subpopulation (15–20%)of trophoblast cells which are binucleate (BNC) and migrate/invade, not as far as the uterine stroma but to fuse with uterine epithelial cells or their derivatives to form fetomaternal hybrid tissue in the placentomes (see Chap. 6). This occurs throughout pregnancy, not transiently, as in the horse.

Before migration the late pregnant BNC in the cow have been shown (Bainbridge et al. 2001) to express both mRNA and antibodies for MHC determinants but because of the unique method of formation of the fetomaternal hybrid trinucleate cell (see Chap. 6.2.4) none of the original potentially allogeneic transcripts on the BNC plasmalemma are presented to the uterine stroma. The contents of the numerous BNC granules, bovine placental lactogen and numerous pregnancy associated glycoproteins (PAGS), are released from the trinucleate cell by exocytosis to the uterine stroma. Since the BNC migrate from implantation to term, these fetal products are ideally positioned to act as local immunomodulators in addition to their metabolic and developmental roles in the placentomes (Wooding et al. 2005b).

A more recent study in the cow (Davies et al. 2006) detected no expression of MHC on any trophoblast in the placentomes, but there was increasing expression of both classical and non classical MHC but only during the last 3 months of a 9 month pregnancy and always restricted to the interplacentomal uninucleate trophoblast cells. This is another small subset of trophoblast, never more than 5% of the total area. Since the MHC expression here is limited toward the end of pregnancy it was suggested to be more relevant to parturition than any general immunoregulatory role. In support of this, the greater the extent of MHC immunoincompatability between sire and dam, the fewer problems there were with placental retention.

In sharp contrast to the cow, the sheep with the same BNC migration, fusion and granule release, shows no MHC positive subpopulations in the trophoblast in any region at any time during pregnancy. The continuous BNC migration forms fetomaternal syncytial plaques rather than trinucleate cells in the placentome but this syncytium is also MHC negative (Gogolin-Ewens et al. 1989). Recent work around implantation has shown an enlargement of subepithelial endometrial fibroblasts in a subepithelial zone in the sheep, with the expression of proteins (osteopontin, desmin) characteristic of decidualisation in rodents but without lymphocyte recruitment (Johnson et al. 2003). This mild reaction could be a consequence of the granule release from the uterine epithelial syncytium produced by BNC migration and fusion.

Common to all ruminants so far investigated is a population of intraepithelial γδT cells of the innate system, restricted to the interplacentomal uterine epithelium with similar lytic capacity to that in human and rodent NK cells. However these cells only increase to significant numbers in late pregnancy and, like interplacentomal MHC expression in the cow, are probably involved in parturition rather than establishing early immunoacommodation between mother and conceptus (Aminoor-Rahman et al. 2002).

9.1.7   Endotheliochorial Placentas

In the endotheliochorial placenta the trophoblast is invasive, transforming into a syncytium before eliminating all of the uterine epithelium and closely apposing the intact maternal capillaries. Mutual growth of maternal and fetal capillaries and the intervening syncytium produce the massive growth in area and blood flow required during the second half of pregnancy.

No investigation of the MHC expression of the endotheliochorial syncytium has been published to our knowledge. Some species such as the cat show a few cells between syncytium and maternal capillaries which have been termed "decidual" but nothing is known of their function. The corpus luteum produces progesterone throughout pregnancy, no hormone /cytokine/growth factor secretion by the syncytium has been identified as yet.

It seems clear that the wide variety of placental structure is matched by the variety of immunological strategies for successful placentation based on both the innate and adaptive systems. Far too little is known as yet to draw useful parallels or clues as to the evolution of the various types.

9.2   Viviparity

The evolution of placentation depends on the advantages of viviparity to the species. The widespread occurrence of viviparity throughout the animal kingdom, both invertebrate and vertebrate, indicates that it is a very successful strategy for survival both in the aquatic and the terrestrial environments (Hogarth 1976; Amoroso 1981; Wourms 1981) In addition, the existence of many reptiles which can be oviparous [egg laying] in some environmental conditions or habitats, and viviparous in others together with the number of times the condition has evolved in the different phyla (Shine and Guillette 1988) indicate that the transition is both advantageous and genetically straightforward (Thompson and Speake 2006). There are two main types of viviparity; ovoviviparity, which involves retention of the shelled, yolky egg inside the mother until birth and has evolved much more frequently than matrotrophic viviparity, where the retained conceptus has much less yolk, a thin or absent shell and has developed specialized apposed uterine and extraembryonic placental structures to access nutrition directly from the mother throughout gestation. It has been estimated that the number of times that ovoviviparity and matrotrophic viviparity have evolved in the different classes are: Fish 28 and 17; Amphibia 5 and 3; Reptiles 100 and 3; Birds 0 and 0 and Mammals 1 and 1 (Blackburn 1993, 2000).

The main advantages of viviparity are the egg economy, and the ability to protect and maintain a suitable environment and nutrition for embryonic development. These plainly outweigh the drawbacks of the demands placed on the maternal structure and life style.

In the marine and terrestrial invertebrates the most usual provision for viviparity is maternal secretion of water and nutrients into a brood pouch in which the young develop, but there are examples of extraembryonic epithelial structures probably adapted for nutrient acquisition by apposition to the maternal tissues such as that found in the tunicate *Salpa* (Bone et al. 1985) and the onychophoran *Plicatoperipatus* (Huebner and Lococo 1994).

In the vertebrates in all cases, the system employed modifies a preexisting maternal organ, usually ovary, oviduct or uterus. The preparation and subsequent modifications are controlled by an interplay of estradiol and progesterone. These steroids are synthesized initially by the tissue from which the oocyte was produced and which becomes the corpus luteum (CL). The hypothalamus initiates the secretion by the pituitary of Luteinising Hormone (LH) which controls the frequency and periodicity of this ovulation.

Hormones and cytokines from the conceptus are also important in maintaining the CL progesterone secretion and preparing the uterus for implantation. Subsequently the placenta is an important source of signal molecules including progesterone for fetomaternal coexistence (see Porter et al. 1982 for a review).

The earliest mode of vertebrate reproduction in the Fish [Anamniotes] was release of numerous eggs and sperm coincidently into the sea producing hundreds of tiny very vulnerable embryos requiring sufficient yolk to survive to independence. Pituitary and ovarian hormones synchronized oocyte development, ovulation and release with mating. Most Teleost fish use this system. The advantages of gamete economy and safety led to the development of internal fertilization and with it the possibility of deposition of a protective shell, within which, after laying, the conceptus relied on the yolk to develop to a much less vulnerable stage before hatching. Ovarian hormones would be required firstly to modify the normal ovarian/oviductal/ uterine mucus secretion into localized shell constituent secretion, secondly to maintain quiescence of the uterus while the shell is deposited and thirdly to match the ovulation rate to shell production. Many Elasmobranchs and most Teleosts are oviparous (see Chap. 3). Retention of the shelled, yolky egg[s] in the uterine tract until the shell ruptures for live birth [ovoviviparity] increases the safety factor but considerably reduces numbers. Once the yolky egg is retained in uterine safety the shell can be reduced in thickness not only to facilitate gas exchange but also for direct uptake by the embryo of maternal nutrients via the epidermis and gut. Uterine glands can be developed to supply nutrients to the uterus together with other uterine/ovarian glandular secretions under ovarian hormonal control. This will allow reduction in the amount of yolk required initially and also the possibility of developing specialized epithelial outgrowths from both uterus and embryo to optimize the nutrient and gas transfer between the two and forming a placental structure if apposed.

In some of these matrotrophic viviparous fish embryos the yolk/belly sac is modified to assume this function in apposition to the uterine epithelium, usually with a very thin remnant of the shell between, (see Chap. 3.1.4). In others trophonemata from the uterine epithelium or trophotaeniae from the embryo hindgut produce similar 1,000-fold increases in the dry weight gain of the conceptus between ovulated egg and neonate whereas ovoviviparous fish relying on yolk alone, show no significant increase (see table 3–1).

Most Amphibians (Anamniotes) return to the aquatic environment to breed but a few have unusual sites for ovoviviparity (see Chap. 3.2). If in the oviduct, the embryos rely initially on the egg yolk but then take in oviductal secretions by mouth. No extraembryonic structures for access to nutrition have been identified.

In the evolution of viviparity and placentation it is the next stage – the development of further extraembryonic membranes, the amnion and allantois – which is least understood. These membranes are the essential prerequisite for the colonization of the land by the Reptiles, Birds and Mammals and the allantois forms the basis for all forms of placentation in these classes. However the living Amphibia, whose ancestors were undoubtedly the first to colonise the land, show neither.

Romer (1966) has suggested that the earliest reptiles had an amphibious lifestyle and range but because of the frequent intermittent droughts developed an egg which could be laid on land. This also had the advantage of avoiding the pressures of predation in the water. The major problems of desiccation, embryonic nutrition and protection and oxygen supply can be met by relying on further elaboration of the developmental flexibility of the extraembryonic membranes seen in the fish as modifications for viviparity.

Expansion of the celomic cavity would allow extension of the headfold seen in the Teleost Heterandria genus (Fig. 3.2) to enclose the embryo in a membranous sac – the amnion. This would provide a considerable barrier to desiccation, and equally important a "private sea" in which the delicate embryo would be protected from mechanical shock during development.

Similarly, the expansion of the celom provides an increased area of [troph-] ectoderm which could be vascularised for respiration by the sort of outgrowth from the hindgut as seen in the Teleost trophotaeniae (Fig. 3.2).This allantoic outgrowth would replace the very limited capacity of the yolk sac membrane for respiration, since it must concentrate on its primary function of yolk mobilization for nutrition.

In parallel with, or prior to, these membrane changes the eggshell needs to evolve to a form strong enough to support the bulk of the egg and to allow sufficient gas exchange(Packard and Seymour 1997).

This is one possible scenario, which over millions of years of stepwise evolution, could produce the first oviparous amniote, forerunner of the reptiles. The advantages of viviparity in the sea are equally valid on land and the extant reptiles have evolved ovoviviparity in at least 100 species and the ease of the transition is emphasized by the fact that some species adopt the viviparous habit when necessary. Several oviparous lizard species rely on this strategy when food is scarce or the environmental conditions unfavourable. It is suggested that these stresses modify the ovarian and or uterine hormone secretion so that the period of uterine quiescence needed for shell deposition is extended to suit developmental need. Once established on land, the rapid spread and diversification of the reptiles was testimony to the efficiency of the reproductive mode culminating in the age of the Dinosaurs, all of which were oviparous.

The allantois vascularises the chorion/trophectoderm in all reptiles forming a large respiratory surface but the yolk remains the primary nutritive source and there is no significant dry weight increase from egg to neonate in the 100 or so ovoviviparous forms as compared to the far more numerous oviparous forms. Only five or six species of skinks have true matrotrophic viviparity with very little yolk but with "placental" elaborations of limited areas of the chorioallantoic apposition to the uterine epithelium designed for nutrient transfer rather than respiration (Fig. 3.3). As in the fish this produces an enormous increase, up to 900,000% in some Mabuya species, in the dry weight gain during pregnancy which can last many months (Thompson and Speake 2006). Again as in the fish, the growth and quiescence of the uterus, and the provision of nutrients to the fetus are maintained by ovarian and uterine hormones [e.g. progesterone and relaxin] which also control parturition.

Birds and mammals are generally considered to have evolved from reptiles. Fossil skeletal evidence indicates many intermediate forms, most of which are now extinct, and suggests that the mammary gland and homeothermy evolved before viviparity. This is shown by the monotremes which lay eggs but suckle their young and have an amnion and allantois pattern of development very similar to that of the birds. The monotremes were the first development from the stem mammalian line, several million years later the marsupial ancestors appeared. Extant species are certainly viviparous, giving birth to very undeveloped neonates just capable of crawling to attach to the teat in the mothers pouch and relying on milk for most of fetal growth and development. The initial very short gestation period is reliant on yolk sac nutrition with a small allantoic sac. There is only one genus, the Peramelids (Figs. 4.1–4.3), which develop a short lived chorioallantoic placenta in the last third of the 15 day gestation, but this does produce a more developed neonate.

It is plain that neither the monotremes nor the marsupials seem likely starting points for the main mammalian line, and there is good molecular and fossil evidence that the eutherian stem genus evolved soon after the marsupials at least 124 mya ago (Ji et al. 2002). These small mammalian ancestral forms then had many millions of years of evolution alongside the dominant reptiles. There is no fossil evidence to indicate which form of reptile viviparity is most likely to have been the basic design for the wide variety of placental structures which today provide eutherian matrotrophic viviparity, relying on the amnion and the allantois.

All four mammalian superorders were present 75 mya, and all available evidence indicates that that the eutherians are monophyletic, sharing a single ancestor. The most complex reptile placenta, such as that in *Mabuya mabuya* (Chap. 3.4.3.2) would seem a suitable immediate precursor for a simple non invasive epitheliochorial eutherian placenta. However most recent molecular and geological evidence indicates that the superorders Xenarthra and Afrotheria, with no extant epitheliochorial members, originated before the Euarchontoglires or Laurasiatheria which show epithelio-,endothelio- and hemo- chorial types (Fig. 9.1). This suggests that the earliest "stem" mammal had an invasive trophoblast with the non invasive as subsequent developments. The viviparous mode is easily achieved in the reptiles and there was 70 million years of evolutionary time available for the protoeutherians to develop endothermy and lactation alongside a variety of forms of viviparity.

This produced the monotremes [ovoviviparous] and marsupials with a very short gestation, simple placenta and the teat replacing the umbilical cord. Most reptiles with a simple placenta have a long gestation, so if a short gestation time became advantageous, as with the marsupials, another way of accessing maternal nutrition rapidly is to develop an invasive placenta, characteristic of all extant mammals with a short gestation. The complete absence of yolk in the eutherian egg also necessitates rapid access to nutrition, although it could also be a consequence. The oldest (124 mya) fossil eutherian found so far is *Eomaia*, a tiny (25 g) furry animal, probably an arboreal insectivore (Ji et al. 2002). With a narrow pelvic girdle and by analogy with extant insectivores it is likely to have produced poorly developed altricial young after a short gestation with an invasive placenta. The oldest fossil metatherian, *Sinodelphys* at 125 mya (Luo et al. 2003), is very similar in size and arboreal skeletal modifications. This animal presumably had a very short gestation with pouch development replacing the need for invasive placentation, by analogy with the current Didelphids.

9.3   Evolution

The study of the evolution of the soft tissue of the eutherian placenta is complicated by the lack of any fossil record. Only the present day wide range of types, epithelio-, endothelio- and hemochorial is available. Comparative morphology of the embryonic membrane development of these types led Mossman (1987) to suggest an epitheliochorial ancestral type with polyphyletic evolution from this state. However the evidence from early fossil bone and tooth morphology indicates a monophyletic eutherian origin and this has been confirmed over the last decade by studies of the DNA sequence data – molecular phylogenetics. This technique relies on mapping changes in the DNA sequence in particular nuclear and mitochondrial genes of extant species. The rate of these changes has been shown to be fairly constant and this provides a molecular clock which can be calibrated against the fossil record. Computer algorithms then allow sequence comparisons from large numbers of individual gene sequences to be aggregated to produce the most likely evolutionary progression or tree topology (cladogram) for the sampled species (Fig. 9.1; Murphy et al. 2001).

The molecular confirms the fossil evidence to a considerable extent (Archibald 2003) but the molecular evidence is becoming ever more extensive. This has for example established a single Cetartiodactyl order from the Cetacea (whales, dolphins) and the Artiodactyla (bovids, camels) now confirmed by fossil evidence of the ankle joints (Carter and Mess 2006). It has also indicated that the Tenrecoidea should be in the Afrotheria superorder and not with other insectivores in the Laurasiatheria. Both methods support four eutherian superorders, Laurasiatheria (bats, carnivores, horses, insectivores, ruminants and whales), Euarchontoglires (lagomorphs, primates, rodents and tree shrews), Afrotheria (aardvarks, elephants, hyraxes, manatees and Tenrecoidea), and Xenarthra (anteaters, armadillos and sloths). See Fig. 9.1. The main problem still unresolved is which is the basal superorder, Xenarthra or Afrotheria (Mess and Carter 2006; Wildman et al. 2006). However since neither contain any species with epitheliochorial placentation, the ancestral species must have been endo- or hemochorial with epitheliochorial as a derived state. One possible tree topology/cladogram is shown in Fig. 9.1.

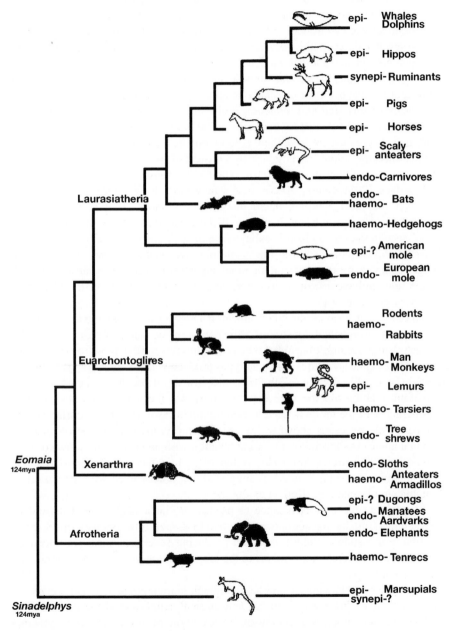

Fig. 9.1 Phylogenetic tree or Cladogram of eutherian mammals, showing the placental types of the various Orders. Modified from Vogel (2005) and Wildman et al. (2006)

Fossil morphological evidence can provide information that the molecular studies cannot, such as the size and possible habitat of the ancestral eutherian. The oldest (125 mya) known fossil eutherian so far is *Eomaia* (Ji et al. 2002), reconstructed as a hairy, tiny (25 g), agile animal with a dentition consistent with an insectivore diet, living in the shadow of the dinosaurs. Present day animals of this type all have endothelio- or hemochorial placentas, reinforcing the conclusion from the molecular and fossil derived evolutionary scenarios (Vogel 2005). It is interesting to note that the earliest fossil marsupial, *Sinadelphys,* also lived 124 mya and was a very similar mammal both in size and habitat (Luo et al. 2003).

It must be remembered that outside the localised zone of an endothelio- or hemochorial placenta there is usually an epitheliochorial type area or paraplacenta. Expansion of this area and reduction in the invasive area would offer a possible evolutionary route to a diffuse epitheliochorial placenta. What is unknown is what evolutionary pressure or drive there is to favour such a change.

The ancestral eutherian is generally considered to have evolved from a milk producing, homeothermic relative of extant viviparous synapsid reptiles (Kumar and Hedges 1998; Benton 2005) all of which have an epitheliochorial placenta. However the most complex lizard placenta does show different regions/zones adapted for gas and nutrient exchange. If the trophoblast of the discoid "placentome" seen in the scincid lizard Mabuya (Ramirez-Pinilla et al. 2006) became more invasive and endotheliochorial the whole placenta would be very similar to that suggested above for the ancestral mammal. However invasion also carries a considerable immunological cost requiring systems to control and limit the allogeneic intruder (see Chap. 9.1).

Since the uterine amniotic environment in a homeothermic mammal is very uniform, the environmental constraints which are a major factor in promoting evolutionary change are minimal. What then is the stimulus for the evolution of the wide range of placental structures, since all types produce viable young? It has been suggested (Haig 1993; Zeh and Zeh 2000) that instead of pregnancy being a mutual symbiotic relationship the fetal drive to maximise nutrient uptake is constantly conflicting with the mothers need to ration limited resources. This conflict is controlled mostly by genes showing sex specific imprinting, where one of a pair of parental alleles is silenced by DNA methylation with no change in sequence (John and Surani 2000; Kiefer 2007).This system of genetic manipulation is not found in fish, amphibia nor reptiles and it is suggested that it coevolved with the variety of mammalian placentation (Reik et al. 2003).

Maintaining the correct balance to produce a viable neonate will depend on developing a suitable barrier between the two to support different patterns of reproduction. For example a short gestation producing several helpless tiny neonates would require different adjustments to the long gestation producing the single run-soon-as-born types.

The invasive hemochorial structure, although immunologically costly, gives the fetus immediate direct access to gas and nutrient supplies, and the possibility of influencing the mothers metabolism by secretion of hormones and cytokines directly into her blood. Also the area of the fetal exchange surface is independent of any maternal tissue growth. All these considerations favour the fetus and many of the genes involved in placental growth have paternal expression only. The mother's alleles of these genes are imprinted and silenced. Genetic imprinting plays an important role in maintaining the maternofetal balance, with the mothers contributions (paternal allele silenced) blunting the more aggressive fetal drives and favouring fetal brain development (Martin 2003; Reik et al. 2003). Hemochorial placentation would favour a high fetal demand which is only sustainable for a short gestation, producing neonates with small brains and bodies. Epitheliochorial placentas, on the other hand, appose a cellular uterine epithelial barrier to the trophoblast, limiting and slowing its ability to manipulate the mother's metabolism. This enables the mother to sustain the lower demand for longer, eventually producing a single large neonate with, in the case of the dolphin, a brain almost as large as the human.

However this scenario doesn't explain how the mother withstands the long hemochorial gestation in the primates, although it does produce the largest brain of all. Wildman et al. (2006) suggest that the villous form of the primate placenta is less metabolically demanding, less efficient than the labyrinthine type in smaller hemochorial placentas, producing a mutually acceptable level of nutrient transfer, but there is no direct evidence for this. This emphasizes the difficulty, in comparative studies, of being able to define and measure the efficiency of a placenta. Baur 1977, pioneered this and measured the area of the placental interface per gram of conceptus at term. This is a better measure than *grams* of placenta per gram of conceptus because all transfers are limited by the area available. The results (see Fig. 2.9) clearly separate the diffuse epitheliochorial placentas (pigs, horses, camels) from the "compact" placentas with a localised area(s) of placental membrane proliferation which include synepitheliochorial (cow, giraffe) endotheliochorial (cat, elephant) and hemochorial (man, rodents, guinea pig) placentas. The diffuse epitheliochorial types provide six to ten times less area per gram of conceptus than any compact placenta so the horse ($4.4\,cm^2\,g^{-1}$) is much more 'efficient' than either the cow ($46\,cm^2$) or the human ($34\,cm^2$). The hemochorial human does seem more efficient on this criterion than the epitheliochorial cow until the different lengths of the pregnancies are factored in: the average weight gain of the conceptus per day in the cow is twice that of the human but equivalent to the horse (Vogel 2005).

Certainly the data indicate that the epitheliochorial placental type seems at least comparable to the other two in terms of gross transfer efficiency, but also that no one criterion is suitable for judging the overall efficiency of an individual placenta in terms of producing viable offspring. Most attempts at generalisations have too many exceptions to be useful because the evolutionary pressures are multifactorial. The maternofetal conflict and imprinting generate numerous potential sources of variability. The lifestyle and reproductive pattern are equally important in shaping the outcome. Close examination of a particular clade can provide some clues to evolutionary mechanisms and directions. A recent study of Cetartiodactyla, for example, was based on an analysis of six morphological placental characters and each produced the same evolutionary sequence (tree topology) (Klisch and Mess 2007). This showed that an increase in placental area per gram (to increase glucose transport) and the evolution of two new trophoblast cell hormone delivery systems (to manipulate the maternal metabolism) could plausibly be interpreted as the fetal initiatives. These were designed to reposition the maternofetal balance. It was suggested this had been tilted towards the maternal side by the change from hemo- or endotheliochorial to the basal Cetartiodactyl epitheliochorial form which interposed a uterine epithelial barrier to fetal invasion and ready access to maternal supplies.

It will be necessary to investigate a much wider range of taxa for molecular, structural and functional information to achieve a better understanding of the pathways of placental and vertebrate evolution.

Chapter 10
Hybridisation, Cloning and Fetal Origins of Adult Disease

10.1 Hybridisation

The range of placental structures found in the different mammalian orders makes the prospect of spontaneous hybridisation very unlikely. By definition the species are non hybridising groups so even though the detailed placental structures here are similar, other postzygotic factors block crossing even between closely related species. There are exceptions, kangaroos will cross fertilise interspecifically (references in O'Neill et al. 1998) and recently guanaco inseminated with camel sperm produced one offspring (Skidmore et al. 1999) but usually the success rate is low, live healthy neonates unusual and fertile crosses even rarer. The development of in vitro fertilisation and embryo transfer techniques has improved the success rates significantly and raised the possibility of increasing the breeding stock of endangered species such as Prezwalski horses raised in normal horses (Allen et al. 1986) and rare bovine species raised in standard cattle (Durrant and Benirschke 1981). The exact nature of the trophoblast genotype is another important factor, as *Mus musculus* will only produce live births of *Mus caroli* if the inner cell mass of *caroli* is camouflaged by injecting it into a *musculus* blastocyst (Rossant and Frels 1980). Other interspecific mouse crosses have revealed that one of critical loci for placental development maps to the proximal part of the X chromosome (Zechner et al. 1996). The X chromosome is imprinted in the extraembryonic tissues, with genes on the maternally inherited X being expressed. Similar results have been produced in sheep and goat crosses (Fehilly et al. 1984). However these are rare examples and it is clear that even with very similar placental structures, immunological and endocrinological imbalances will rapidly terminate gestation in interspecific crosses (Allen et al. 1987; Crepeau et al. 1989).

Recent molecular biological investigation of this hybrid incompatibility has indicated that the basic stability of the genome is also compromised. Up to 45% of the mammalian DNA is non functional and needs to be inactivated by methylation to ensure normal replication (Yoder et al. 1997). Studies on mouse (Shi et al. 2005), *Peromyscus* (Vrana et al. 2000) and kangaroo (O'Neill et al. 1998) hybrids show that a common feature is the disruption of the demethylation then de novo remethylation of DNA which occurs between gamete formation and early implantation stages in mice and cows (Dean et al. 2003). Imprinted genes are exempt from this genome wide demethylation at this time (Armstrong et al. 2006), but recent results show that significant losses of imprinting result from standard embryo transfer and tissue cultured embryos lose even more (Rivera et al. 2008).Whether this was an effect on methylation status was not investigated.

Identification of the genes which are directly involved in placental growth and development has relied on molecular biological manipulation of the DNA of the mouse embryo before transfer. Genes can be added (knock in) silenced (knockout) or multiplied and if the resulting embryo implants, phenotype examination with in situ hybridisation and/or immunocytochemistry can identify the stage at which the modified gene is expressed and the structure(s) it affects. Such studies have produced a wealth of information about mouse placentation and although the mouse is uniquely amenable to and suitable for these manipulations many useful parallels to gene function in other species are now becoming apparent (Georgiades et al. 2002; Cross et al. 2003; Cross 2006; Soares et al. 2006).

10.2 Cloning

Each of the blastomeres at the four cell stage of development can produce genetically identical neonates (or 'clones') if cultured separately to blastocysts and transferred to separate recipients. However the differentiation inherent in the subsequent divisions of the four cell stage blocks this possibility. Further manipulations of mouse zygotes showed that each cell of the 16 cell stage could produce viable blastocysts but only if they were fused (with chemical or electrical stimuli) with enucleated oocytes. Willadsen (1986) then extended the technique to sheep. Subsequently it was shown using the same method, that a non embryonic standard diploid adult somatic cell could be fused with an enucleated oocyte to produce a viable fertile sheep – Dolly, the first animal produced by Somatic Cell Nuclear Transfer (SCNT) (Wilmut et al. 1997).

This cloning technique has enormous potential. Not only could an unlimited number of genetic replicates of any commercially valuable or endangered animal be produced but similar molecular biological manipulations as used on the mouse could be adapted to the somatic cell nucleus before fusion. Clones could be engineered to produce human compatible antibodies in their milk or blood; antibody modification of, say, pig heart tissue, could produce an organ suitable for transplants into humans; much better animal models of human diseases could be developed in pig (atherosclerosis) and sheep (cystic fibrosis); and bulls could be designed to produce spermatozoa which produced only female calves (Vajta and Gjerrie 2006).

Unfortunately there is a major problem with SCNT in that the success rate of pregnancies carrying SCNT derived conceptuses is usually less than 5% and many of the offspring display abnormal growth and developmental defects (Farin et al. 2006; Vajta and Gjerrie 2006; Keefer 2007). Most of the work with SCNT has been with cows or sheep and there are many examples of successful completely normal progeny but at present the outcome of an SCNT based trial is entirely unpredictable. One of the reasons is because the placenta shows such a variety of defects. In the cow the placentomes are usually fewer and larger but the modifications to the histology are again variable. Some report a doubling in the frequency of the characteristic trophoblast binucleate cell (Ravelich et al. 2004), others a decrease (Heyman et al. 2002), yet others, normal levels (Miglino et al. 2007). There are reports of vascular changes (Miles et al. 2004; Farin et al. 2006), but nothing consistent to suggest possible remedies. There is now good evidence that a major part of the problem lies in the insufficiently accurate reprogramming of the genome of the donor cell by methylation postfusion and in the early division stages (Dean et al. 2003; Armstrong et al. 2006; Farin et al. 2006). Since many genes controlling placental development are imprinted (that is, differentially methylated) it is not surprising that there are so many problems throughout pregnancy. In the cloned mouse, with a 2% success rate, most SCNT pregnancies are lost before 10 dpc, and most of the remainder show abnormally large spongiotrophoblast and giant cell populations and a severely disrupted labyrinth zone (Tanaka et al. 2001). There is also evidence of abnormal gene expression in such pregnancies (Humpherys et al. 2002; Han et al. 2003).

The technique DOES have potential, but until the controls on gene expression are better understood, and the methylation status is surely only one of many factors here, that potential will remain unfulfilled.

10.3   Fetal Origins of Adult Disease

There is now a considerable weight of evidence that stresses on the mother during gestation not only modify placental structure, but such modification can also produce deleterious long term effects on the health of the offspring much later in life. The neonatal birthweight is closely correlated to that of the placenta in a wide variety of mammals (Baur 1977). Stresses, whether they are nutritional or from exposure to heat or altitude, or from restriction of uterine size, usually decrease the placental weight and nutrient transfer function and tend to produce smaller/thinner neonates. Human epidemiological studies on adults who were in utero during the Dutch famine of 1944 show that the smaller babies born to the starved mothers suffered an increased incidence in later life of diseases, notably heart attacks, hypertension, glucose intolerance, diabetes type II, obesity and reproductive disorders (collectively referred to as the 'metabolic syndrome' Barker 2001; Bertram and Hanson 2001). Similar results have now been described in a wide variety of human populations and can also be mimicked exactly in animal experiments. Such experiments also show that nutritional stress during gestation can produce long term deleterious effects even if the birthweight is normal (Bertram and Hanson 2001; Fowden et al. 2005, 2006a,b; Symonds et al. 2007).

The Dutch famine results also show that the timing of the nutritional stress is critical to the outcome. First trimester stress produced cardiovascular and brain defects, second trimester stress kidney problems and stress in the third trimester insulin resistance and eventually, diabetes II. This pattern of results is also seen in nutritionally or heat stressed sheep. Here nutrient, usually protein, restriction during the first 30 days of pregnancy (term is 147 days) produces cognitive defects and hypertension, during 30–90 dpc reduced nephron numbers and compromised renal function and during the final 90–147 dpc undernutrition results in normal adult cardiovascular function but increased adiposity, insulin resistance and diabetes II (Symonds et al. 2007).

The effects of maternal nutrient restriction on placental development is only beginning to be quantified accurately, most studies relying only on placental weight as a proxy marker. Comparing studies is also made difficult by the various diets employed, and the extent of the nutrient deprivation. However, a reduction in villous surface area appears to be a common feature, with effects on the fetal placental vasculature being more variable (Schlabritz-Loutsevitch et al. 2007 for a recent review). There are also marked species differences. For example 50% less protein in the isocaloric diets fed for the first 3.5 days of pregnancy with normal levels for the rest of the 19 days produces larger neonates in mice (Watkins et al. 2008) but smaller pups in rats (Kwong et al. 2000). However in both species the offspring show sustained hypertension and increased anxiety related behaviour as adults. The changes are seen even if the 3.5 day mouse blastocysts are transferred to a normally fed mother, the effect is definitely on the conceptus. In this case the function of the yolk sac, the sole nutrient uptake route at this stage, was shown to be altered (Watkins et al. 2008). At the blastocyst stage the embryonic founder cell lineages for embryonic and extraembryonic tissues are established so it is a particularly sensitive stage, but equally the subsequent regulation of the development of the various organ systems can be critically affected at specific times by the chorioallantoic placenta dependent, nutrient availability. The complex cascade(s) of specific gene expression necessary for normal conceptus development will depend upon the placenta's ability to deliver suitable levels of nutrients, notably protein, at each stage. There is good evidence that the placenta can alter its size, structure and transporter abundance and type in response to fetal nutrient demand but how this signalling is accomplished is unknown (Angeolini et al. 2006; Fowden et al. 2006b). The SCNT results indicate how important the methylation levels are. Since many differentially methylated, that is imprinted, genes are involved in placental growth it is possible that changes in gene methylation could be affected by suboptimal uterine conditions but there is no direct evidence for this as yet. It seems more likely that normal development relies on a critical level of nutrient supplies at particular times. Normally this is ensured, and the whole process orchestrated, by the maternal and fetal endocrine systems. In support of this scenario, evidence is accumulating to indicate that many of the adult problems are accompanied by abnormal hormone concentrations, and these would be the result of irreversible fetal reprogramming, in response to placental stresses, of the basal levels and sensitivities of a wide range of fetal endocrine functions (Fowden et al. 2005). This is a very active area of investigation currently and it is perhaps possible to see some parts of the complex hierarchy of controls that ensures the remarkable progression from zygote to neonate.

References

Abd-Elnaeim, M.M.M., Leiser, R., Wilsher, Allen, W.R. (2006) Structural and haemovascular aspects of placental growth throughout gestation in young and aged mares. Placenta 27: 1103–1113

Adamson, S.L., Lu, Y., Whiteley, K.J., Holmyard, D., Hemberger, M., Pfarrer, C., Cross, J.C. (2002) Interactions between trophoblast cells and the maternal and fetal circulation in the mouse placenta. Developmental Biology 250: 358–373

Ain, R., Canham, L.N., Soares, M.J. (2003) Gestation stage dependent intrauterine trophoblast cell invasion in the rat and mouse: novel endocrine phenotype and regulation. Developmental Biology 260: 176–190

Aitken, R.J. (1974) Delayed implantation in roe deer. Journal of Reproduction and Fertility 39: 225–233

Al-Lamki, R.S., Skepper, J.N., Burton, G.J. (1999) Are placental bed giant cells merely aggregates of small mononuclear trophoblast cells? An ultrastructural and immunocytochemical study. Human Reproduction: 496–504

Al-Timimi, A., Fox, H. (1986) Immunohistochemical localisation of FSH, LH, GH, ACTH and PRL in the human placenta. Placenta 7: 163–172

Allen, W.R. (1975) Endocrine functions of the placenta pp 214–267. In Comparative Placentation (D.H. Steven, Editor) Academic Press London

Allen, W.R. (1982) Immunological aspects of the endometrial cup reaction and the effect of xenogeneic pregnancy in horses and donkeys. Journal of Reproduction and Fertility 31: 57–94

Allen, W.R., Moor, R.M. (1972) The origin of the equine endometrial cups. I. Production of PMSG by fetal trophoblast cells. Journal of Reproduction and Fertility 29: 313–316

Allen, W.R., Stewart, F. (2001) Equine placentation. Reproduction Fertility and Development 13: 623–634

Allen, W.R., Hamilton, D.W., Moor, R.M. (1973) Origin of equine endometrial cups. II. Invasion of the endometrium by trophoblast. Anatomical Record 177: 485–502

Allen, W.R., Kydd, J.H., Antczak, D.F. (1986) Successful applications of immunotherapeutics to a model of pregnancy failure in equids, in Reproductive Immunology (eds. D.A. Clark, B.A. Croy), Elsevier Science, Amsterdam, pp. 253–260

Allen, W.R., Kydd, J.H., Boyle, M.S., Antczak, D.F. (1987) Extraspecific donkey in horse pregnancy as a model for early fetal death. Journal of Reproduction and Fertility Supplement 35: 197–209

Allen, W.R., Wilsher, S., Turnbull, C., Stewart, F., Ousey, J., Rossdale, P.D., Fowden, A.F. (2002) Influence of maternal size on placental, fetal and neonatal growth in the horse I. development in utero. Reproduction 123: 445–453

Aminoor-Rahman, A.N.D., Meeusen, E.N.T., Lee, C.S. (2002) Peripartum changes in the intraepithelial lymphocyte population of the sheep (uterine) interplacentomal epithelium. American Journal of Reproductive Immunology 47: 132–141

Aminoor-Rahman, A.N.D., Snibson, K.J., Lee, C.S., Meeusen, E.N.T. (2004) Effects of implantation and early pregnancy on the expression of cytokines and vascular surface molecules in the sheep endometrium. Journal of Reproductive Immunology 64: 45–58

Amoroso, E.C. (1952) Placentation, in Marshall's Physiology of Reproduction, Vol. II, Chap. 15 (ed. A. Parkesed), Longman, Green and Co., London, pp. 127–311

Amoroso, E.C. (1960) Viviparity in fishes. Symposia of the Zoological Society of London 1: 153–181

Amoroso, E.C. (1981) Viviparity, in Cellular and Molecular Aspects of Implantation (eds. S.R. Glasser, D.H. Bullock), Plenum Press, New York, pp. 3–25

Amoroso, E.C., Perry, J.S. (1975) The existence during gestation of an immunological buffer zone at the interface between maternal and fetal tisues. Phlosophical Transactions of the Royal Society. B. 271: 343–361

Amoroso, E.C., Heap, R.B., Renfree, M.B. (1979) Hormones and the evolution of viviparity, in Hormones and Evolution (ed. E.J.W. Barrington), Academic Press, New York

Anderson, J.W. (1969) Ultrastructure of the placenta and fetal membranes of the dog. I. The placental labyrinth. Anatomical Record 165: 1536

Anderson, D., Faber, J.J. (1982) Regulation of fetal placental blood flow in the lamb. American Journal of Physiology 247: R567–R574

Anderson, T.L., Hoffman, L.B. (1984) Alterations in epithelial glycocalyx of rabbit uteri during early pseudopregnancy and pregnancy and following ovariectomy. American Journal of Anatomy 171: 321–334

Anderson, T.L., Olson, G.E., Hoffman, L.B. (1986) Stage specific alterations in the apical membrane glycoproteins of endometrial epithelial cells related to implantation in rabbits. Biology of Reproduction 34: 701–720

Angeolini, E., Fowden, A.L., Coan, P., Sandovici, I., Smith, P., Dean, W., Burton, G., Tycko, B., Reik, W., Sibley, C., Constancia, M. (2006) Regulation of placental efficiency for nutrient transport by imprinted genes. Placenta 27 Supplement A Trophoblast Research 20: S98–S102

Anson-Cartwright, L., Dawson, K., Holmyard, D., Fisher, S.J., Lazzarini, R.A., Cross, J.C. (2000) The glial cells missing-1 protein is essential for branching morphogenesis in the chorioallantoic placenta. Nature Genetics 25: 311–314

Aoki, A., Metz, J., Forssman, W.G. (1978) Studies on the ultrastructural and permeability of the haemotrichorial placenta. II. Fetal capillaries and tracer administration into the fetal blood circulation. Cell and Tissue Research 192: 409–422

Archibald, J.D. (2003) Timing and biogeography of the eutherian radiation: fossils and molecules compared. Molecular Phylogenetics and Evolution 28: 350–359

Arck, P., Hansen, P.J., Jericevic, B.M., Piccinni, M.P., Szekeres-Bartho, J. (2007) Progesterone during pregnancy: endocrine–immune cross talk in mammalian species and the role of stress. American Journal of Reproductive Immunology 58: 268–279

Armstrong, L., Lako, M., Dean, W., Stojkovic, M. (2006) Epigenetic modification is central to genome reprogramming in SCNT. Stem Cells 24: 805–814

Assheton, R. (1906) The morphology of the ungulate placenta, particularly the development of that organ in the sheep and notes upon the placenta of the elephant and hyrax. Philosophical Transactions of the Royal Society B 198: 143–315

Atkinson, D.E., Boyd, R.D.H., Sibley, C.P. (2006) Placental transfer, Chap. 52, in Physiology of Reproduction, 3rd edn (eds. E. Knobil, J.D. Neill), Academic Press, New York, pp. 2787–2875

Ayuk, P.T., Theophanous, D., D'Souza, S.W., Sibley, C., Glazier, J.D. (2002) L-Arginine transport by the microvillous plasma membrane of the syncytiotrophoblast from human placenta in relation to nitric oxide production: effects of gestation, preeclampsia and IUGR. Journal of Clinical Endocrinology and Metabolism 87: 747–751

Badwaik, N.K., Rasweiler, J.J. (2000) Pregnancy, in Reproductive Biology of Bats, Chap. 6 (eds. E.G. Chrichton, P.H. Krutzsch), Academic Press, London, pp. 221–294

Badwaik, N.K., Rasweiler, J.J., Muradali, F. (1998) Coexpression of cytokeratin and vimentin by highly invasive trophoblast in the white winged vampire bat, *Diaemus youngii* and the black mastiff bat, *Molossus ater*. Journal of Reproduction and Fertility 114: 307–325

Bainbridge, D.R., Sargent, I.L., Ellis, S.A. (2001) Increased expression of major histocompatibility complex (MHC) class I transplantation antigens in bovine trophoblast cells before fusion with maternal cells. Reproduction 122: 907–913

Bainter, K. (1986) Intestinal Absorption of Macromolecules and Immune Transmission from Mother to Young. CRC Press, Florida

Bao, L., Tessier, C., Prigent-Tessier, A., Li, F., Buzzio, O.L., Calegari, E.A., Horseman., E.A., Gibori, G. (2007) Decidual prolactin silences the expression of genes detrimental to pregnancy. Endocrinology 148: 2326–2334

Barcroft, J., Barron, D.H. (1946) Observations on the form and relations of the maternal and fetal vessels in the placenta of the sheep. Anatomical Record 94: 569–595

Barker, D.J.P. (2001) The malnourished baby and infant. British Medical Journal 60: 69–88

Barlow, P.W., Sherman, M.L. (1972) The bio- chemistry of differentiation of mouse trophoblast: studies on polyploidy. Journal of Experimental Embroyology and Medicine 27: 447–465

Barnes, R.J., Comline, R.S., Silver, M., Steven, D.H. (1976) Ultrastructural changes in the placenta of the ewe after fetal hypophysectomy or adrenalectomy. Journal of Physiology 263: 173–174

Barrau, M.D., Abel, J.H., Torbit, E.A., Tietz, W.J. (1975) Development of the implantation chamber in the pregnant bitch. American Journal of Anatomy 143: 115–130

Basir, G.S., O, W.S., So, W.W., Ng, E.H., Ho, P.C. (2002) Evaluation of cycle-to-cycle variation of endometrial responsiveness using transvaginal sonography in women undergoing assisted reproduction. Ultrasound in Obstetrics and Gynaecology 19: 484–489

Bass, F., Krane, E.J., Mallon, K, Nathanielz, P.W., Steven, D.H. (1977) Mobilisation of binucleate cells in the placenta of the ewe after foetal pituitary stalk section. Journal of Endocrinology 73: 36–37

Bassett, J.M. (1991) Current perspectives on placental development and its integration with fetal growth. Proceedings of the Nutrition Society 50: 311–319

Battaglia, F.E., Meschia, G. (1978) Principal substrates of fetal metabolism. Physiological Reviews 58: 499–527

Battaglia, F.E., Meschia, G. (1986) An Introduction to Fetal Physiology. Academic Press, Orlando, FL

Battaglia, F.C., Regnault, T.R.H. (2001) Placental transport and metabolism of amino acids. Placenta 22: 145–161

Battista, P.J., Bell, A.W., Deaver, D,R., Currie, W.B. (1990) Differential control of the placental lactogen release and progesterone production by ovine tissue in vitro. Placenta 11: 337–348

Baur, R. (1972) Quantitative analyse des wachstums der Zottenoberflache bei der Placenta des Rindes und des Menschen. Zeitschrift fur Anatomie Entwicklungsgeschichte 136: 87–97

Baur, R. (1977) Morphometry of the placental exchange area. Advances in Anatomy Embryology and Cell Biology 53: 1–63

Bazer, F.W., First, N.L. (1983) Pregnancy and parturition. Journal of Animal Science 57 (Supplement 2): 425–460

Bazer, F.W., Vallet, J.L., Roberts, R.M., Sharp, D.C., Thatcher, W.W. (1986) Role of conceptus secretory products in establishment of pregnancy. Journal of Reproduction and Fertility 76: 841–850

Beaulaton, J., Lockshin, R.A. (1982) The relation of programmed cell death to development and reproduction: comparative studies and an attempt at classification. International Review of Cytology 79: 215–235

Becher, H. (1931) Placenta und uterus schleirnhaut von Tamandua tetradactyla (Myrmecophaga). Morphologisches Jahrbucher 67: 381–458

Beck, F., Lowe, J.R. (1972) The stimulation of a maternal pregnancy reaction in the ferret. Journal of Anatomy 111: 333

Beckers, J.F., DeCoster, R., Wouters-ballman, P., Froment-Lienhard, Ch., Van der Zwalmen, P., Ectors, F. (1982) Dosage radioimmunologique de l'hormone placentaire somatotrope et mammotrope bovine

Belkacemi, L., Gariepy, G., Mounier, C., Simoneau, L., Lafond, J. (2004) Calbindin-D9k localization and levels of expression in trophoblast cells from human term placenta. Cell and Tissue Research 315: 107–117

Bell, C. (1972) Autonomic nervous control of reproduction: circulatory and other factors. Pharmacological Reviews 24: 657–736

Bell, C. (1974) Control of uterine blood flow in pregnancy. Medical Biology 52: 219–228

Bell, S.C. (1983) Decidualisation: regional differentiation and associated functions. Oxford Review of Reproductive Biology 5: 220–271

Bell, A.W. (1984) Factors controlling placental and fetal growth and their effects on future production, in Reproduction in Sheep (eds. D.H. Lindsay, D.T. Pearce), Cambridge University Press, Cambridge, pp. 144–152

Bell, A.W. (1991) The whole animal – pregnancy and fetal metabolism, in Quantitative Aspects of Ruminant Digestion and Metabolism (eds. J.M. Forbes, J. France), Butterworths, London, pp. 100–145

Bell, S.C., Patel, S.R., Jackson, J.A., Waites, G.T. (1988) Major secretory protein of human decidualised endometrium is an IGF binding protein. Journal of Endocrinology 118: 317–328

Bell, A.W., Hay, W.W., Ehrhardt, R.A. (1999) Placental transport of nutrients and its implications for fetal growth. Journal of Reproduction and Fertility Supplement 54: 401–410

Ben David, E., Shemesh, M. (1990) Localisation of cytoplasmic side chain cleavage enzyme in the bovine placentome using the protein A gold technique. Biology of Reproduction 42: 131–138

Benton, M.J. (2005) Vertebrate Palaeontology, 3rd edn. Blackwell, London

Bertram, C.E., Hanson, M.A. (2001) Animal models and programming of metabolic syndrome. British medical Bulletin 60: 103–121

Betteridge, K.J. (1989) The structure and function of the equine capsule in relation to embryo manipulation and transfer. Equine Veterinary Journal. Supplement 8: 91–100

Bevilacqua, E.M.A.F., Abrahamsohn, P.A. (1988) Trophoblast giant cell formation during the invasive stage of implantation of the mouse embryo. Journal of Morphology 198: 341–351

Bhavnani, B.R., Woolever, C.A. (1978) Formation of steroids by the pregnant mare. VI. Metabolism of [3H] dehydroepiandrosterone injected into the fetus. Endocrinology 103: 2291–2298

Bhiwgade, D.A. (1990) Comparative electron microscopy of the chorioallantoic placental barrier in some Indian Chiroptera. Acta Anatomica 138: 302–317

Billingsley, S.A., Wooding, F.B.P. (1990) An immunogold cryoultrastructural study of sites of synthesis and storage of chorionic gonadotrophin and placental lactogen in human syncytiotrophoblast. Cell and Tissue Research 261: 375–382

Billington, W.D., Bell, S.C. (1983) Immunology of mouse trophoblast, in Biology of Trophoblast (eds. C. Loke, A. Whyte), Elsevier, Amsterdam, pp. 571–594

Bissonnette, J.M. (1981) Studies in vivo of glucose transfer across the guinea-pig placenta. Placenta Supplement 2: 155–162

Bissonnette, J.M., Farrell, R.C. (1973) Pressure-flow and pressure-volume relationships in the fetal placental circulation. Journal of Applied Physiology 35: 355–360

Bjorkman, N.H. (1968) Fine structure of cryptal and trophoblastic giant cells in the bovine placentome. Journal of Ultrastructural Research 24: 249–258

Bjorkman, N.H. (1970) An Atlas of Placental Fine Structure. Williams & Wilkins, Baltimore, USA

Bjorkman, N.H., Bloom, G. (1957) On the fine structure of the foetal-maternal junction in the bovine placentome. Zeitschrift Zellforschung 45: 649–659

Bjorkman, N.H., Wimsatt, W.A. (1968) The allantoic placenta of the vampire bat (Desmodus rotundus marinus): a reinterpretation of its structure based on electron microscopic observations. Anatomical Record 162: 83–98

Black, S., Kadyrov, M., Kaufmann, P., Ugele, B., Emans, N., et al. (2004) Syncytial fusion of human trophoblast depends on caspase 8. Cell Death and Differentiation 11: 90–98

Blackburn, D.G. (1993) Chorioallantoic placentation in squamate reptiles: structure, function, development and evolution. Journal of Experimental Zoology 266: 414–430

Blackburn, D.G. (2000) Reptilian viviparity: past research, future directions and appropriate models. Comparative Biochemistry and Physiology A 127: 391–409

Blackburn, D.G., Callard, I.P. (1997) Morphogenesis of placental membranes in the viviparous, placentotrophic lizard Chalcides chalcides. Journal of Morphology 232: 35–45

Blackburn, D.G., Lorenz, R.L. (2003a) Placentation in garter snakes. II. Transmission EM of the chorioallantoic placenta of Thamnophis radix and T. sirtalis. Journal of Morphology 256: 171–186

Blackburn, D.G., Lorenz, R.L. (2003b) Placentation in garter snakes. III. Transmission EM of the omphalallantoic placenta of *Thamnophis radix* and *T. sirtalis*. Journal of Morphology 256: 187–204

Blackburn, D.G., Vitt, L.J. (2002) Specialisations of the chorioallantoic placenta in the Brazilian scincid lizard, *Mabuya heathi*: a new placental morphotype for reptiles. Journal of Morphology 254: 121–131

Blackburn, D.G., Vitt, L.J., Beuchat, C.A. (1984) Eutherian like reproductive specialisation in a viviparous reptile. Proceedings of the National Academy of Sciences of the United States of America 81: 4860–4863

Blackburn, D.G., Osteen, K.G., Winfrey, V.P., Hoffman, L.H. (1989) Obplacental giant cells of the domestic rabbit. Development, morphology and intermediate filament composition. Journal of Morphology 202: 185–203

Blankenship, T.N., Enders, A.C. (2003) Modification of uterine vasculature during pregnancy in macaques. Microscopy Research and Technique 60: 390–401

Blankenship, T.N., Given, R.L. (1992) Penetration of the uterine epithelial basement membrane during blastocyst implantation in the mouse. Anatomical Record 233: 196–204

Blankenship, T.N., Given, R.L., Parkening, T.A. (1990) Blastocyst implantation in the chinese hamster (*Cricetulus griseus*). American Journal of Anatomy 187: 137–157

Blankenship, T.N., Enders, A.C., King, B.F. (1992) Distribution of laminin, type IV collagen and fibronectin in the cell columns and trophoblast shell of early macaque placentas. Cell and Tissue Research 270: 241–248

Blois, S.M., Kammerer, U., Arck, P.C. (2007) Dendritic cells: key to fetal tolerance? Biology of Reproduction 77: 590–598

Blomberg, L.A., Garret, W.M., Guillomot, M., Miles, J.R., Sonstergard, T.S., VanTassell, C.P., Zuellse, K.A. (2006) Transcriptome profiling of the tubular porcine conceptus identifies the differential regulation of growth and developmentally associated genes. Molecular Reproduction and Development 73: 1491–1502

Bogic, L.V., Yamamoto, S.Y., Millar, L.K., Bryant-Greenwood, G.D. (1997) Developmental regulation of the human relaxin genes in the decidua and the placenta: overexpression in the preterm premature rupture of the fetal membranes. Biology of Reproduction 57: 908–920

Bone, Q., Pulsford, A.L., Amoroso, E.C. (1985) The placenta of the salp (Tunicata: Thaliacea). Placenta 6: 53–64

Boomsma, R.A., Mavrogianis, P.A., Verhage, H.G. (1997) Immunocytochemical colocalisation of TGFa, EGF and EGFR in the cat endotheliochorial placenta. Histochemical Journal 29: 495–504

Borke, J.L., Caride, A., Verma, A.K., Kelley, L.K., Smith, C.H., Penniston, J.T., Kumar, R. (1989) Calcium pump epitopes in the placental trophoblast basal membranes. American Journal of Physiology 257: C341–C346

Borowicz, P.P., Arnold, D.R., Johnson, M.L., Grazul-Bilska, A.T., Redmer, D.A., Reynolds, L.P. (2007) Placental growth throughout the last two thirds of pregnancy in sheep: vascular development and angiogenic factor expression. Biology of Reproduction 76: 259–267

Boshier, D.P. (1969) A histological and histochemical examination of implantation and early placentome formation in sheep. Journal of Reproduction and Fertility 19: 51–61

Boshier, D.P. (1970) The pontamine blue reaction in pregnant sheep uteri. Journal of Reproduction and Fertility 22: 595–596

Boshier, D.P., Holloway, H. (1977) The sheep trophoblast and placental function: an ultrastructural study. Journal of Anatomy 124: 287–298

Bouillot, S., Rampon, C., Tillet, E., Huber, P. (2006) Tracing the glycogen cells with protocadherin 12 during mouse placenta development. Placenta 27: 882–888

Boving, B. (1971) Biomechanism of implantation, in Biology of the Blastocyst (ed. R.J. Blandau), University of Chicago Press, Chicago

Bowen, J.A., Burghardt, R.C. (2000) Cellular mechanisms of implantation in domestic farm animals. Seminars in Cell and Developmental Biology 11: 93–104

Bowen, J.A., Bazer, F.W., Burghardt, R.C. (1996) Spatial and temporal analysis of integrin and Muc-1 expression in porcine uterine epithelium and trophoblast in vivo. Biology of Reproduction 55: 1098–1106

Bower, D.J. (1987) Chromosome organisation in polyploid mouse trophoblast nuclei. Chromosoma 95: 76–80

Boyd, J.D., Hamilton, W.J. (1970) The Human Placenta. Heffers, Cambridge

Bracher, U., Mathias, S., Allen, W.R. (1996) Influence of chronic degenerative endometritis (endometrosis) on placental development in the mare. Equine Veterinary Journal 28: 180–188

Bradbury, S., Billington, W.O., Kirby, D.RS. (1965) A histochemical and electron microscopical study of the fibrinoid of the mouse placenta. Journal of the Royal Microscopical Society 84: 199–211

Brambel, C.E. (1933) Allantochorionic differentiations of the pig studied morphologically and histochemically. American Journal of Anatomy 51: 397–459

Brambell, F.W.R. (1966) The transmission of passive immunity from mother to young and the catabolism of immunoglobulins. Lancet ii: 1087–1093

Brambell, F.W.R. (1970) The Transmission of Passive Immunity from Mother to Young. Frontiers of Biology, Vol. 18. North Holland Press, Amsterdam

Braun, T., Li, S., Moss, T.J.M., Newnham, J.P., Challis, J.R.G. Gluckman, P., Sloboda, D,M. (2007) Maternal betamethasone administration reduces binucleate cell number and placental lactogen in sheep. Journal of Endocrinology 194: 337–347

Braverman, M.B., Bagni, A., DeZiegler, D., Den, T., Gurpide, E. (1984) Isolation of PRL producing cells from first and second trimester decidua. Journal of Clinical Endocrinology and Medicine 58: 521–528

Brent, R.L., Fawcett, L.B. (1998) Nutritional studies of the embryo during early organogenesis with normal embryos and embryos exhibiting yolk sac dysfunction. Journal of Pediatrics 132: S6–S16

Brierley, J., Clark, D.A. (1987) Characterisation of hormone dependent suppressor cells in the uterus of mated and pseudopregnant mice. Journal of Reproductive Immunology 10: 201–217

Brosens, J.J., Pijnenborg, R., Brosens, I.A. (2002) The myometrial junctional zone spiral arteries in normal and abnormal pregnancies. American Journal of Obstetrics and Gynecology 187: 1416–1423

Brunette, M.G. (1988) Calcium transport through the placenta. Canadian Journal of Physiology and Pharmacology 66: 1261–1269

Bruns, M.E., Overpeck, J.G., Smith, G.C., Hirsch, G.N., Mills, S.E., Bruns, D.E. (1988) Vitamin D dependent Ca binding protein in rat uterus: differential effects of estrogen, tamoxifen, progesterone and pregnancy. Endocrinology 122: 2371–2378

Bryant Greenwood, G.D., Rees, M.C.P., Turnbull, A.C. (1987) Immunohistochemical localisation of relaxin, prolactin and prostaglandin synthesase in human amnion, chorion and decidua. Journal of Endocrinology 114: 491–496

Bryden, M.M., Evans, H.E., Binns, W. (1972) Embryology of the sheep. I. Extra embryonic membranes and the development of body form. Journal of Morphology 138: 160–186

Bucher, K. Leiser, R.,Tiemann, U.,Pfarrer, C. (2006) Platelet activating factor and acetyl hydrolase are expressed in immature bovine trophoblast giant cells throughout gestation, but not at parturition. Prostaglandins and other lipid mediators 79: 74–83

Bulmer, J.N., Billington, W.D., Johnson, P.M. (1984) Immunohistological identification of trophoblast populations in early human pregnancy with the use of monoclonal antibodies. American Journal of Obstetrics and Gynecology 148: 19–26

Bulmer, J.N., Morrison, L., Johnson, P.M., Meager, A. (1990) Immunohistochemical localisation of interferons in human placental tissue in normal ectopic and molar pregnancy. American Journal of Reproductive Immunology 22: 109–116

Burd, L.I., Jones Jr., M.D., Simmons, M.A., Makowski, E.L., Meschia, G., Battaglia, F.C. (1975) Placental production and foetal utilization of lactate and pyruvate. Nature 254: 710–711

Burgess, T.L., Kelly, R.B. (1987) Constitutive and regulated secretion of proteins. Annual Review of Cell Biology 3: 243–293

Burke, L., Rice, G.E., Ralph, M.M., Thorburn, G.D. (1989) Effects of calcium availability on the release of ovine choriomammotropin from cotyledonary cells incubated in vitro. Comparative Biochemistry and Physiology 93A: 489–492

Burton, G. (1982) Review article. Placental uptake of maternal erythrocytes: a comparative study. Placenta 3: 407–434

Burton, G. (1987) The fine structure of the human placental villus as revealed by scanning electron microscopy. Scanning Microscopy 1: 1811–1828

Burton, G. (1990) On the varied appearances of the human placenta villous surface visualised by scanning electron microscopy. Scanning Microscopy 4: 501–507

Burton, G.J., Tham, S.W. (1992) Formation of vasculosyncytial membrane in the human placenta. Journal of Developmental Physiology 18: 43–47

Burton, G.J., Jauniaux, E. (1995) Sonographic, stereological and Doppler flow velocimetric assessments of placental maturity. British Journal of Obstetrics and Gynaecology 102: 818–825

Burton, G.J., Jauniaux, E., Watson, A.L. (1999) Maternal arterial connections to the placental intervillous space during the first trimester of human pregnancy; the Boyd Collection revisited. American Journal of Obstetrics and Gynecology 181: 718–724

Burton, G.J., Watson, A.L., Hempstock, J., Skepper, J.N., Jauniaux, E. (2002) Uterine glands provide histiotrophic nutrition for the human fetus during the first trimester of pregnancy. Journal of Clinical Endocrinology and Metabolism 87: 2954–2959

Burton, G.J., Jauniaux, E., Charnock-Jones, D.S. (2007) Human early placental development: potential roles of the endometrial glands. Placenta 28 Supplement A: S64–S69

Busch, L.C., Winterhager, E., Fischer, B. (1986) Regeneration of the uterine epithelium in later stages of pseudopregnancy in the rabbit. An ultrastructural study. Anatomy and Embryology 174: 97–104

Butler, H. (1967) The giant cell trophoblast of the Senegal galago (*Galago senegalensis*) and its bearing on the evolution of the primate placenta. Journal of Zoology 152: 195–207

Byatt, J.C., Shimomura, K., Duello, T.M., Bremel E. (1986) Isloation and characterisation of multiple forms of bovine placental lactogen from secretory granules of the fetal cotyledon. Endocrinology 119: 1343–1350

Byatt, J.C., Eppard, P.J., Veenhuizen, J.J., Sorbet, R.H., Buonomo, F.C., Curran,D.F., Collier, R.J. (1992) Srerum half life and in vivo actions of recombinant bovine placental lactogen in the cow. Journal of Endocrinology 132: 185–193

Caluwaerts, S., Vercruysse, L., Luyten, C., Pijnenborg, R. (2005) Endovascular trophoblast invasion and associated changes in the spiral arteries of the pregnant rat. Placenta 26: 574–584

Cardell, R.R., Hisaw, F.L., Dawson, A.B. (1969) The fine structure of granular cells in the uterine endometrium of the rhesus monkey *Macaca mulatta* with a discussion of the possible function of these cells in relaxin secretion. American Journal of Anatomy 124: 307–340

Carpenter, S.J. (1972) Light and electron microscope observations on the morphology of the chorioallantoic placenta of the golden hamster (Cricetusa (ureatus)) days 7 through 9 of gestation. American Journal of Anatomy 135: 445–476

Carpenter, S.J. (1975) Ultrastructural observations on the maturation of the placental labyrinth of the golden hamster (days 10–16 (term) of gestation). American Journal of Anatomy 143: 315–348

Carpenter, S.J. (1982) Tropoblast invasion and alteration of maternal arteries in the pregnant hamster. Placenta 3: 219–242

Carstensen, M., Leichtweiss, H.-P., Molsen, G., Schrcer, H. (1977) Evidence for a specific transport of D-hexoses across the human term placenta in vitro. Archiv fUr Gynakologie 222: 187–196

Carter, A.M. (1975) Placental circulation, in Comparative Placentation: Essays in Structure and Function (ed. D.H. Steven), Academic Press, New York, pp. 108–160

Carter, A.M. (2005) Classics revisited. Placentation in an American mole, *Scalopus aquaticus*. Placenta 26: 597–600

Carter, A.M. (2007) Animal models of human pregnancy – a review. Placenta 28 Supplement 1: S41–S47

Carter, A.M., Gothlin, J., Olin, T. (1971) An angiographic study of the structure and function of uterine and maternal placental vasculature in the rabbit. Journal of Reproduction and Fertility 25: 201–210

Carter, A.M., Enders, A.C., Künzle, H., Oduor-Okelo, D., Vogel, P. (2004a) Placentation in species of phylogenetic importance: the Afrotheria. Animal Reproduction Science 82/83: 35–48

Carter, A.M., Blankenship, T.N., Künzle, H., Enders, A.C. (2004b) Structure of the definitive placenta of the tenrec, *Echinops telfairi*. Placenta 25: 218–232

Carter, A.M., Blankenship, T.N., Künzle, H., Enders, A.C. (2005) Development of the haemophagous region and labyrinth of the placenta of the tenrec, *Echinops telfairi*. Placenta 26: 251–261

Carter, A.M., Blankenship, T.N., Enders, A.C., Vogel, P. (2006) The fetal membranes of the otter shrews and a synapomorphy for Afrotheria. Placenta 27: 258–268

Casey, M.L., Winkel, E.A., Porter, J.E., MacDonald, P.E. (1983) Endocrine regulation of the initiation of parturition. Clinics in Perinatology 10: 709–712

Castellucci, M., Scheper, M., Scheffen, I., Celora, A., Kaufmann, P. (1990) The development of the human placental villous tree. Anatomy and Embryology 181: 117–128

Cate-Hoedemaker, N.J. (1933) Beitrage zur kenntnis der plazentation bei Haien und Reptilien. ZeitschriJt Zellforschung & Mikroskopisches Anatomie 18: 299–345

Cateni, C., Paulesu, L., Bigliardi, E., Hamlett, W.C. (2003) The interleukin system in the uteroplacental complex of the cartilaginous fish the smoothhound shark *Mustela canis*. Reproductive Biology and Endocrinology 1: 25

Challis, J.R.G., Dilley, S.R., Robinson, J.S., Thorburn, G.D. (1976) Prostaglandins in the circulation of the foetal lamb. Prostaglandins 11: 1041–1052

Champion, E.E., Glazier, J.D., Greenwood, S.L., Mann, S.J., Rawlings, J.M., Sibley, C.P., Jones, C.J.P. (2003) Localization of alkaline phosphatase and Ca^{2+}-ATPase in the cat placenta. Placenta 24: 453–461

Champion, E.E., Mann, S.J., Glazier, J.D., Jones, C.J.P., Rawlings, J.M., Sibley, C.P., Greenwood, S.L. (2004) System β and system A amino acid transporters in the feline endotheliochorial placenta. American Journal of Physiology 287: R1369–R1379

Chan, J.S.D., Robertson, H.A., Friesen, H.G. (1978) Maternal and fetal concentrations of OPL measured by radioimmunoassay. Endocrinology 102: 1606–1613

Chan, J.S.D., Nie, Z.-R., Pang, S.E. (1990) Cellular localisation of ovine placental lactogen using monoclonal antibodies. Animal Reproduction Science 23: 33–40

Chaouat, G. (1987) Placental immunoregulatory factors. Journal of Reproductive Immunology 10: 179–188

Chatterjee, S., Hasrouni, S., Lala, P.R. (1982) Localisation of paternal H2K antigens on murine trophoblast cells in vivo. Journal of Experimental Medicine 155: 1679–1689

Chavez, D.J., Enders, A.E. (1982) Lectin binding of mouse blastocysts: appearance of *Dolichos biflorus* binding sites on the trophoblast during delayed implantation and their subsequent disappearance during implantation. Biology of Reproduction 26: 545–552

Chen, T.T., Bazer, F.W., Gebhardt, B.M., Roberts, R.M. (1975) Uterine secretion in mammals. Biology of Reproduction 13: 304–313

Chene, N., Martal, J., Charrier, J. (1988) Ovine chorionic somatomammotropin and fetal growth. Reproduction Nutrition and Development 28: 1707–1730

Cherny, R.A., Findlay, J.K. (1990) Separation and culture of endometrial epithelial and stromal cells: evidence of morphological and functional polarity. Biology of Reproduction 43: 241–250

Christofferson, R.H., Nillson, B.O. (1988) Morphology of the endometrial vasculature during early placentation in the rat. Cell and Tissue Research 253: 209–220

Chung, M., Teng, C., Timmerman, M., Meschia, G., Battaglia, F.C. (1998) Production and utilisation of aminoacids by the ovine placenta. American Journal of Physiology 274: E1–E22

Clark, D.A. (1985a) Maternofetal relations. Immunology Letters 9: 239–247

Clark, D.A. (1985b) Prostaglandins and immunoregulation during pregnancy. American Journal of Reproduction and Immunological Microbiology 9: 111–112

Clark, D.A., Manuel, J., Chaouat, G., Gorczynski, R.M., Levy, G.A. (2004) Ecology of danger-dependent cytokine boosted spontaneous abortion in the CBA X DBA/2 mouse model. I. Synergistic effect of LPS and (TNFa + INFg) on pregnancy loss American Journal of Reproductive Immunology 52: 370–378

Coan, P.M., Ferguson-Smith, A.C., Burton, G.J. (2004) Developmental dynamics of the definitive mouse placenta assessed by stereology. Biology of Reproduction 70: 1806–1813

Coan, P.M., Ferguson-Smith, A.C., Burton, G.J. (2005) Ultrastructural changes in the interhaemal membrane and junctional zone of the murine chorioallantoic placenta across gestation. Journal of Anatomy 207: 783–796

Coan, P.M., Conroy, N., Burton, G.J., Ferguson-Smith, A.C. (2006) Origin and characteristics of glycogen cells in the developing murine placenta Developmental Dynamics 235: 3280–3294

Coceani, F., Olley, P.M., Bisha, I, Bodach, E., White, K.P. (1978) Significance of the prostaglandin system to the control of muscle tone of the ductus arteriosus. Advances in Prostaglandin and Thromboxane Research 4: 325–333

Colbern, E.T., Main, E.K. (1991) Immunology of the maternal placental interface in normal pregnancy. Seminars in Perinatology 15: 196–205

Cole, W.E. Garfield, R.E. (1989) Ultrastructure of the myometrium, in Biology of the Uterus, 2nd edn, Chap. 15 (eds. R.M. Wynn, W.P. Jollie), Plenum, New York, pp. 455–504

Comline, R.S., Silver, M. (1970) pO_2, pCO_2 and pH levels in the umbilical and uterine blood of the mare and ewe. Journal of Physiology 209: 587–608

Comline, R.S., Silver, M. (1974) A comparative survey of blood gas tensions, oxygen affinity and red cell 2,3-DPG concentrations in the fetal and maternal blood in the mare, cow and sow. Journal of Physiology 242: 805–826

Concannon, P.W., McCann, J.P., Hemp, K.M. (1989) Biology and endocrinology of ovulation pregnancy and parturition in the dog. Journal of Reproduction and Fertility Supplement 39: 3–25

Corbacho, A.M., Martínez De La Escalera, G., Clapp, C. (2002) Roles of prolactin and related members of the prolactin/growth hormone/placental lactogen family in angiogenesis. Journal of Endocrinology 173: 219–38

Covone, A.K, Johnson, P.M., Mutton, D., Adinolfi, M. (1984) Trophoblast cells in peripheral blood from pregnant women. Lancet ii: 841–843

Crane, L.H, Martin, L. (1991) In vivo myometrial activity during early pregnancy and pseudopregnancy in the rat. Reproduction Fertility and Development 3: 233–244

Crawford, R.J., Tregear, G.W., Niall, H.D. (1986) The nucleotide sequences of baboon chorionic gonadotropin f3-subunit genes have diverged from the human. Gene 46: 161–169

Crepeau, M.A., Yamashito, S., Croy, B.A. (1989) Morphological demonstration of the failure of *Mus caroli* trophoblast in the *Mus musculus uterus*. Journal of Reproduction and Fertility 86: 277–288.

Critchley, G.R., Burton, G.J. (1987) Intralobular variation in barrier thickness in the mature human placenta. Placenta 8: 185–194

Cronier, L., Frendo, J-L., Defamie, N., Pidoux, G., Bertin, G., et al. (2003) Requirement of gap junctional intercellular communication for human villous trophoblast differentiation. Biology of Reproduction 69: 1472–1480

Cross, J.C. (2006) Placental function in development and disease Reproduction Fertility and Development 18: 71–76

Cross, J.C., Roberts, R.M. (1989) Porcine conceptuses secrete an interferon during the preattachment period of early pregnancy. Biology of Reproduction 40: 1109–1118

Cross, J.C., Hemberger, M., Lu, Y., Nozaki, T., Whiteley, K., Masutani, M., Adamson, S.L. (2002) Trophoblast functions, angiogenesis and remodeling of the maternal vasculature in the placenta. Molecular and Cellular Endocrinology 187: 207–212

Cross, J.C., Baczyk, D., Dobric, N., Hemberger, M., Hughes, M., Simmons, D.G., Yamamoto, H., Kingdom, J.C.P. (2003) Genes development and evolution of the placenta. Placenta 24: 123–130

Croy, B.A., Chantakru, S., Essadeg, S., Ashkar, A.A., Weiss, Q. (2002) Decidual NK cells: key regulators of placental development. Journal of Reproductive Immunology 57: 151–168

Croy, B.A., Van den Heuvel, M.J., Borzychowski, A.M., Tayade, C. (2006) uNK cells: a specialized differentiation regulated by hormones. Immunological Reviews 214: 161–185

Crump, A., Donaldson, W.L., Miller, J., Kydd, J.H., Allen, W.R., Antczak, D.F. (1987) Expression of MHC antigens on horse trophoblast. Journal of Reproduction and Fertility Supplement 35: 379–388

Cukierski, M.A. (1987) Synthesis and transport studies of the intra syncytial lamina: an unusual placental basement membrane in the little brown bat *Myotis lucifugus*. American Journal of Anatomy 178: 387–409

Currie, M.J., Bassett, M.J., Gluckman, P.D. (1997) Ovine Glut-1 and -3: cDNA partial sequences and developing gene expression in the placenta. Placenta 18: 393–401

Dallenbach-Hellweg, G. (1967) Endometrial granulocytes and implantation. Excerpta Medica International Congress, Series 133: 411–418

Dantzer, V. (1985) Electron microscopy of the initial stages of placentation in the pig. Anatomy and Embroyology 172: 281–293

Dantzer, V., Leiser, R. (1994) Initial vascularisation in the pig placenta. Demonstration of non-glandular areas by histology and corrosion casts. Anatomical Record 238: 177–190

Dantzer, V., Bjorkman, N., Hasselager, E. (1981) An electron microscopic study of histotrophe in the interareolar part of the porcine placenta. Placenta 21: 19–28

Dantzer, V., Leiser, R., Kaufmann, P., Luckhardt, M. (1988) Comparative morphological aspects of placental vascularisation. Trophoblast Research 3: 235–260

Das, S.K., Das, N., Wang, J.H. Lim, H., Schryver, B., Plowman, G.D., Dey, S.K. (1997) Expression of betacellulin and epiregulin genes in the mouse uterus temporally by the blastocyst solely at the site of its apposition is coincident with the "window" of implantation. Developmental Biology 190: 178–190

Davies, J., Davenport, G.R. (1979) Symplasma formation and decidualisation in the pseudopregnant rabbit after intraluminal instillation of tricaprylin. Journal of Reproduction and Fertility 55: 141–145

Davies, J., Glasser, S.R. (1968) Histological and fine structural observations on the placenta of the rat. I. Organisation of the normal placenta. II. Changes after surgical removal of the fetus. Acta Anatomica 69: 542–608

Davies, J., Hoffman, L.H. (1975) Studies on the progestational endometrium of the rabbit. II. Electron microscopy day 0 to day 13 of gonadotrophin induced pseudopregnancy. American Journal of Anatomy 142: 335–366

Davies, J., Dempsey, E.W., Amoroso, E.C. (1961a) The subplacenta of the guinea pig: an electron-microscope study. Journal of Anatomy 95: 311–324

Davies, J., Dempsey, E.W., Amoroso, E.C. (1961b) The subplacenta of the guinea pig: development, histology and histochemistry. Journal of Anatomy 95: 457–473

Davies, C.J., Hill, J.R., Edwards, J.L., Shrick, F.N., Fisher, P.J., Eldridge, J.A., Schlafer, D.H. (2004) MHC gene expression on the bovine placenta: its relationship to abnormal pregnancies and retained placenta. Animal Reproduction Science 82/83: 267–280

Davies, C.J., Eldridge, J.A., Fisher, P.J., Schlafer, D.H. (2006) Evidence for expression of both classical and non classical MHC I genes in bovine trophoblast cells. American Journal of Reproductive Immunology 55: 188–200

Dawes, G.S. (1968) Foetal and neonatal physiology: a comparative study of the changes at birth. Year Book Medical Publishers, Chicago

Daya, S., Clark, D.A., Derlin, C., Javrell, J. (1985) Preliminary characterisation of two types of suppressor cells in the human uterus. Fertility and Sterility 44: 778–785

Dean, W., Santos, F., Reik, W. (2003) Epigenetic reprogramming in early mammalian development and following SCNT. Seminars in Cell and Developmental Biology 1: 193–100

Deane, H.W., Rubin, B.L., Driks, E.C, Lobel, B.L., Leipsner, G. (1962) Trophoblast giant cells in placentas of rats and mice and their probable role in steroid hormone production. Endocrinology 70: 407–419

Del Pino, E.M., Galarza, M.L., DeAlbuja, C.M., Humphries, A.A. (1975) The maternal pouch and development in the marsupial frog *Gastrotheca riobambae*. Biological Bulletin 149: 480–491

Denker, H.W. (1978) Role of trophoblastic factors in implantation, in Novel Aspects of Reproductive Physiology (eds. C.H. Spilman, J.W. Wilks), Spectrum, New York. 181–182

Denker, H.W. (1993) Implantation: a cell biological paradox. Journal of Experimental Zoology 266: 541–558

Denker, H.W., Tyndale-Biscoe, C.H. (1986) Embryonic implantation and proteinase activity in a marsupial (*Macropus eugenii*) histochemical patterns. Cell and Tissue Research 246: 279–291

Denker, H.W., Eng, L.W., Hamner, C.E. (1978) Studies on the early development and implantation in the cat. II. Implantation processes. Anatomy and Embryology 54: 39–54

Desmarais, J.A., Bordignon, V., Lopes, F.L., Smith, L.C., Murphy, B.D. (2004) The escape of the mink embryo from obligate diapause. Biology of Reproduction 70: 662–670

DeSouza, M.M., Gulnar, A., Surveyor, G.A., Price, R.E., JoAnne Julian, J., Kardon, R., Zhou, H., Gendler, S., Hilkens, J., Carson, D.D. (1999) MUC1/episialin: a critical barrier in the female reproductive tract. Journal of Reproductive Immunology 45: 127–158

Dey, S.K., Lim, H., Reese, J.R., Paria, R.C., Daikoku, T., Wang, H. (2004) Molecular cues to implantation. Endocrine Reviews 25: 341–373

Dietl, J., Hönig, A., Kämmerer, U., Rieger, L. (2006) Natural killer cells and dendritic cells at the human feto-maternal interface: an effective cooperation? Placenta 26: 341–347

Dimova, T., Mihaylova, A., Spassover, P., Georgieva, R. (2007) Establishment of the porcine epitheliochorial placenta is associated with endometrial T-cell recruitment. American Journal of Reproductive Immunology 57: 250–261

Donaldson, W.L., Zhang, G.B., Oriol, J.G., Antczak, D.F. (1990) Invasive equine trophoblast expresses conventional class I MHC antigens. Development 110: 63–71

Donaldson, W.L., Oriol, J.G., Plavin, A., Antczak, D.F. (1992) Developmental regulation of class I MHC antigen expression by equine trophoblast cells. Differentiation 52: 69–78

Drieux, B., Thiery, G. (1951) La placentation chez Mammiferes domestiques. m. Placenta des Bovides. Recueil de Midecin Veterinaire 127: 5–25

Durrant, B., Benirschke, K. (1981) Embryo transfer in exotic animals. Theriogenology 15: 7783

Dziuk, P.J., Polge, C., Rowson, L.E. (1964) Intrauterine migration and mixing of embryos in swine following egg transfer. Journal of Animal Science 23: 37–42

Edson, J.L., Hudson, D.G., Hull, D. (1975) Evidence for increased fatty acid transfer across the placenta during a maternal fast in rabbits. Biology of the Neonate 27: 50–55

Ehrhardt, R.A., Bell, A.W. (1997) Developmental increases in glucose transporters in the sheep placenta. American Journal of Physiology 273: R1132–R1141

Elbrink, J., Bihler, I. (1975) Membrane transport: its relation to cellular metabolic rates. Science 188: 1177–1184

Elcock, J.M., Searle, R.F. (1985) Antigen presenting capacity of mouse decidual tissue and placenta. American Journal of Reproductive Immunology 7: 99–103

Ellington, S.K.L. (1985) A morphological study of the development of the allantois of rat embryos in vivo. Journal of Anatomy 142: 1–11

Ellington, S.K.L. (1987) A morphological study of the development of the chorion of rat embryos. Journal of Anatomy 150: 247–263

Enders, A.C. (1965) A comparative study of the fine structure of the trophoblast in hemochorial placentas. American Journal of Anatomy 116: 2968

Enders, A.C. (1982) Whither studies of placental morphology. Journal of Reproduction and Fertility 31: 9–15

Enders, A.C. (1989) Trophoblast differentiation during the transition from trophoblast plate to lacunar stage of implantation in the rhesus monkey and human. American Journal of Anatomy 186: 85–98

Enders, A.C. (1991) Structural responses of the primate endometrium to implantation. Placenta 12: 309–325

Enders, A.C. (2007) Implantation in the macaque: expansion of the implantation site during the first week of implantation. Placenta 28: 794–802

Enders, A.C. (2008) Chapter 31b Implantation of the blastocyst ii. Implantation in Primates, in "The Endometrium", 2nd edn (eds. J.D. Aplin, A. Fazleabas), Taylor and Francis, London and New York

Enders, A.C., Carter, A.M. (2006) Comparative placentation: some interesting modifications for histotrophic nutrition – a review. Placenta 27 Supplement 1: S11–S16

Enders, A.C., Enders, R.K. (1969) The placenta of the four eyed opossum (*Philander opossum*). Anatomical Record 165: 431–450

Enders, A.C., Given, R.L. (1977) The endometrium of delayed and early implantation, in Biology of the Uterus (ed. R.M. Wynn), Plenum, New York, pp. 203–238

Enders, A.C., King, B.F. (1991) Early stages of trophoblast invasion of the maternal vascular system during implantation in the macaque and baboon. American Journal of Anatomy 192: 329–346

Enders, A.C., Liu, L.K.M. (1991a) Lodgement of the equine blastocyst in the uterus from fixation through endometrial cup formation. Journal of Reproduction and Fertility Supplement 44: 427–438

Enders, A.C., Liu, L.K.M. (1991b) Trophoblast-uterine interactions during equine chorionic girdle cell maturation migration and transformation. American Journal of Anatomy 192: 366–381

Enders, A. C., King, B.F. (1993) Development of the yolk sac. In The human yolk sac and yolk sac tumours. Nogales , F.F., Editor. Springer, London.

Enders, A.C., Lopata, A. (1999) Implantation in the marmoset monkey: expansion of the early implantation site. Anatomical Record 256: 279–99

Enders, A.C., Schlafke, S. (1969) Cytological aspects of trophoblast-uterine interaction in early implantation. American Journal of Anatomy 125: 1–30

Enders, A.C., Schlafke, S. (1971a) Implantation in the ferret: epithelial penetration. American Journal of Anatomy: 291–316

Enders, A.C., Schlafke, S. (1971b) Penetration of the uterine epithelium during implantation in the rabbit. American Journal of Anatomy 132: 219–240

Enders, A.C., Welsh, A.O. (1993) Structural interactions of trophoblast and uterus during hemochorial placenta formation. Journal of Experimental Zoology 266: 578–587

Enders, A.C., Wimsatt, W.A. (1968) Formation and structure of the hemodichorial chorioallantoic placenta of the bat (*Myotis lucifugus* Inc.). American Journal of Anatomy 122: 453–463

Enders, A.C., Wimsatt, W.A. (1971) Transport and barrier function in the chorioallantoic placenta of the bat, *Myotis lucifugus*. Anatomical Record 170: 381–400

Enders, A.C., Given, R.L. (1977) The endometrium of delayed and early implantation. pp 203–244. In Biology of the Uterus. Wynn, R.M., Editor. Plenum Press New York.

Enders, A.C., Schlafke, S., Welsh, A.O. (1980) Trophoblastic and uterine luminal epithelial surfaces at the time of blastocyst adhesion in the rat. American Journal of Anatomy 159: 59–72

Enders, A.C., Hendrickx, A.G., Schlafke, S. (1983) Implantation in the rhesus monkey: initial penetration of the endometrium. American Journal of Anatomy 167: 275–298

Enders, A.C., Welsh, A.O., Schlafke, S. (1985) Implantation in the rhesus monkey: endometrial responses. American Journal of Anatomy 173: 147–169

Enders, A.C., Lantz, K.C., Schlafke, S. (1996) Preference of invasive cytotrophoblast for maternal vessels in early implantation in the macaque. Acta Anatomica 155: 145–162

Enders, A.C., Blankenship, T.N., Lantz, K.C., Enders, S.S. (1998) Morphological variation in the interhemal areas of chorioallantoic placentae. Trophoblast Research 12: 1–19

Enders, A.C., Blankenship, T.N., Conley, A.J., Jones, C.J.P. (2006) Structure of the midterm placenta of the spotted hyena, *Crocuta crocuta*, with emphasis on the diverse hemophagous regions. Cells Tissues and Organs 183: 141–155

Enders, A.C., Blankenship, T.N., Goodman, S.M., Soarimalala, V., Carter, A.M. (2007) Placental diversity in malagasy tenrecs: placentation in shrew tenrecs (*Microgale* spp.), the mole-like rice tenrec (*Oryzorictes hova*) and the web-footed tenrec (*Limnogale mergulus*). Placenta 28: 748–749

Esterman, A., Greco, M.A., Mitani, Y., Finlay, T.H., Ismail-Beigi, F., et al. (1997) The effect of hypoxia on human trophoblast in culture: morphology, glucose transport and metabolism. Placenta 18: 129–136

Farin, P.W., Piedrahita, J.A., Farin, C.E. (2006) Errors in development of fetuses and placentas from in vitro-produced bovine embryos. Theriogenology 65: 178–191

Farkash, Y., Timberg, R., Orly, J. (1986) Preparation of an antiserum to rat cytochrome P450 cholesterol side chain cleavage enzyme and its use for ultrastructural localisation of the immunoreactive enzyme by protein A gold technique. Endocrinology 118: 1353–1365

Fatemi, S.H. (1987) The role of secretory granules in the transport of basement membrane components: ARG studies of rat parietal yolk sac employing 3H proline as a precursor of type IV collagen. Connective Tissue Research 16: 1–14

Faulk, W.P., Hsi, B.L. (1983) Immunobiology of human trophoblast membrane antigens, in Biology of Trophoblast (eds. C. Loke, A. Whyte), Elsevier, Amsterdam, pp. 535–565

Faulk, W.P., McIntyre, J.A. (1983) Immunological studies of human trophoblast: markers, subsets and functions. Immunological Reviews 75: 139–175

Fazleabas, A.T., Kim, J.J., Strakova, Z. (2004) Implantation: embryonic signals and the modulation of the uterine environment – a review. Placenta 25 Supplement A: S26–S31

Fehilly, C.B., Willadsen, S.M., Tucker, E.M. (1984) Interspecific chimaerism between sheep and goat. Nature 307: 634–636

Ferry, B.L., Starkey, P.M., Sargent, I.L., Watt, G.M.O., Jackson, M., Redman, C.W.G. (1990) Cell populations in the human early pregnancy decidua: NK activity and response to interleukin 2 of CD56 positive large granular lymphocytes. Immunology 70: 446–452

Fincher, K.B., Bazer, F.W., Hansen, P.J., Thatcher, W.W., Roberts, K.M. (1986) Proteins secreted by the sheep conceptus suppress induction of uterine PGF_{2a} release by oestradiol and oxytocin. Journal of Reproduction and Fertility 76: 425–433

Finn, C.A. (1986) Implantation, menstruation and inflammation. Biological Reviews 61: 313–328

Finn, C.A., Bredl, I.C.S. (1973) Studies on the development of the implantation reaction in the mouse uterus: influence of actinomycin D. Journal of Reproduction and Fertility 34: 247–253

First, N.L., Bosc, M.J. (1979) Proposed mechanisms controlling parturition and the induction of parturition in swine. Journal of Animal Science 48: 1407–1421

Firth, J.A., Sibley, C.P. Ward, B.S. (1986a) Histochemical localisation of phosphatases in the pig placenta. I. Nonspecific phosphatases and their relation to uteroferrin. Placenta 7: 1725

Firth, J.A., Sibley, C.P., Ward, B.S. (1986b) Histochemical localisation of phosphatases in the pig placenta. II. Potassium dependent and potassium independent paranitrophenylphosphatases at high pH: relationship to sodium dependent ATP'ase. Placenta 7: 26–36

Fischer, T.V. (1971) Placentation in the American beaver (*Castor canadensis*). American Journal of Anatomy 131: 159–184

Flaminio, M.J.B.F., Antczak, D.F. (2005) Inhibition of lymphocyte proliferation and activation: a mechanism used by equine invasive trophoblast to escape from the maternal immune response Placenta 26: 148–159

Flint, A.P.F., Henville, A., Christie, W.B. (1979) Presence of placental lactogen in bovine conceptuses before attachment. Journal of Reproduction and Fertility 56: 305–308

Flint, A.P.F., Burton, R.D., Heap, R.B. (1983) Sources of progesterone during gestation in Barbary sheep (*Ammotragus lervia*). Journal of Endocrinology 98: 283–288

Folkart, G.R., Dancis, J., Money, W.L. (1960) Transfer of carbohydrates across guinea-pig placenta. American Journal of Obstetrics and Gynecology 80: 221–223

Ford, S.P., Weber, L.J., Stormshak, F. (1977) Role of estradiol 17beta and progesterone in regulating constriction of ovine uterine arteries. Biology of Reproduction 17: 480–483

Forsyth, I.A. (1991) The biology of the placental PRL/GH family. Oxford Reviews of Reproductive Biology 13: 97–148

Fowden, A.L. (1997) Comparative aspects of fetal carbohydrate metabolism. Equine Veterinary Journal 24: 19–25

Fowden, A.L., Silver, M. (1995) The effects of thyroid hormone on oxygen and glucose metabolism in the sheep fetus during late gestation. Journal of Physiology 482: 203–213

Fowden, A.L., Forhead, A.J, White, K.L., Taylor, P.M. (2000) Equine uteroplacental metabolism at mid and late gestation. Experimental Physiology 85: 539–540

Fowden, A.L., Giussani, D.A., Forhead, A.J. (2005) Endocrine and metabolic programming during intrauterine development. Early Human Development 81: 723–734

Fowden, A.L., Ward, J.W. Forhead, A.F. (2006a) Control of fetal metabolism: relevance to developmental origins of heath and disease, in Developmental Origins of Health and Disease (eds. P.D. Gluckman, M.A. Hanson), Cambridge University Press, Cambridge

Fowden, A.L., Ward, J.W., Wooding, F.B.P., Forhead, A.F. (2006b) Programming placental nutrient transfer capacity. Journal of Physiology 572: 5–16

Fox, H. (1964) The villous cytotrophoblast as an index of placental ischaemia. Journal of Obstetrics and Gynaecology of the British Commonwealth 71: 885–893

Fraser, E.A., Renton, R.M. (1940) Observations on the breeding and development of the viviparous fish *Heterandria formosa*. Quarterly Journal of Microscopical Science 81: 479–520

Freemark, M., Handwerger, S. (1986) The glycogenic effects of PL and GH in ovine fetal liver are mediated through binding to specific fetal OPL receptors. Endocrinology 118: 613–618

Freemark, M., Cromer, M., Korner, G., Handwerger, S. (1987) A unique placental lactogen receptor: implications for fetal growth. Endocrinology 120: 1865–1872

Frendo, J.L., Olivier, D., Cheynet, V., Blond, J.L., Bouton, O., et al. (2003) Direct involvement of HERV-W Env glycoprotein in human trophoblast cell fusion and differentiation. Molecular Cell Biology 23: 3566–3574

Freyer, C., Zeller, U., Renfree, M.B. (2003) The marsupial placenta: a phylogenetic analysis. Journal of Experimental Zoology 299A: 59–77

Freyer, C., Zeller, U., Renfree, M.B. (2007) Placental function in two distantly related marsupials. Placenta 28: 249–257

Friess, A.E., Sinowatz, F., Skolek-Winnischen, K., Trautner, W. (1981) The placenta of the pig. II. The ultrastructure of the areolae. Anatomy and Embryology 163: 43–53

Fuchs, A.R., Helmer, H., Chang, S.M., Fields, M.J. (1992) Concentration of OT receptors in the placenta and fetal membranes of cows during pregnancy and labour. Journal of Reproduction and Fertility 96: 775–783

Galton, M. (1962) DNA content of placental nuclei. Journal of Cell Biology 13: 183–191

Gambel, P., Rossant, J., Hunziker, K.D., Wegmann, T. (1985) Origin of decidual cells in murine pregnancy and pseudopregnancy. Transplantation 39: 443–445

Gardner, K. (1975) Origins and properties of trophoblast, in Immunology of Trophoblast (eds. RG. Edwards, C.W.S. Howe, M.H. Johnson), Cambridge University Press, Cambridge, pp. 43–61

Garfield, K.F., Robideau, S., Challis, J.K.G., Daniel, E.E. (1979) Hormonal control of gap junction formation in sheep myometrium during parturition. Biology of Reproduction 21: 999–1007

Garfield, R.E., Blennerhassett, M.G., Miller, S.M. (1988) Control of myometrial contractility: role and regulation of gap junctions. Oxford Review of Reproductive Biology 10: 436–490

Geier, G., Schuhmann, R., Kraus, H. (1975) Regionale unterschiedliche Zellproliferation innerhalb der Plazentome reifer menschlicher Plazenten. Archiv GylUlkologie 218: 31–37

Geisert, R.D., Brookbank, I.W., Roberts, R.M., Bazer, F.W. (1982b) Establishment of pregnancy in the pig. II. Cellular remodelling of the porcine blastocyst during elongation on day 12 of pregnancy. Biology of Reproduction 27: 941–955

Geisert, R.D., Lee, C.L., Simmen, F.A., Zavi, M.T., Fliss, A.B., Bazer, F.W., Simmen, R.C.M. (1991) Expression of mRNAs encoding IGFI and II and IGFBP2 in bovine endometrium during the estrus cycle and early pregnancy. Biology of Reproduction 45: 975–983

Georgiades, P., Ferguson-Smith, A.C., Burton, G.J. (2002) Comparative developmental anatomy of the murine and human definitive placentae. Placenta 23: 3–19

Ghosh, D., Bell, S.C., Sengupta, J. (2004) Immunohistological localization of insulin-like growth factor binding protein-1 in primary implantation sites and trauma-induced deciduomal tissues of the rhesus monkey. Placenta 25: 197–207

Gibb, W., Challis, J.R. (2002) Mechanisms of term and preterm birth. Journal of Obstetrics and Gynaecology Canada 24(11): 874–883

Gibori, G., Ralison, B., Basuray, R, Rowe, M.C, Hunzicker, B., Dunn, M. (1984) Endocrine role of the decidual tissue: decidual and luteotrophic 0 regulation of luteal and cyclase activity, luteal hormone receptors and steroidogenesis. Endocrinology 115: 1153–1157

Gill, T.A., Wegmann, T.G. (1987) Immunoregulation and Fetal Survival. Oxford University Press, New York

Ginther, O.I. (1979) Reproductive Biology of the Mare (privately published book). 4343 Garfoot Rd Cross Plains, Wisconsin 53528, USA

Ginther, O.I. (1983) Mobility of early equine conceptus. Theriogenology 19: 603–610

Ginther, O.I. (1984a) Mobility of twin embryonic vesicles in mares. Theriogenology 22: 83–95

Ginther, O.I. (1984b) Intrauterine movement of the' early conceptus in barren and postpartum mares. Theriogenology 21: 633–644

Ginther, O.I. (1989) The nature of embryo reduction in mares with twin conceptuses: deprivation hypothesis. American Journal of Veterinary Research 50: 45–52

Gipson, I.K., Blalock, T., Tisdale, A., Spurr-Michaud, S., Allcorn, S. Stavreus-Evers, A., Gemzell, K. (2008) MUC16 is lost from the uterodome surface of the receptive human endometrium: in vitro evidence that MUC 16 is a barrier to trophoblast adherence. Biology of Reproduction 78: 134–142

Gitlin D, Perricelli, A. (1970) Synthesis of serum albumin, prealbumin, alphafetoprotein, alpha1-antitrypsin and transferrin by the human yolk sac. Nature 228: 995–997

Given, R.L., Enders, A.C. (1989) The endometrium of delayed and early implantation, in Biology of the Uterus, 2nd edn, Chap. 8 (eds. R.M. Wynn, W.P. Iollie), Plenum Medical, New York, pp. 175–231

Glasser, S.R. (1990) Biochemical and structural changes in uterine endometrial cell types following natural or artificial deciduogenic stimuli. Trophoblast Research 4: 377–416

Glasser, S.R., McCormack, S.A. (1980) Functional development of rat trophoblast and decidual cells during establishment of the hemochorial placenta. Advances in Bioscience 25: 165–198

Glazier, J.D., Atkinson, D.E., Thornburg, K.L., Sharpe, P.T., Edwards, D., Boyd, R.D., Sibley, C.P. (1992) Gestational changes in Ca2 + transport across rat placenta and mRNA for calbindin9K and Ca(2+)-ATPase. American Journal of Physiology 263: 930–935

Gluckman, P.D., Hanson, M. (2005) The Fetal Matrix. Cambridge University Press, Cambridge

Gluckman, P.D., Hanson, M. (eds.) (2006) Developmental Origins of Health and Disease. Cambridge University Press, Cambridge

Gluckman, P., Breier, B.H., Davis, S.R. (1987) Physiology of the somatotrophic axis with particular reference to the ruminant. Journal of Dairy Science 70: 442–466

Goff, A.K. (2002) Embryonic signals and survival. Reproduction in Domestic Animals 37: 133–139

Goffin, F., Munaut, C., Malassiné, A., Evain-Brion, D., Frankenne, F., Fridman, V., Dubois, M., Uzan, S., Merviel, P., Foidart, J.M. (2003) Evidence of a limited contribution of feto–maternal interactions to trophoblast differentiation along the invasive pathway. Tissue Antigens 62: 104–116

Gogolin-Ewens, K.J., Lee, C.S., Mercer, W.R., Brandon, M.R. (1989) Site directed differences in the immune response to the fetus. Immunology 66: 312–317

Goldstein, S.R. (1926) A note on the vascular relations and areolae in the placenta of the pig. Anatomical Record 34: 25–33

Grandis, A.S., Handwerger, S. (1983) Differential effects of ornithine on placental lactogen and GH secretion. Journal of Endocrinology 97: 175–178

Gray, A.P. (1971) Mammalian Hybrids. Commonwealth Agricultural Bureau. Farnham Royal, Slough, UK, pp. 262

Greenstein, I.S., Murray, R.W., Foley, R.C. (1958) Observations on the morphogenesis and histochemistry of the bovine preattachment placenta between 16 and 33 days of gestation. Anatomical Record 132: 321–325

Greiss, F.E. (1966) Pressure-flow relationship in the gravid uterine vascular bed. American Journal of Obstetrics and Gynecology 96: 41–47

Greiss, F.C., Anderson, S.G. (1970) Effect of ovarian hormones on the uterine vascular bed. American Journal of Obstetrics and Gynecology 107: 829–836

Gross, T.S., Williams, W.F., Russek-Cohen, E. (1991) Cellular changes in the peripartum bovine fetal placenta related to placental separation. Placenta 12: 27–35

Grove, B.D., Wourms, J.P. (1991) The follicular placenta of the viviparous fish Heterandria formosa. 1. Ultrastructure and development of the embryonic absorptive surface. Journal of Morphology 209: 265–284

Gruenwald, P. (1973) Lobular structure of hemochorial primate placentas and its relationship to maternal vessels. American Journal of Anatomy 136: 133–152

Grunert, E. (1986) Etiology and pathology of retained bovine placenta, in Current Therapy in Theriogenology, 2 edn (ed. D.A. Morrow), W.B. Saunders, London, pp. 237–242

Guillette, L.J. (1991) The evolution of viviparity in aminote vertebrates – new insights, new questions. Journal of Zoology 223: 521–526

Guillette, L.J., Lavia, L.A., Walker, N.J., Roberts, D.K. (1984) Luteolysis induced by prostaglandin F2a in the lizard, *Anolis carolinensis*. General and Comparative Endocrinology 56: 271–277

Guillette, L.J., Herman, C.A., Dickey, D.A. (1988) Synthesis of prostaglandin by tissues of the viviparous lizard, *Sceloporus aeneus*. Journal of Herpetology 22: 180–185

Guillomot, M., Guay, P. (1982) Ultrastructural features of the cell surfaces of uterine and trophoblastic epithelium during embryonic attachment in the cow. Anatomical Record 204: 315–322

Guillomot, M., Flechon, E., Wintenberger-Torres, S. (1982) Cytochemical studies of the uterus and trophoblastic surface during blastocyst attachment in the ewe. Journal of Reproduction and Fertility 65: 1–8

Habashi, S., Burton, G.J., Steven, D.H. (1983) Morphological study of the fetal vasculature of the human term placenta: scanning electron microscopy of corrosion casts. Placenta 4: 41–56

Hafez, E.S.E., Tsutsumi, Y. (1966) Changes in endometrial vascularity during implantation and pregnancy in the rabbit. American Journal of Anatomy 118: 249–282

Hagan, H. (1951) Embryology of the Viviparous Insects. Ronald Press, New York

Hahn, T., Hartmann, M., Blaschitz, A., Skofitsch, G., Graf, R., Dohr, G., Desoye, G. (1995) Localisation of the high affinity facilitative glucose transporter protein GLUT 1 in the placenta of human, marmoset monkey (*Callithrix jacchus*) and rat at different developmental stages. Cell and Tissue Research 280: 49–57

Haig, D. (1993) Genetic conflicts in human pregnancy. Quarterly Reviews in Biology 68: 495–532

Haines, A.N., Flajnik, M.F., Wourms, J.P. (2006) Histology and immunology of the placenta in the atlantic sharpnose shark, *Rhizoprionodon terraenovae*. Placenta 27: 1114–1121

Halestrap, A.P., Price, N.T. (1999) The proton linked monocarboxylate transporter family: structure, function and regulation. Biochemical Journal 343: 281–299

Hall J., Talamantes, F. (1984) Immunocytochemical Localisation of Mouse Placental Lactogen. Journal of Histochemistry and Cytochemistry 32: 379–382

Haluska, G.J., Currie, W.B. (1988) Variation in plasma concentrations of oestradiol 1713 and their relationship to those of progesterone, PGF2a and oxytocin across pregnancy and at parturition in pony mares. Journal of Reproduction and Fertility 84: 635–644

Hama, K, Aok, J., Inoue, A., Endo, T., Amano, T., Motoki, R., Kanai, M., Ye, X., Chun, J., Matsuki, N., Suzuki, H., Shibasaki, M., Arai, H. (2007) Embryo spacing and implantation timing are differentially regulated by LPA3-mediated lysophosphatidic acid signaling in mice. Biology of Reproduction 77: 954–959

Hamilton, W.J., Harrison, R.J., Young, B.A. (1960) Aspects of placentation in certain Cervidae. Journal of Anatomy 94: 1–33

Hamilton, D.W., Allen, W.R., Moor, R.M. (1973) Origin of endometrial cups. III. Light and electron microscopic study of fully developed equine endometrial cups. Anatomical Record 177: 503–518

Hamlett, W.C. (1989) Evolution and morphology of the placenta in sharks. Journal of Experimental Zoology Supplement 2 35–52

Hamlett, W.C., Wourms, J.P. (1984) Ultrastructure of the preimplantation shark yolk sac placenta. Tissue and Cell Research 16: 613–625

Hamlett, W.C., Wourms, J.P., Hudson, J.S. (1985) Ultrastructure of the full term shark yolk sac placenta. 1. Morphology and cell transport at the fetal attachment site. Journal of Ultrastructure Research 91: 192–206

Hamlett, W.C., Eulitt, A.M., Jarrell, R.L., Kelly, M.A. (1993) Utero-gestation and placentation in Elasmobranchs. Journal of Experimental Zoology 266: 347–367

Hamon, M., Heap, R.B., Morgan, G., Wango, E., Wooding, F.B.P. (1985) Steroid production by isolated binucleate cells from ruminant placenta. Journal of Physiology 371: 218P

Han, Y.M., Kang, Y.K., Koo, D.B., Lee, K.K. (2003) Nuclear reprogramming of cloned embryos produced in vitro. Theriogenology 59: 33–44

Hanna, J., Goldman-Wohl, D., Hamani, Y., Avraham, I., Greenfield, C., et al. (2006) Decidual cells regulate key developmental processes at the human fetal-maternal interface. Nature Medicine 12: 1065–1074

Hannan, N.J., Jones, R.L., White, C.A., Salamonsen, L.A. (2006) The chemokines CX3CL1, CCL14, and CCL4 promote human trophoblast migration at the feto-maternal interface. Biology of Reproduction 74: 896–904

Harder, J.D., Stonebrook, M.J., Pondy, J. (1993) Gestation and placentation in two new world opossums: *Didelphis virginiana* and *Monodelphis domestica*. Journal of Experimental Zoology 266: 463–479

Harding, J.E., Jones, C.T., Robinson, J.S. (1985) Studies on experimental growth retardation in sheep. The effects of a small placenta in restricting transport to and growth of the fetus. Journal of Developmental Physiology 7: 427–442

Harris, J.W.S., Ramsey, E.M. (1966) The morphology of human uteroplacental vasculature. Contributions to Embryology 38: 43–58

Harrison, R.J., Hyett, A.R. (1954) The development and growth of the placentomes in the fallow deer (*Dama dama*). Journal of Anatomy 88: 338–355

Hasselager, E. (1985) Surface exchange area of the porcine placenta: morphometry of anisotropic interdigitating microvilli. Journal of Microscopy 141: 91–100

Hauguel, S., Leturque, A., Kande, J., Girard, J. (1988) Glucose utilization by the placenta and fetal tissues in fed and fasted pregnant rabbits. Pediatric Research 23: 480–483

Hay, W.W. (1995) Regulation of placental metabolism by glucose supply. Reproduction Fertility and Development 7: 365–385

Head, J.R., Seelig, L.L. (1984) Lymphatic vessels in the uterine endometrium of virgin rats. Journal of Reproductive Immunology 6: 157–166

Heap, R.B., Flint, A.P.F., Staples, L.D. (1983) Endocrinology of trophoblast in farm animals, in Biology of Trophoblast (eds. C. Loke, A. Whyte), Elsevier, Amsterdam, pp. 353–400

Heap, R.B., Fleet, L.R., Hamon, M. (1985) PGF$_{2a}$ is transferred from the uterus to the ovary in the sheep by lymphatic and blood vascular pathways. Journal of Reproduction and Fertility 74: 645–656

Heasman, L., Clarke, L., Stephenson, T.J., Symons, M.E. (1999) The influence of maternal nutrient restriction in early to mid pregnancy on placental and fetal development in sheep. Proceedings of the Nutrition Society 58: 283–288

Hees, H., Moll, W., Wrobel, K.H., Hees, Z. (1987) Pregnancy induced structural changes and trophoblast invasion in the segmental mesometrial arteries of the guinea pig. Placenta 8: 609–626 (in rat and mouse)

Hemberger, M., Nozaki, T., Masutani, M., Cross, J.C. (2003) Differential expression of angiogenic and vasodilatory factors by invasive trophoblast cells depending on depth of invasion. Developmental Dynamics 227: 185–191

Hemmings, W.A. (1957) Protein selection in yolk sac splanchnopleur of the rabbit: the total uptake as estimated as loss from the uterus. Proceedings of the Royal Society B 148: 76–83

Hempstock, J., Bao, Y-P., Bar-Issac, M., Segaren, N., Watson, A.L., et al. (2003) Intralobular differences in antioxidant enzyme expression and activity reflect oxygen gradients within the human placenta. Placenta 24: 517–523

Hempstock, J., Cindrova-Davies, T., Jauniaux, E., Burton, G.J. (2004) Endometrial glands as a source of nutrients, growth factors and cytokines during the first trimester of human pregnancy; a morphological and immunohistochemical study. Reproductive Biology and Endocrinology 2: 58

Herington, J.L., Bany, B.M. (2007) Effect of the conceptus on uNK cell numbers and function in the mouse uterus during decidualisation. Biology of Reproduction 76: 579–588

Hess, A.P., Hamilton, S., Talbi, C., Giudice, L.C. (2007) Decidual stromal cell response to paracrine signals from the trophoblast: amplification of immune and angiogenic modulators. Biology of Reproduction 76: 102–117

Heyman, Y., Chavatte-Palmer, P., LeBourhis, D., Camous, S., Vignon, X., J.P., Renard, J.P. (2002) Frequency and occurrence of late-gestation losses from cattle cloned embryos. Biology of Reproduction 66: 6–13

Hill, J.P. (1932) The developmental history of the primates. Philosophical Transactions of the Royal Society 221: 45–178

Hill, P.M.M., Young, M. (1973) Net placental transfer of free amino acids against varying concentration gradients. Journal of Physiology 235: 409–422

Hill, D.J., Clemmons, D.R., Riley, S.C, Bassett, N., Challis, J.RG. (1993) Immunohisto-chemical localisation of IGFs and IGFl/2/3 binding proteins in the human placenta and fetal membranes. Placenta 14: 1–12

Hoenderop, J.G., Nilius, B., Bindels, R.J.M. (2005) Calcium absorption across epithelia. Physiological Reviews 85: 373–422

Hofmann, G.E., Scott, R.T., Bergh, P.A., Deligdisch, L. (1991) Immunohistochemical localisation of EGF in human endometrium decidua and placenta. Journal of Clinical Endocrinology and Medicine 73: 882–887

Hoffman, L.H. (1970) Placentation in the garter snake, *Thamnophis sirtalis*. Journal of Morphology 131: 57–88

Hoffmann, L.H., Wooding, F.B.P. (1993) Giant and binucleate trophoblastic cells of mammals. Journal of Experimental Zoology 266: 559–577

Hoffman, L.H., Winfrey, V.P., Anderson, T.Z., Olson, G.E. (1990a) Uterine receptivity to implantation in the rabbit: evidence for a 42 kDa glycoprotein as a marker of receptivity. Trophoblast Research 4: 243–258

Hogarth, P.J. (1976) Viviparity, Studies in Biology No. 75. Arnold, London

Hradecky, P, Mossman, H.W., Stott, G.G. (1988) Comparative histology of antelope placentomes. Theriogenology 29: 693–714

Huebner, E, Lococo, D.J. (1994) Oogenesis in a placental viviparous onychophoran. Tissue Cell 26: 867–889

Hughes, A.L., Green, J.A., Garbayo, J.M., Roberts, R.M. (2000) Adaptive Diversification within a large family of recently duplicated placentally expressed genes. Proceedings of the National Academy of Sciences Washington USA 97: 3319–3323

Hughes, R.L. (1993) Monotreme development with particular reference to extraembryonic membranes. Journal of Experimental Zoology 266: 480–488

Humblot, P., De Montigny, G., Jeanguyot, N., Tetodoie, F., Payen, B., Thibier, M., Sasser, R.G. (1990) PSPB and progesterone concentrations in French alpine goats throughout gestation. Journal of Reproduction and Fertility 89: 205–212

Humpherys, D., Eggan, K., Akutsu, H., Friedman, A., Hochedlinger, K., Yanagimachi, R., Lander, E.S., Golub, T.R., Jaenisch, R. (2002) Abnormal gene expression in cloned mice derived from embryonic stem cell and cumulus cell nuclei. Proceedings of the National Academy of Sciences (Wash) 99: 12889–12894

Hunt, J.S., Manning, L.S., Wood, G.W. (1984) Macrophages in murine uterus are immunosuppressive. Cellular Immunology 85: 499–510

Hunt, J.S, Morales, P.J, Pace, J.L, Fazleabas, A.T., Langat, D.K. (2007) A commentary on gestational programming and functions of HLA-G in pregnancy. Placenta 28 Supplement 857–863

Huppertz, B., Kertschanska, S., Frank, H.G., Gaus, G., Funayama, H., et al. (1996) Extracellular matrix components of the placental extravillous trophoblast: immunocytochemistry and ultrastructural distribution. Histochemistry and Cell Biology 106: 291–301

Huppertz, B., Kertschanska, S., Demir, A.Y., Frank, H.G., Kaufmann, P. (1998) Immunohistochemistry of matrix metalloproteinases, their substrates, and their inhibitors during trophoblast invasion in the human placenta. Cell and Tissue Research 291: 133–148

Huppertz, B., Kadyrov, M., Kingdom, J.C.P. (2006) Apoptosis and its role in the trophoblast. American Journal of Obstetrics and Gynecology 195: 29–39

Ilgren, E.B. (1983) Review article: control of trophoblast growth. Placenta 4: 307–328

Illsley, N.P. (2000) Glucose transporters in the human placenta. Placenta 21: 14–22

Ingermann, K.L. (1987) Control of placental glucose transfer. Placenta 8: 557–571

Jackson, M.R., Mayhew, T.M., Boyd, P.A. (1992) Quantitative description of the elaboration and maturation of villi from 10 weeks of gestation to term. Placenta 13: 357–370

Jansson, T. (2001) Amino acid transporters in the human placenta. Pediatric Research 49: 141–147

Jauniaux, E., Gulbis, B. (2000) Fluid compartments of the embryonic environment. Human Reproduction Update 6: 268–278

Jauniaux, E., Hempstock, J., Greenwold, N., Burton, G.J. (2003) Trophoblastic oxidative stress in relation to temporal and regional differences in maternal placental blood flow in normal and abnormal early pregnancies. American Journal of Pathology 162: 115–125

Jauniaux, E., Cindrova-Davies, T., Johns, J., Dunster, C., Hempstock, J., et al. (2004) Distribution and transfer pathways of antioxidant molecules inside the first trimester human gestational sac. Journal of Clinical Endocrinology and Metabolism 89: 1452–1459

Jerez, A., Ramirez-Pinilla, M.P. (2001) The allantoplacenta of *Mabouya mabouya*. Journal of Morphology 249: 132–146

Jerez, A., Ramirez-Pinilla, M.P. (2003) Morphogenesis of extraembryonic membranes and placentation in *Mabouya mabouya* (Squamata, Scincidae). Journal of Morphology 258: 158–178

Ji, Q., Luo, Z.X., Yuan, C.X., Wible, J.R., Zhang, J.P., Georgi, J.A. (2002) The earliest known eutherian mammal. Nature 416: 816–822

John, R.M., Surani, M.A. (2000) Genomic imprinting, mammalian evolution and the mystery of egg laying mammals. Cell 101: 585–588

Johns, K., Renegar, K.H. (1990) Ultrastructural morphology and relaxin immunolocalisation in giant trophoblast cells of the golden hamster placenta. American Journal of Anatomy 189: 167–178

Johnson, S., Wooding, F.B.P. (1988) Synthesis and storage of chorionic gonadotrophin and placental lactogen in human syncytiotrophoblast. Journal of Physiology 403: 29P

Johnson, G.A., Burghardt, R.C., Joyce, M.M., Spencer, T.E., Bazer, F.W., Pfarrer, C., Gray, A.C. (2003) Osteopontin expression in uterine stroma indicates a decidualisation-like differentiation during ovine pregnancy Biology of Reproduction 68: 1951–1958

Jollie, W.P. (1969) Nuclear and cytoplasmic fine structure in the trophectoderm giant cells of rat placenta. Anatomical Record 65: 1–14

Jollie, W.P., Jollie, L.G. (1967) Electron microscopic observations on the yolk sac of the spiny dogfish *Squalus acanthias*. Journal of Ultrastructure Research 18: 102–126

Jollie, W.P., Craig, S.S. (1979) The fine structure of the placental junctional zone cells during prolonged pregnancy in rats. Acta Anatomica 105: 386–400

Jones, C.J.P. (1997) The life and death of the embryonic yolk sac, in Embryonic Medicine and Therapy (eds. E. Jauniaux, E.R. Barnea, R.G. Edwards), Oxford University Press, Oxford, pp. 180–196

Jones, K.E., Guillette, L.J. (1982) Hormonal control of oviposition and parturition in lizards. Herpetologica 38: 80–93

Jones, C.J.P., Fox, H. (1991) Ulltrastructure of the normal human placenta. Electron Microscopy Reviews 4: 129–178

Jones, C.J.P., Fazleabas, A.T. (2001) Ultrastructure of epithelial plaque formation and stromal cell transformation by post ovulatory chorionic gonadotropin treatment in the baboon (*Papio anubis*). Human Reproduction 16: 2680–2690

Jones, C.J.P., Hamlett, W.C. (2004) Structure and glycosylation of the term yolk sac placenta and uterine attachment site in the viviparous shark *Mustelus canis*. Placenta 25: 820–829

Jones, R.L., Stoikos, C., Findlay, J.K., Salamonsen, L.A. (2006a) TGFβ superfamily: expression and actions on the endometrium and placenta. Reproduction 132: 217–232

Jones, R.L., Kaitu'u-Lino, T.J., Nie, G., Gabriel Sanchez-Partida, L.G., Findlay, J.K., Salamonsen, L.A. (2006b) Complex expression patterns support potential roles for maternally derived activins in the establishment of pregnancy in mouse. Reproduction 132: 799–810

Jones, H.N., Powell, T.L., Jansson, T. (2007) Regulation of placental transport: a review. Placenta 28: 763–774

Jones, C.J.P., Harris, L.K., Whittingham, J., Aplin, J.D., Mayhew, T.M. (2008) A reappraisal of the morphophenotype and the basal lamina coverage of cytotrophoblasts in human term placenta. Placenta 29: 215–219

Joost, H.G., Thorens, B. (2001) The extended GLUT family of sugar/polyol transport facilitators. Molecular Membrane Biology 18: 247–256

Kalter, S.S., Helmke, R.J., Panigel, M., Heberling, K.C., Felsburg, P.J., Axelrod, L.K. (1973) Observation of apparent C type particles in Baboon (*Papio cynocephalaus*) placentas. Science (Washington) 179: 1332–1333

Kanbour, A., Ho, H.N., Misra, D.N., Macpherson, T.A., Kinz, H.W., Gill, T.O. (1987) Differential expression of MHC class I antigens on the placenta of the rat. Journal of Experimental Medicine 166: 1882–1887

Kappes, S.M., Warren, W.C., Pratt, S.L., Liang, K., Anthony, K.V. (1992) Quantification and cell localisation of OPL mRNA expression during mid and late gestation. Endocrinology 131: 2829–2838

Karimu, A.L., Burton, G.J. (1994) The effects of maternal vascular pressure on the dimensions of the placental capillaries. British Journal of Obstetrics and Gynaecology 101: 57–63

Kashiwagi, A., DiGirolamo, C.M., Kanda, Y., Niikura, Y., Esmon, C.T., Hansen, T.R., Shioda, T., Pru, J.T. (2007) The postimplantation embryo differentially regulates endometrial gene expression and decidualisation. Endocrinology 148: 4173–4184

Kaufmann, P. (1981a) Electron microscopy of the guinea pig placental membranes. Placenta 2: 310–320

Kaufmann, P. (1981b) Entwicklung der Plazenta,. in Die Plazenta des Menschen (eds. V. Becker, T.H. Schiebler, F. Kubli), Thieme, New York; pp. 13–50

Kaufmann, P. (1982) Development and differentiation of the human placental villous tree. Bibliotheca Anatomica 22: 29–39

Kaufmann, M.H. (1983) Origin, properties and fate of trophoblast in the mouse, in Biology of Trophoblast (eds. C. Loke, A. Whyte), Elsevier, Amsterdam, pp. 23–70

Kaufmann, P. (1985) Basic morphology of the fetal and maternal circuits in the human placenta. Contributions to Gynecology and Obstetrics 13: 517

Kaufmann, P., Bruns, V., Leiser, K., Luckhard, M., Winterhager, E. (1985) The fetal vasculature of term human placental villi. II. Intermediate and terminal villi. Anatomy and Embryology 173: 203–214

Kaufmann, P., Mayhew, T.M., Charnock Jones, D.S. (2004) Aspects of human fetoplacental vasculogenesis and angiogenesis. II. Changes during normal pregnancy. Placenta 25: 114–126

Keams, M., Lala, P.K. (1983) Life history of decidual cells: a review. American Journal of Reproduction and Immunological Microbiology 3: 78–82

Keefer, C.L. (2007) Lessons learned from nuclear transfer (cloning). Theriogenology 69: 48–54

Kellas, L.M. (1966) The placenta and foetal membranes of the antelope Ourebia ourebi (Zimmerman). Acta Anatomica 64: 390–445

Kemp, B., Kertschanka, S., Kadyrov, M., Rath, W., Kaufmann, P., Huppertz, B. (2002) Invasive depth of the extravascular trophoblast correlates with cellular phenotype: a comparison of intra- and exra-uterine sites. Histochemistry and Cell Biology 117: 401–414

Kennedy, T.G. (1983) Embryonic signals and the initiation of blastocyst implantation. Australian Journal of Biological Science 36: 531–543

Kennedy, T.G., Lukash, L.A. (1982) Induction of decidualisation in rats by the intrauterine infusion of prostaglandins. Biology of Reproduction 27: 253–260

Kennedy, T.G., Gillio-Meina, C., Phang, S.G. (2007) Prostaglandins and the initiation of blastocyst implantation and decidualisation. Reproduction 134: 635–643

Kevorkova, O., Ethier-Chiasson, M., Lafond, J. (2007) Differential expression of glucose transporters in rabbit placenta: effect of hypercholesterolemia in dams. Biology of Reproduction 76: 487–495

Keys, J.L., King, G.J. (1988) Morphological evidence for increased uterine vascular permeability at the time of embryonic attachment in the pig. Biology of Reproduction 39: 473–487

Keys, J.L., King, G.J. (1989) Structural changes in the luminal epithelium of the porcine uterus between days 10 and 19 of the estrous cycle. American Journal of Anatomy 185: 42–57

Keys, J.L., King, G.J. (1990) Microscopic examination of procine conceptus-maternal interface between days 10 and 19 of pregnancy. American Journal of Anatomy 188: 221–228

Keys, J.L., King, C.J., Kennedy, T.G. (1986) Increased uterine vascular permeability at the time of embryonic attachment in the pig. Biology of Reproduction 34: 405–411

Keys, J.L., King, G.J., Laforest, J.P. (1989) Autofluorescence of. the porcine endometrium during early pregnancy. Biology of Reproduction 40: 220–222

Kiefer J.C. (2007) Epigenetics in development. Developmental Dynamics 236: 1144–1146

Kim, Y.M., Bujold, E., Chaiworapongsa, T., Gomez, R., Yoon, B.H., et al. (2003) Failure of physiologic transformation of the spiral arteries in patients with preterm labor and intact membranes. American Journal of Obstetrics and Gynecology 189(4): 1063–1069

Kimber, S.J. (2005) Leukemia inhibition factor (LIF) in implantation and uterine biology. Reproduction 130: 131–145

Kimber, S.J., Lindenberg, S. (1990) Hormonal control of. a carbohydrate epitope involved in implantation in mice. Journal of Reproduction and Fertility 89: 13–21

Kimberley, D.J., Thornburg, K.L. (1989) Electron microscopical tracers in the uterine epithelium of the pregnant guinea pig. Placenta 10: 531–543

Kimura, K., Uchida, T.A. (1983) Electron microscopical observations of delayed implantation in the Japanese long fingered bat Miniopterus schreibersii fuliginosus. Journal of Reproduction and Fertility 69: 187–193

Kimura, K., Uchida, T. (1984) Development of the main and accessory placentae in the Japanese long fingered bat *Miniopterus schreibersii fuliginosus*. Journal of Reproduction and Fertility 71: 119–126

Kimura, J., Sasaki, M., Endo, H., Fukuta, K. (2004) Anatomical and histological characterisation of the female reproductive organs of the mouse deer (Tragulidae) Placenta 25: 705–711

King, B.F. (1976) Localization of transferrin on the surface of the human placenta by electron microscopic immunocytochemistry. Anatomical Record 186(2): 151–159

King, B.F. (1977) An electronmicroscopic study of absorption of peroxidase conjugated immunoglobulin G by guinea pig visceral yolk sac in vitro. American Journal of Anatomy 148: 447–456

King, B.F. (1984) The fine structure of the placenta and chorionic vesicles of the bush baby *Galago crassicaudata*. American Journal of Anatomy 169: 101–117

King, B.F. (1986) Morphology of the placenta and fetal membranes, in Comparative Primate Biology, Vol. 3 (eds. W.R Dukelow, J.E. Erwin), Wiley, New York, Reproduction and Development, pp. 311–331

King, B.F. (1992) Ultrastructural evidence for transtrophoblastic channels in the hemomonochorial placenta of the Degu (*Octodon degus*). Placenta: 13: 35–41

King, G.J., Atkinson, B.A., Robertson, H.A. (1979) Development of the bovine placentome during the second month of gestation. Journal of Reproduction and Fertility 55: 173–180

King, G.J., Atkinson, B.A. (1987) The bovine intercaruncular placenta throughout gestation. Animal Reproduction Science 12: 241–254

King, B.F., Enders, A.C. (1971) Protein absorption by the guinea pig chorioallantoic placenta. American Journal of Anatomy 130: 409–430

King, B.F., Hastings, R.A. (1977) The comparative fine structure of the interhemal membrane of chorioallantoic placentas from six genera of myomorph rodents. American Journal of Anatomy 149: 165–180

King, A., Loke, Y.W. (1993) Effect of IFN and a on killing of human trophoblast by decidual LAK cells. Journal of Reproductive Immunology 23: 51–62

King, B.F., Mossman, H.W. (1974) The fetal membranes and unusual giant cell placenta of the jerboa Uaculus) and jumping mouse (Zapus). American Journal of Anatomy 140: 405–432

King, B.F., Tibbitts, F.D. (1969) The ultrastructure of the placental labyrinth in the kangaroo rat Dipodomys. Anatomical Record 163: 543–554

King, B.F., Tibbitts, F.D. (1976) The fine structure of the chinchilla placenta. American Journal of Anatomy 145: 33–56

King, B.F., Enders, A.C., Wimsatt, W.A. (1978) The annular hematoma of the shrew yolk sac placenta. American Journal of Anatomy 152: 45–58

King, G.J., Atkinson, B.A., Robertson, H.A. (1979) Development of the bovine placentome during the second month of gestation. Journal of Reproduction and Fertility 55: 173–180

King, G.J., Atkinson, B.A., Robertson, H.A. (1981) Development of the intercaruncular areas during early gestation and the establishment of the bovine placenta. Journal of Reproduction and Fertility 61: 469–474

King, B.F., Pinheiro, P.B.N., Hunter, R.L. (1982) The form of the placental labyrinth in the sloth *Bradypus tridactylis*. Anatomical Record 202: 15–22

King, A.E., Kelly, R.W., Sallenave, J-M., Bocking, A.D., Challis, J.R.G. (2007) Innate immune defences in the human uterus during pregnancy. Placenta 28: 1099–1106

Kirby, D.R.S. (1963) The development of mouse blastocysts transplanted to the scrotal and cryptorchid testis. Journal of Anatomy 97: 119–130

Kirby, D.R.S. (ed.) (1965) The 'invasiveness' of the trophoblast, in The early Conceptus, normal and abnormal (ed. W. Wallace Park), Livingstone, London, pp. 68–73

Kisalus, L.L., Hew, J.C., Little, C.D. (1987) Immunolocalisation of extracellular matrix proteins and collagen synthesis in first trimester human decidua. Anatomical Record 218: 402–415

Kiso, Y., Yasufuku, K., Matsuda, H., Yamavchi, S. (1990) Existence of an endothelio-endothelial placenta in the insectivore *Suncus murinus*. Cell and Tissue Research 262: 195–197

Kitchen, H., Bunn, H.F. (1975) Ontogeny of equine haemoglobins. Journal of Reproduction and Fertility Supplement 23: 595–598

Kliman, H.J., Nestler, J.E., Sermasi, E., Sanger, J.M., Strauss, J.F. (1986) Purification, characterisation and in vitro differentiation of cytotrophoblast from human term placenta. Endocrinology 118: 1567–1582

Klisch, K., Mess, A. (2007) Evolutionary differentiation of Cetoardactyl placentae in the light of the viviparity- driven conflict hypothesis. Placenta 28: 353–360

Klisch, K., Schuler, G., Miglino, M.R., Leiser, R. (2000) Genome multiplication in trophoblast giant cells of sheep, goat, water buffalo and deer: an image cytophotometric study. Reproduction in Domestic Animals 35: 145–148

Klisch, K., Thomsen, P.D., Dantzer, V., Leiser, R. (2004) Genome multiplication is a generalized phenomenon in placentomal and interplacentomal trophoblast giant cells in cattle. Reproduction Fertility and Development 16: 301–306

Klisch, K., Bevilacqua, E., Olivera, L.V. (2005) Mitotic polyploidization in trophoblast giant cells of the alpaca. Cells Tissues Organs. 181: 103–108

Klonisch, T., Wolf, P., Hombach-Klonisch, S., Vogt, S., Kuechenhoff, A., Tetens, F., Fischer, B. (2001) Epidermal growth factor-like ligands and erbB genes in the peri-implantation rabbit uterus and blastocyst. Biology of Reproduction 64: 1835–1844

Knight, P.M., Lombardi, J., Wourms, J.P., Burns, J.R. (1985) Follicular placenta and embryonic growth of the viviparous four eyed fish (Anableps). Journal of Morphology 185: 131–142

Kovacs, C.S., Chafe, L.L., Woodland, M.L., McDonald, K.R., Fudge, N.J., Wookey, P.J. (2002) Calcitropic gene expression suggests a role for the intraplacental yolk sac in maternofetal calcium exchange. American Journal of Physiology 282: E721–E732

Krcek, J., Clark, D.A. (1985) Selective localisation of a bone marrow cell subpopulation at the implantation site in murine decidua. American Journal of Reproduction and Immunological Microbiology 7: 95–98

Krebs, C., Winther, H., Dantzer, V., Leiser, R. (1997) Vascular interrelationships of near term mink placenta: LM combined with SEM of corrosion casts. Microscopy Research and Technique 38: 125–136

Kumar, S., Hedges, S.B. (1998) A molecular timescale for vertebrate evolution. Nature 392: 917–920

Kunz, Y.W. (1971) Histological study of the greatly enlarged pericardial sac in the embryo of the viviparous telost – *Lebistes reticulatus*. Revue Suisse Zoologie 78: 187–207

Kurman, R.J., Main, C.S., Chen, H.C. (1984) Intermediate trophoblast: a distinctive form of trophoblast with, specific morphological biochemical and functional features. Placenta 5: 349–370

Kwong, W.Y., Wild, A.E., Roberts, P., Willis, A.C., Fleming, T.P. (2000) Maternal undernutrition during the preimplantation period of rat development causes blastocyst abnormalities and programming of postnatal hypertension Development 127: 4195–4202

Lala, P.K. (1989) Similarities between immunoregulation in pregnancy and malignancy: the role of prostaglandin E2. American Journal of Reproductive Immunology 20: 147–152

Lala, P., Chatterjee, G., Hasrouni, L.S., Kearns, M., Montgomery, B., Colavicenzo, V. (1983) Immunobiology of the fetomaternal interface. Immunological Reviews 75: 87–116

Lambert, R.T., Ashworth, C.J., Beattie, L., Gebbie, F.E., Hutchinson, J.S., Kyle, D.J., Racey, P.J. (2001) Temporal changes in reproductive hormones and conceptus-endometrial interactions during embryonic diapause and reactivation of the blastocyst in European roe deer (*Capreolus capreolus*) Reproduction 121: 863–871

Lanier, L.K. (2006) NK cells: roundup. Immunological Reviews 214: 5–8

Larkin, L.H., Flickinger, C.J. (1969) Ultrastructure of the metrial gland cell in the pregnant rat. American Journal of Anatomy 126: 337–354

Larsen, J.F. (1961) Electron microscopy of the implantation site in the rabbit. American Journal of Anatomy 109: 319–334

Lawn, A.M., Chiquoine, A.D., Amoroso, E.C. (1969) The development of the placenta in the sheep and goat: an electron microscopic study. Journal of Anatomy lOS: 557–578

Leco, K.J., Edwards, D.R., Schultz, G.A. (1996) TIMP-3 is the major inhibitor in the decidualising mouse uterus. Molecular Reproduction and Development 45: 458–465

Ledger, W.L., Ellwood, D.A., Taylor, M.J. (1983) Cervical softening in late pregnant sheep by infusion of PGEz into a cervical artery. Journal of Reproduction and Fertility 69: 511–523

Lee, K.Y., Demayo, F.J. (2004) Animal models of implantation Reproduction 128: 679–695

Lee, S.Y., Mossman, H.W., Mossman, A.S., DelPino, G. (1977) Evidence for a specific implantation site in ruminants. American Journal of Anatomy 150: 631–640

Lee, S.Y., Anderson, J.W., Scott, G.L., Mossman, H.W. (1983) Ultrastructure of the placenta and fetal membranes of the dog. II. The yolk sac. American Journal of Anatomy 166: 313–327

Lee, C.S., Gogolin-Ewens, K., White, T.R., Brandon, M.R. (1985) Studies on the distribution of binucleate cells in the placenta of sheep with a monoclonal antibody SBU-3. Journal of Anatomy 140: 565–576

Lee, C.S., Gogolin-Ewens, K., Brandon, M. (1986a) Comparative studies on the distribution of binucleate cells in the placentae of the deer and cow using the monoclonal antibody SBU3. Journal of Anatomy 147: 163–180

Lee, C.S., Wooding, F.B.P., Brandon, M.R. (1986b) Light and electron microscope immunocytochemical studies on the role of binucleate cells in villus growth in goat placentomes. Journal of Submicroscopical Cytology 18: 661–672

Lee, C.S., Wooding, F.B.P., Brandon, M.R. (1986c) Immunogold co-localization of ovine placental lactogen and the antigen recognized by the SBU-3 monoclonal antibody in sheep placental granules. Journal of Reproduction and Fertility 78: 653–662

Lee, C.S., Wooding, F.B.P., Brandon, M.R. (1986d) Ultrastructural immunogold investigations of the function and diversity of binucleate cells in the sheep placenta using a monoclonal antibody. Placenta 7: 495–504

Lee, C.S., Ralph, M.M., Gogolin Ewens, K.J., Brandon, M.R. (1990) Monoclonal antibody identification of cells dissociated from the sheep placentomal trophoblast. Journal of Histochemistry and Cytochemistry 38: 649–652

Lee, C.S., Meeusen, E., Gogolin Ewens, K., Brandon, M.R. (1992) Quantitative and qualitative changes in the intraepithelial lymphocyte population in the uterus of nonpregnant and pregnant sheep. American Journal of Reproductive Immunology 28: 90–96

Lee, V.H., Zhang, S.J., Chang, S.M., Fields, M.J., Fields, P.A. (1995) In vitro transformation of rabbit cytotrophoblast cells into syncytiotrophoblast: stimulation of hormone secretion by progesterone and dibutyryl cyclic 3',5'-adenosine monophosphate. Biology of Reproduction 52: 868–877

Leiser, R. (1975) Kontaktaufnahme zwischen trophoblast und uterus epithel wahrend der When implantation beim rind. Anatomy Histology Embryology 4: 63–86

Leiser, R. (1979) Blastocysten implantation bei der Hauskatze. Licht und Elektronen optische untersuchungen. Anatomy Histology Embryology 8: 79–96

Leiser, R. (1987) Mikrovaskularisation der Ziegenplazenta, dargestellt mit Rasterelektronischunterschten Gefassausgussen. Schweizwer Archiv fur Tierheilkunde: 129: 59–74

Leiser, R., Dantzer, V. (1988) Structural and functional aspects of porcine placental vasculature. Anatomy and Embryology 177: 409–419

Leiser, R., Enders, A.C. (1980a) Light and electron microscopy study of the near term paraplacenta of the domestic cat. 1. Polar zone and paraplacental functional areas. Acta Anatomica 106: 293–311

Leiser, R., Enders, A.C. (1980b) Light and electron microscopy study of the near term paraplacenta of the domestic cat. 2. Paraplacental hematoma. Acta Anatomica106: 312–326

Leiser, R., Kaufmann, P. (1994) Placental structure: in a comparative aspect. Experimental and Clinical Endocrinology 102: 122–134

Leiser, R., Kohler, T. (1983) The blood vessels of the cat girdle placenta. Observations on corrosion casts. SEM and histological studies. I. Maternal vasculature. Anatomy and Embryology 167: 85–93

Leiser, R., Kohler, T. (1984) The blood vessels of the cat girdle placenta. observations on corrosion casts, scanning electron microscopical and histological studies. II. Fetal vasculature. Anatomy and Embryology 170: 209–216

Leiser, R., Krebs, C., Klisch, K., Ebert, B., Dantzer, V.B., Schuler, G., Hoffmann, B. (1997) Fetal villosity and microvasculature of the bovine placentome in the second half of gestation. Journal of Anatomy 191: 517–527

Lemons, J.A., Adcock III, E.W., Jones J.I., M.D., Naughton, M.A., Meschia, G., Battaglia, F.C. (1976) Umbilical uptake of amino acids in the unstressed fetal lamb. Journal of Clinical Investigation 58: 1428–1434

Liggins, G.E. (1981) Endocrinology of parturition, in Fetal Endocrinology (eds. M.J. Novy, J.A. Resko), Academic Press, New York, pp. 211–237

Lindenberg, S., Hyttel, P., Sjogren, A., Greve, T. (1989) A comparative study of attachment of human bovine and mouse blastocysts to uterine epithelial monolayers. Human Reproduction 4: 446–456

Loke, Y.W., King, A. (1989) Immunology of pregnancy: quo vadis? Human Reproduction 4: 613–615

Loke, Y.W., Burland, K. and Butterworth, B. (1986) Antigen bearing trophoblast in human placental implantation site, in Reproduction and Immunology (eds. D.A. Clark, B.A. Croy), Elsevier, London, pp. 53–59

Lombardi, J., Wourms, J.P. (1985a) The trophotaenial placenta of a viviparous goodeid Fish. 1. Ultrastructure of the internal ovarian epithelium: the maternal component. Journal of Morphology 184: 277–292

Lombardi, J., Wourms, J.P. (1985b) The trophotaenial placenta of a viviparous goodeid fish. II. Ultrastructure of trophotaeniae: the embryonic component. Journal of Morphology 184: 293–309

Lombardi, J., Wourms, J.P. (1985c) The trophotaenial placenta of a viviparous goodeid fish. III. Protein uptake by trophotaeniae: the embryonic component. Journal of Experimental Zoology 236: 165–179

Lopata, A., Sibson, M.C., Enders, A.C., Bloomfield, K.L., Gregory, M.S., Trapani, G.D., Perkins, A.V., Tonissen, K.F., Clarke, F.M. (2001) Expression and localization of thioredoxin during early implantation in the marmoset monkey. Molecular Human Reproduction 7: 1159–65

Lopes, F.L., Desmarais, J., Gévry, N.Y., Ledoux, S., Murphy, B.D. (2003) Expression of vascular endothelial growth factor isoforms and receptors Flt-1 and KDR during the peri-implantation period in the mink, Mustela vison. Biology of Reproduction 68: 1926–1933

Lopes, F.L., Desmarais, J.A., Murphy, B.D. (2004) Embryonic diapause and its regulation. Reproduction 128: 669–678

Lowe, K.C., Beck, N.F.G., McNaughton, D.C., Jansen, C.A.M., Thomas, A.L., Nathanielz, P.W., Mallon, K., Steven., D.H. (1979) Ultrastructural changes in the placenta of the ewe after long term intravascular infusion of 2bromo-13-ergocryptine (CBI54) into mother or fetus. Quarterly Journal of Experimental Physiology 64: 253–262

Lowndes, K., Amano, A., Yamamoto, S.Y., Bryant-Greenwood, G.D. (2006) The human relaxin receptor (LGR7): expression in the fetal membranes and placenta. Placenta 27: 610–618

Luckett, W.P. (1976) Cladistic relationship among primate higher categories: evidence from fetal membranes and placenta. Folia Primatologia 25: 245–276

Luckett, W.P. (1977) Ontogeny of amniote fetal membranes and their application to phylogeny, in Major Patterns in Vertebrate Evolution (eds. M.K. Hecht, P.C. Goody, B.M. Hecht), Plenum Press, New York, pp. 439–516

Luckett, W.P. (1978) Origin and differentiation of the yolk sac and embryonic mesoderm in presomite human and rhesus monkey embryos. American Journal of Anatomy 152: 59–98

Luckhardt, M., Kaufmann, P., Elger, W. (1985) The structure of the tupaia placenta. L Histology and vascularisation. Anatomy and Embryology 171: 201–210

Ludwig, K.S. (1962) Zur feinstruktur der maternofetalen verbindung in placentom des schafes (Ovis aries. L.). Experientia 18: 212–213

Luo, Z.X., Ji, Q., Wible, J.R., Yuan, C.X. (2003) An early cretaceous tribosphenic mammal and metatherian evolution. Science 302: 1934–1940

Luu, K.C., Nie, G.Y., Salamonsen, L.A. (2004) Endometrial calbindins are critical for embryo implantation: evidence from in vivo use of morpholino antisense oligonucleotides. Proceedings of the National Academy of Sciences 101: 8028–8033

MacDonald, A.A. (1975) Uterine vasculature of the pregnant pig: a scanning electron microscopy study. Anatomical Record 184: 689–698

MacDonald, A.A. (1981) The vascular anatomy of the pig placenta: a scanning electron microscope study. Acta Morphologica Neerlandica Scandinavica 19: 171–172

MacDonald, A.A., Chavatte, P., Fowden, A.L. (2000) Scanning electron microscopy of the microcotyledonary placenta of the horse in the latter half of gestation. Placenta 21: 565–574

Magness, R.R., Mitchell, M.D., Rosenfeld, C.R. (1990) Uteroplacental production of eicosanoids in ovine pregnancy. Prostaglandins 39: 7588

Makowski, E.L. (1968) Maternal and fetal vascular nets in placentas of sheep and goats. American Journal of Obstetrics and Gynecology 100: 283–288

Malassine, A. (1974) Evolution ultrastructurale du labyrinthe de placenta de chatte. Anatomy and Embryology 146: 1–20

Malassine, A., Leiser, R. (1984) Morphogenesis and fine structure of the near term placenta of Talpa europea. 1. Endotheliochorial labyrinth. Placenta 5: 145–158

Malassiné, A, Frendo, J.L., Evain-Brion, D. (2003) A comparison of placental development and endocrine functions between the human and mouse model. Human Reproduction Update 9: 531–539

Martal, J., Lacroix, M.C. (1978) Production of chorionic somatomammotrophin (OCS): fetal growth of the placenta and corpus luteum in ewes treated with 2-bromo-l3-ergocryptine (BC). Endocrinology 103: 193–199

Martal, L, Djiane, J., Dubois, M.P. (1977) Immunofluorescent localisation of ovine placental lactogen. Cell and Tissue Research 184: 427–433

Martin, C.B. (1981) Models of placental blood flow. Placenta Supplement 1: 65–80

Martin, L. (1979) Early cellular changes and circular muscle contractions associated with the induction of decidualisation by intruterine oil in mice. Journal of Reproduction and Fertility 55: 135–139

Martin, R.D. (2003) Human reproduction: a comparative background for medical hypotheses. Journal of Reproductive Immunology 59: 111–135

Martinek, J.J. (1970) Fibrinoid and the fetal-maternal interface of the rat placenta. Anatomical Record 166: 587–604

Maruo, T., Matsuo, H., Murata, K., Mochizuki, M. (1992) Gestational age-dependent dual action of epidermal growth factor on human placenta early in gestation. Journal of Clinical Endocrinology Metabolism 75: 1362–1367

Maslar, L.A., Powers, J., Craddock, P., Ansbacher, K. (1986) Decidual prolactin production by organ cultures of human endometrium: effects of intermittent progesterone treatment. Biology of Reproduction 34: 741–750

Mathieu, C.L., Burnett, S.H., Mills, S.E., Overpeck, J.G., Bruns, D.E., Bruns, M.E. (1989) Gestational changes in Calbindin-D$_{9k}$ in rat uterus, yolk sac and placenta: implications for maternal–fetal calcium transport and uterine muscle function. Proceedings of the National Academy of Sciences of America 86: 3433–3437

Matthews, J.C., Beveridge, M.J., Malandro, M.S., Rothstei, J.D., Campbell-Thompson, M., Verlander, J., Kilberg, M.S., Novak, D.A. (1998) Activity and protein localisation of multiple glutamate transporters in gestation day 14 vs. day 20 rat placenta. American Journal of Physiology 274: C603–C614

Mattson, B.A., Overstrom, E.W., Albertini, D.E. (1990) Transitions in trophectodermal cellular shape and cytoskeletal organisation in the elongating pig blastocyst. Biology of Reproduction 42: 195–205

Mayhew, T.M., Joy, C.F., Haas, J.D. (1984) Structure-function correlation in the human placenta: the morphometric diffusing capacity for oxygen at full term. Journal of Anatomy 139: 691–708

Mayhew, T.M., Jackson, M.R., Haas, J.D. (1986) Microscopical morphology of the human placenta and its effects on oxygen diffusion: a morphometric model. Placenta 7: 121–131

Mazariegos, M.R, Leblond, C.P., van der Rest, M. (1987) Radioautographic tracing of 3H proline in the endodermal cells of the parietal yolk sac as an indicator of the biogenesis of basement membrane components. American Journal of Anatomy 179: 79–93

Mead, R.A. (1993) Embryonic diapause in vertebrates. Journal of Experimental Zoology 266: 629–641

Merker, H.J., Bremer, D., Barach, H.J., Gossrau, R. (1987) The basement membrane of the persisting maternal blood vessels in the placenta of *Callithrix jacchus*. Anatomy and Embryology 176: 87–97

Meseguer, M., Pellicier, A., Simon, C. (1998) MUC1 and endometrial receptivity. Molecular Human Reproduction 4: 1089–1098

Meseguer, M., Aplin, J.D., Caballero-Campo, P, O'Connor, J.E., Martín, J.C., Remohí, J., Pellicer, A., Simón, C. (2001) Human endometrial mucin MUC1 is up-regulated by progesterone and down-regulated in vitro by the human blastocyst. Biology of Reproduction 64: 590–601

Mess, A., Carter, A.M. (2006) Evolutionary transformations of fetal membrane characters in Eutheria with special reference to Afrotheria. Journal of Experimental Zoology (Molecular and Developmental Evolution) 306B: 140–163

Metz, J. (1980) On the development of the rat placenta 1. Differentiation and functional alterations of labyrinthine layers II and III. Anatomy and Embryology 159: 289–305

Metz, J., Heinrich, D., Forssmann, W.G. (1976) Gap junctions in haemodi-, and trichorial placentas. Cell and Tissue Research 171: 305–316

Metz, J., Aoki, A., Forssman, W.G. (1978) Studies on the ultrastructure and permeability of the haemotrichorial placenta 1. Intercellular junctions of layer I and tracer administration into the maternal compartment. Cell and Tissue Research 192: 391–407

Meuris, S., Soumenkoff, G., Malengrean, A., Robyn, C. (1980) Immunoenzymological localisation of prolactin like immunoreactivity in decidual cells of the endometrium from pregnant and non pregnant women. Journal of Histochemistry and Cytochemistry 28: 1347–1350

Michael, K., Ward, B.S., Moore, W.M.O. (1985) In vitro permeability of the pig placenta in the last third of gestation. Biology of the Neonate 47: 170–178

Miglino, M.A., Carter, A.M., Ambrosio, C.E., Bonatelli, B., De Oliveira, M.F., Dos Santos Ferraz, R.H., Rodrigues, R.F., Santos, T.C. (2004) Vascular organization of the hystricomorph placenta: a comparative study in the agouti, capybara, guinea pig, paca and rock cavy. Placenta 25: 438–448

Miglino, M.A., Pereira, F.T., Visintin, J.A., Garcia, J.M., Meirelles, F.V., Rumpf, R., Ambrósio, C.E., Papa, P.C., Santos, T.C., Carvalho, A.F., Leiser, R., Carter, A.M. (2007) Placentation in cloned cattle: structure and microvascular architecture. Theriogenology 68: 604–617

Miles, J.R., Farin, C.E., Rodriguez, K.F., Alexander, J.E., Farin, P.W. (2004) Angiogenesis and morphometry of bovine placentas in late gestation from embryos produced in vivo or in vitro. Biology of Reproduction 71: 1919–1926

Milosavljevic, M., Duello, T.M., Schuler, L.A. (1989) In situ localisation of the prolactin related mRNAs to binuleate cells of bovine placentomes. Endocrinology 125: 883–889

Mincheva-Nilsson, L. (2003) Pregnancy and γδ T cells: talking the hard questions. Reproductive Biology and Endocrinology 2: 120

Mitchell, M.D., Flint, A, P.F. (1978) Prostaglandin production by intrauterine tissues from peri-parturient sheep: use of a superfusion technique. Journal of Endocrinology 76: 111–121

Miyazaki, H., Imai, M., Hirayama, T., Saburi, S., Tanaka, M., Maruyama, M., Matsuo, C., Meguro, C., Nishibashi, K., Inoue, J., Djiane, J., Gertler, G., Tachi, S., Imakawa, K., Tachi, C. (2002) Establishment of feeder-independent cloned caprine trophoblast cell line which expresses placental lactogen and interferon tau. Placenta 23: 613–630

Moffett-King, A. (2002) Natural killer cells and pregnancy. Nature Reviews in Immunology 2: 656–663

Moffett-King, A., Loke, C. (2006) Immunology of placentation in eutherian mammals. Nature Reviews in Immunology 6: 584–594

Moll, W. (1972) Gas exchange in concurrent, countercurrent and crosscurrent flow systems. The concept of the fetoplacental flow unit, in Respiratory Gas Exchange and Blood Flow in the Placenta (eds. L.D. Longo, H. Bartels), US DHEW Publication No. (NIH), pp. 73–361

Moll, W. (1985) Physiological aspects of placental ontogeny and phylogeny. Placenta 6: 141–154

Moll, W. (2003) Structure adaptation and blood flow control in the uterine arterial system after hemochorial placentation. European Journal of Obstetrics, Gynecology, and Reproduction Biology 110: S19–S27

Moore, H.D.M., Gems, S., Hearn, J.P. (1985) Early implantation stages in the marmoset monkey (*Callithrix jacchus*). American Journal of Anatomy 172: 265–278

Mor, G., Abrahams, D. (2003) Potential role of macrophages as immunoregulators of pregnancy. Reproductive Biology and Endocrinology 2: 119

Morgan, G., Wooding, F.B.P. (1983) Cell migration in the ruminant placenta, a freeze fracture study. Journal of Ultrastructure Research 83: 148–160

Morgan, G., Wooding, F.B.P., Brandon, M.R. (1987) Immunogold localisation of placental lactogen and the SBD3 antigen by ultracryomicrotomy at implantation in the sheep. Journal of Cell Science 88: 503–512

Morgan, G., Whyte, A., Wooding, F.B.P. (1990) Characterisation of the synthetic capacities of isolated binucleate cells from sheep and goats. Anatomical Record 226: 27–36

Morrish, D.W., Marusyk, H., Siy, O. (1987) Demonstration of specific secretory granules for human choriogonadotrophin in the placenta. Journal of Histochemistry and Cytochemistry 35: 93–101

Morrish, D.W., Marusyk, H., Bhardwaj, D. (1988) Ultrastructural localisation of hPL in distinctive granules in human term placenta: comparison with granules containing hCG. Journal of Histochemistry and Cytochemistry 36: 193–197

Morriss, F.H., Boyd, R.D.H. (1988) Placental transport, in Physiology of Reproduction, Chap. 50 (eds. E. Knobil, J. Neill), Raven Press, New York

Morton, W.R.M. (1961) Observations on the full term foetal membranes of the three members of the Camelidae (*Camelus dromedarius* L., *C. bactrianus* L., *Lama glama* L.). Journal of Anatomy 95: 200–209

Mossman, H.W. (1926) The rabbit placenta and the problem of placental transmission. American Journal of Anatomy 37: 433–497

Mossman, H.W. (1937) Comparative morphogenesis of the fetal membranes and accessory uterine structures. Carnegie Institute Contributions to Embryology 26: 129–246

Mossman, H.W. (1965) The principal interchange vessels of the chorioallantoic placenta of mammals, in Organogenesis (eds. R.L. De Haan, H. Ursprung), Holt Rinehart Wilson, New York, pp. 771–786

Mossman, H.W. (1987) Vertebrate Fetal Membranes. Rutgers University Press, New Brunswick, NJ

Moxon, L.A., Wildt, A.E., Slade, B.S. (1976) Localization of proteins in coated micropinocytotic vesicles during transport across rabbit yolk sac endoderm. Cell and Tissue Research 171: 175–193

Murphy, C.R. (2004) Uterine receptivity and the plasma membrane transformation Cell Research 14: 259–267

Murphy, W.J., Eizirik, E., Johnson, W.E., Zhang, Y.P., Ryder, O.A., Obrien, S.J. (2001) Molecular phylogenetics and the origins of placental mammals. Nature 409: 614–618

Musah, A.I., Schwabe, C, Willham, R.L., Anderson, L.L. (1987) Induction of parturition, progesterone secretion and delivery of placenta in beef heifers given relaxin with cloprostenol or dexamethasone. Biology of Reproduction 37: 797–803

Nakano, H., Takahashi, T., Imai, K., Hashizume, K. (2001) Expression of placental lactogen and cytokeratin in bovine placental binucleate cells in culture. Cell Tissue Research 303: 263–270

Nakano, H., Shimada, A., Imai, K., Takezawa, T., Takahashi, T., Hashizume, K. (2002) Bovine trophoblastic differentiation on collagen substrata: formation of binucleate cells expressing placental lactogen. Cell Tissue Research 307: 225–235

Nakano, H., Shimada, A., Imai, K., Takahashi, T., Hashizume, K. (2005) The cytoplasmic expression of E-cadherin and β-catenin in bovine trophoblasts during binucleate cell differentiation. Placenta 26: 393–401

Nelson, D.M., Ortman-Nabi, J., Curran, E.M. (1986) Morphology and glycoprotein synthesis of uterine gland epithelium in human basal plate at term: an ultrastructural and autoradiographic study. Anatomical Record 216: 146–153

Ness, R.B., Sibai, B.M. (2006) Shared and disparate components of the pathophysiologies of fetal growth restriction and preeclampsia. American Journal of Obstetrics and Gynecology 195(1): 40–49

Nikitenko, L., Morgan, G., Kolesnikov, S.I., Wooding, F.B.P. (1998) Immunocytochemical and in situ hybridisation studies of the distribution of calbindin D9K in the bovine placenta throughout pregnancy. Journal of Histochemistry and Cytochemistry 46: 679–688

Njogu, A., Owiti, G.O., Persson, E., Oduor-Okelo, D. (2006) Ultrastructure of the chorioallantoic placenta and chorionic vesicles of the lesser bush baby (*Galago senegalensis*). Placenta 27: 771–9

Ogura, Y., Takakura, N., Yoshida, H., Nishikawa, C.S. (1998) Essential role of PDGF receptor in the development of the intraplacental yolk sac/sinus of Duval in the mouse placenta. Biology of Reproduction 58: 65–72

O'Shea, J.D., Kleinfeld, F.G., Morrow, H.A. (1983) Ultrastructure of decidualisation in the pseudopregnant rat. American Journal of Anatomy 166: 271–298

Ockleford, C.D., Menon, G. (1977) Differentiated regions of human placental cell surface associated with exchange of materials between maternal and fetal blood: a new organelle and the binding of iron. Journal of Cell Science 25: 279–291

Oduor-Okelo, D. (1984a) Histology of chorioallantoic placenta of the golden rumped elephant shrew (*Rhyncocyon chrysopygus*). Anatomische Anzeiger 157: 395–407

Oduor-Okelo, D. (1984b) An electron microscopy study of the chorioallantoic placenta and subplacenta of the cane rat (*Thryonomys swinderianus Temminck*). Placenta 5: 433–442

Oduor-Okelo, D., Neaves, W.B. (1982) The chorioallantoic placenta of the spotted hyena (Crocuta, crocuta Erxleben): an electron microscopy study. Anatomical Record 204: 215–222

Oduor-Okelo, D., Musewe, D.O., Gombe, S. (1983) Electron microscopic study of the chorioallantoic placenta of the rock hyrax (*Heterohyrax brunei*). Journal of Reproduction and Fertility 68: 311–316

Ogura, A., Nishida, T., Hayashi, Y., Mochida, K. (1991) The development of the uteroplacental vascular system in the golden hamster *Mesocricetus auratus*. Journal of Anatomy 175: 65–77

Ohlsson, R., Falck, P., Hellström, M., Lindahl, P., Boström, H., Franklin, G., Ährlund-Richter, L., Pollard, J., Soriano, P., Betsholtz, C. (1999) PDGFB regulates the development of the labyrinthine layer of the mouse fetal placenta. Developmental Biology 212: 124–136

Oksenberg, J.R., Moryosef, S., Persitz, E., Schenker, Y., Mazes, E., Brantbar, C. (1986) Antigen presenting cells in human decidual tissue. American Journal of Reproduction and Immunological Microbiology 11: 82–88

Oliveira, S.F., Rasweiler, J.J., Badwaik, N.K. (2000) Advanced oviductal development, transport to the preferred implantation site, and attachment of the blastocyst in captive-bred, short-tailed fruit bats, *Carollia perspicillata*. Anatomy and Embryology 201: 357–381

O'Neill, R.J.W., Oneill, M.J., Graves, J.A.M. (1998) Undermethylation associated with retroelement activation and chromosome remodeling in an interspecific mammalian hybrid. Nature 393: 68–72

Ousey, J.C., Forhead, A.J., Rossdale, P.D., Grainger, L., Houghton, E., Fowden, A.L. (2003) Ontogeny of uteroplacental progestagens production in pregnant mares during the second half of gestation. Biology of Reproduction 69: 540–548

Owens, J.A. (1991) Endocrine and substance control of fetal growth: placental and maternal influences and IGFs. Reproduction, Fertility and Development 3: 501–518.

Owers, N.O. (1960) The endothelia-endothelial placenta of the Indian musk shrew, *Suncus murinus* – a new interpretation. American Journal of Anatomy 106: 1–25

Owiti, G.E.O., Cukierski, M., Tarara, R.P., Enders, A.C., Hendrick, A.G. (1986) Early placentation in the African green monkey (*Cercopithecus aethiops*). Acta Anatomica 127: 184–194

Owiti, G.E.O., Tarava, R.P., Hendrickx, A.G. (1989) Fetal membranes and placenta of the African green monkey *Cercopithecus daethiops*. Anatomy and Embryology 179: 591–604

Padykula, H.A., Taylor, J.M. (1976) Ultrastructural evidence for loss of the trophoblast layer in the chorioallantoic placenta of Australian bandicoots (Marsupialia; Peramelidae). Anatomical Record 186: 357–386

Panigel, M. (1956) Contribution a l'etude de l'ovoviviparite chez les reptiles: gestation et parturition chez Ie lezard Zootoca vivipara. Annales des Sciences Naturelles Zoologie XI 18: 569–668

Parham, P. (2004) NK cells and trophoblast: partners in pregnancy. Journal of Experimental Medicine 200: 951–955

Paria, B.C., Reese, J.R., Das, S.K., Dey, S.K. (2002) Deciphering the cross talk of implantation: advances and challenges. Science 296: 2185–2190

Parr, M.B., Parr, E.L. (1986) Permeability of the primary decidual zone (PDZ) in the rat uterus: studies using fluorescent labelled proteins and dextrans. Biology of Reproduction 34: 393–403

Parr, M.B., Parr, E.L. (1989) The implantation reaction in, Biology of Uterus, 2nd edn (eds. RM. Wyrm, W.P. Jollie), Plenum Press, New York, pp. 233–278

Parr, M.B., Tung, H.N., Parr, E.L. (1986) The ultrastructure of the rat primary decidual zone. American Journal of Anatomy 176: 423–436

Pattillo, R.A., Hussa, R.O., Yorde, D.E., Cole, L.A. (1983) Hormone synthesis by normal and neoplastic human trophoblast. In Biology of the Trophoblast, Loke, C., White A. Editors. Elsevier London

Peel, S. (1989) Granulated metrial gland cells. Advances in Anatomy, Embryology and Cell Biology 115: 1–112

Peel, S., Stewart, 1. (1979) Ultrastructural changes in the rat metrial gland in the latter half of pregnancy. Anatomy and Embryology 155: 209–219

Peel, S., Stewart, J. (1986) Oestrogen and the differentiation of granulated metrial gland cells in chimeric mice. Journal of Anatomy 144: 181–187

Peel, S., Stewart, L.G., Bulmer, D. (1979) Metrial gland cells in deciduomata of pseudopregnancy. Journal of Anatomy 129: 21–30

Peel, S., Stewart, L.G., Bulmer, D. (1983) Experimental evidence for the bone marrow origin of granulated metrial gland cells of the mouse uterus. Cell and Tissue Research 233: 647–656

Perry, J.S. (1974) Implantation of fetal membranes and early placentation of the African elephant *Loxodonta africana*. Philosophical Transactions of the Royal Society, B 269: 109–135

Perry, J.S. (1981) The mammalian fetal membranes. Journal of Reproduction and Fertility 62: 321–335

Perry, J.S, Rowlands, I.W. (1962) Early pregnancy in the pig. Journal of Reproduction and Fertility 4: 175–188

Perry, J.S., Heap, R.B., Ackland, N. (1975) The ultrastructure of the sheep placenta around the time of parturition. Journal of Anatomy 120: 561–570

Pfarrer, C., Ebert, B., Miglino, M.A., Klisch, K., Leiser, R. (2001) the three dimensional feto-maternal vascular interrelationship during early bovine placental development: an SEM study. Journal of Anatomy 198: 591–602

Pfarrer, C., Hirsch, P., Guillomot, M., Leiser, R. (2003) Interaction of integrin receptors with extracellular matrix is involved in trophoblast giant cell migration in bovine placentomes. Placenta 24: 588–597

Pfarrer, C.D., Weise, S., Berisha, B., Schams, D, Leiser, R., Hoffmann, B., Schuler, G. (2006a) Fibroblast growth factor FGF-1,2, 7 and FGF receptors are uniformly expressed in trophoblast giant cells during restricted trophoblast invasion in cows. Placenta 27: 758–770

Pfarrer, C.D., Ruziwa, S.D., Winther, H., Callesen, H., Leiser, R., Schams, D., Dantzer, V. (2006b) Localisation of VEGF and its receptors VEGFR-1 and 2 in bovine placentomes from implantation to term. Placenta 27: 889–898

Phillips, A.F., Holtzman, I.R, Teng, C., Battaglia, F.C. (1978) Tissue concentrations of free amino acids in human term placentas. American Journal of Obstetrics and Gynecology 131: 881–887

Pijnenborg, R, Dixon, G., Robertson, W.B., Brosens, I. (1980) Trophoblast invasion of human decidua from 8–18 weeks of pregnancy. Placenta 1: 3–19

Pijnenborg, R, Robertson, W.B., Brosens, I., Dixon, G. (1983) Uteroplacental arterial changes related to interstitial trophoblast migration in early human pregnancy. Placenta 4: 397–414

Pijnenborg, R., Vercruysse, L., Hanssens, M. (2006) The uterine spiral arteries in human pregnancy: facts and controversies. Placenta 27: 939–958

Polge, E., Rowson, L.E.A., Chang, M.E. (1966) The effect of reducing the number of embryos during the early stages of gestation on the maintenance of pregnancy in the pig. Journal of Reproduction and Fertility 12: 395–397

Porter, D.G. (1967) Observations on the development of mouse blastocysts transferred to the testis and kidney. American Journal of Anatomy 121: 73–85

Porter, D.G., Heap, R.B., Flint, A.P.F. (1982) Endocrinology of the placenta and the evolution of viviparity. Journal of Reproduction and Fertility Supplement 31: 113–138

Potgens, A.J., Schmitz, U., Bose, P., Versmold, A., Kaufmann, P., et al. (2002) Mechanisms of syncytial fusion: a review. Placenta 23 Supplement A: S107–S113

Potts, D.M. (1968) The ultrastructure of implantation in the mouse. Journal of Anatomy 103: 77–90

Power, G.G., Gilbert, R.E. (1977) Umbilical vascular compliance in sheep. American Journal of Physiology 233: H660–H664

Power, G.G., Longo, L.D. (1973) Sluice flow in placenta: maternal vascular pressure effects on fetal circulation. American Journal of Physiology 225: 1490–1496

Prasad, M.R.N., Mossman, H.W., Scott, G.L. (1979) Morphogenesis of the fetal membranes of an American mole Scalopus aquaticus. American Journal of Anatomy ISS: 31–38

Pusateri, A.E., Rothschild, M.F., Warner, C.M., Ford, S.P. (1990) Changes in morphology, cell number, cell size and cellular estrogen content of individual littermate pig conceptuses on days 9–13 of gestation. Journal of Animal Science 68: 3727–3735

Pusey, J., Kelly, W.A., Bradshaw, J.M., Porter, D.G. (1980) Myometrial activity and the distribution of blastocysts in the uterus of the rat: interference by relaxin. Biology of Reproduction 23: 394–397

Ralph, M.M., Lee, E.S., Thorburn, G.D. (1989) Identification and characterisation of monolayer cultures of sheep trophoblast cells maintained in bicameral culture chambers. Biology of Reproduction 41: 481–489

Ramirez-Pinilla, M.P. (2006) Placental transfer of nutrients during gestation in an Andean population of the highly matrotrophic lizard genus Mabuya. Herpetological Monographs 20: 194–204

Ramirez-Pinilla, M.P., De Perez, G., Carreno-Escobar, J.F. (2006) Allantoplacental ultrastructure of an Andean population of Mabuya. Journal of Morphology 267: 1227–1247

Ramsey, E.M. (1967) Vascular anatomy, in Biology of the Uterus (ed. R.M. Wynn), Plenum Press, New York, pp. 59–76

Ramsey, E.M. (1982) The Placenta, Human and Animal. Prager Pub (CBS), New York

Ramsey, E.M., Donner, M.W. (1980) Placental Vasculature and Circulation. Thieme, Stuttgart

Ramsey, E.M., Houston, M.L., Harris, J.W.S. (1976) Interactions of the trophoblast and maternal tissues in three closely related primate species. American Journal of Obstetrics and Gynecology 124: 647–652

Ramsoondar, J.J., Christopherson, R.J., Guilbert, L.J., Dixon, W.T., Ghahary, A., Ellis, S., Wegmann, T.G., Piedrahita, J.A. (1999) Lack of class I major histocompatibility antigens on trophoblast of periimplantation blastocysts and term placenta in the pig. Biology of Reproduction 1999 60: 387–397

Rankin, J.H.G. (1978) Role of prostaglandins in the maintenance of the placental circulation. Advances in Prostaglandin and Thromboxane Research 4: 261–269

Rankin, J.H.G., McLaughlin, M.K. (1979) The regulation of the placental blood flows. Journal of Developmental Physiology 1: 3–30

Rankin, J.H.G., Phernetton, T.M. (1976) Effect of prostaglandin E$_2$ on ovine maternal placental blood flow. American Journal of Physiology 231: 754–765

Rankin, J.H.G., Schneider, J.M. (1975) Effect of surgical stress on the distribution of placental, blood flows. Respiration Physiology 24: 373–383

Rankin, J.H.G., Meschia, G., Makowski, E.L., Battaglia, F.C. (1970) Macroscopic distribution of blood flow in the sheep placenta. American Journal of Physiology 219: 9–16

Rasweiler, J.J. (1979) Early embryonic development and implantation in bats. Journal of Reproduction and Fertility 56: 403–416

Rasweiler, J.J. (1993) Pregnancy in chiroptera. Journal of Experimental Zoology Z66: 495–513

Rattray, P.V., Garrett, W.N., East, N.E., Hinman, N. (1974) Growth, development and composition of the ovine conceptus and mammary gland during pregnancy. Journal of Animal Science 38: 613–626

Raub, T.J., Bazer, F.W., Roberts, R.M. (1985) Localisation of the iron transport protein uteroferrin in the porcine endometrium and placenta using immunocolloid gold. Anatomy and Embryology 171: 253–268

Ravelich, S.R., Bernhard, H., Breier, B.H., Reddy, S., Jeffrey, A., Keelan, J.A., Wells, D.N., Peterson, A.J., Lee, R.S.F. (2004) Insulin-like growth factor-I and binding proteins 1, 2, and 3 in bovine nuclear transfer pregnancies. Biology of Reproduction 70: 430–438

Redman, C.W.G., Sargent, I.L. (2007) Microparticles and immunomodulation in pregnancy and preeclampsia. Journal of Reproductive Immunology 76: 61–67

Regnault, T.R.H., Marconi, A.M., Smith, C.H., Glazier, J.D., Novak, D.A., Sibley, C., Jansson, T. (2005) Placental amino acid transport systems and fetal growth restriction – a workshop report. Placenta 26(Supplement 1): S76–S80

Reik, W., Costancia, M., Anderson, N., Dean, W., Ferguson-Smith, A., Tycko, B., Sibley, C.P. (2003) Regulation of the supply and demand for maternal nutrients in mammals by imprinted genes. Journal of Physiology 547: 35–44

Reimers, T.J., Ullman, M.B., Hansel, W. (1985) Progesterone and prostanoid production by bovine binucleate trophoblastic cells. Biology of Reproduction 33: 1227–1236

Reinius, S. (1967) Ultrastructure of blastocyst attachment in the mouse. Zeitschrift fur Zellforschung Mikroskopie Anatomie 77L: 257–266

Renegar, R.H., Southard, J.N., Talamantes, F. (1990) Immunohistochemical colocalisation of placental lactogen II and relaxin in the golden hamster (*Mesocricetus auratus*). Journal of Histochemistry and Cytochemistry 38: 935–940

Renfree, M.B. (2006) Life in the pouch: womb with a view. Reproduction Fertility and Development 19: 721–734

Reng, R. (1977) Die placenta von *Microcebus murinus*. Miller. Zeitschrift fur Saugetierkunde 42: 201–214

Rennie, M.Y., Whiteley, K.J., Kulandavelu, S., Adamson, S.L., Sled, J.G. (2007) 3D visualisation and quantification by microcomputed tomography of late gestational changes in the arterial and venous feto-placental vasculature of the mouse. Placenta 28: 833–840

Reponen, P., Leivo, I., Sahlberg, C., Apte, S.S., Thesleff, I., Olsen, B.R., Tryggvason, K. (1995) 92-kDa type IV collagenase and TIMP-3, but not 72-kDa type IV collagenase or TIMP-1 or TIMP-2, are highly expressed during mouse embryo implantation. Developmental Dynamics 202: 388–396

Reynolds, L.P., Ferrell, C.L. (1987) Transplacental clearance and blood flows of bovine gravid uterus at several stages of gestation. American Journal of Physiology 253: R735–R739

Reynolds, L.P., Redmer, D.A. (1992) Growth and microvascular development of the uterus during early pregnancy in ewes. Biology of Reproduction 47: 698–708

Reynolds, L.P., Caton, J.S., Redmer, D.A., Grazul-Bilska, A.T., Vonnahme, K.A., Borowicz, P.P., Luther, J.S., Wallace, J.M., Wu, G., Spencer, T.E. (2006) Evidence for altered placental blood flow and vascularity in compromised pregnancies. Journal of Physiology 572: 51–58

Reznick, D.N., Mateos, M., Springer, M.S. (2002) Independent origins and rapid evolution of the placenta in the fish genus Poeciliopsis. Science 298: 1018–1020

Rhodes, L., Cowan, R.G., Nathanielz, P.W., Reimers, T.J. (1986) Production of progesterone and placental lactogen by ovine binucleate trophoblast cells. Biology of Reproduction 34(Supplement 1) abstract no. 84

Rice, G.E., Thorburn, G.D. (1986) Characterisation of particle associated choriosomatomammotropin and progesterone in ovine placentomes. Journal of Endocrinology 111: 217–223.

Riley, S.C., Walton, J.C., Herlick, J.M., Challis, J.R.G. (1991) The localisation and distribution of CRH in the human placenta and fetal membranes throughout gestation. Journal of Clinical Endocrinology and Medicine 72: 1001–1007

Riley, P., Anson-Cartwright, L., Cross, J.C. (1998) The Hand1 bHLH transcription factor is essential for placentation and cardiac morphogenesis. Nature Genetics 18: 271–275

Rivera, R.M., Stein, P., Weave, J.R., Mager, J., Schultz, R.M., Bartolomei, M.S. (2008) Manipulations of mouse embryos prior to implantation result in aberrant expression of imprinted genes on day 9.5 of development. Human Molecular Genetics 17: 1–14

Roberts, R.M. (1989) Conceptus interferons and maternal recognition of pregnancy. Biology of Reproduction 40: 449–452

Roberts, R.M. (1992) Interferons as hormones of pregnancy. Endocrinological Reviews 13: 432–442

Roberts, C.T., Breed, W.G. (1994) Placentation in the *Dasyurid marsupial, Sminthopsis crassicaudata*, the fat tailed dunnart, and notes on the placenta of the didelphid, *Monodelphis domestica*. Journal of Reproduction and Fertility 100: 105–113

Robertson, W.B., Khong, T.Y., Brosens, C.I., Dewolf, F., Sheppard, B.L., Bonnar, J. (1986) The placental bed biopsy: review from three European centres. American Journal of Obstetrics and Gynecology 155: 401–412

Robinson, J.S., Kingston, E.J., Jones, C.T., Thorburn, G.D. (1979) Studies on experimental growth retardation in sheep. The effect of removal of endometrial caruncles on fetal size and metabolism. Journal of Developmental Physiology 1: 379–398

Rodewald, R., Kraehenbuhl, J.P. (1984) Receptor mediated transport of IgG. Journal of Cell Biology 99: 159–164

Rodrigues, R.F., Carter, A.M., Ambrosio, C.E., dos Santos, T.C., Miglino, M.A. (2006) The subplacenta of the red-rumped agouti (*Dasyprocta leporina* L). Reproductive Biology and Endocrinology 4: 31

Ross, J.W., Malayer, J.R., Ritchie, J.W., Geisert, R.D. (2003) Characterisation of the IL1β system during porcine elongation and early placental development. Biology of Reproduction 69: 1251–1259

Rossant, J., Cross, J.C. (2001) Placental development: lessons from mouse mutants. Nature Review Genetics 2: 538–548

Rossant, J, Frels, W.I. (1980) Interspecific chimaeras in mammals: successful production of live chimaeras between *Mus musculus* and *Mus caroli*. Science 208: 419–421

Rossdale, P.D., Ousey, J.C. (2002) Fetal programming for athletic performance in the horse: potential effects of IUGR. Equine Veterinary Education 14: 98–112

Rossman, I. (1940) The deciduomal reaction in the rhesus monkey (*Macaca mulatta*). American Journal of Anatomy 66: 277–366

Rowson, L.E-A., Moor, R.M. (1966) Development of the sheep conceptus during the first 14 days. Journal of Anatomy 100: 777–785

Rugh, R. (1975) The Mouse: Its Reproduction and Development. Burgess Publishing, Minneapolis, USA

Sacks, G., Sargent, I., Redman, C. (1999) An innate view of human pregnancy. Immunology Today 20: 114–118

Saito, S. (2000) Cytokine network at the feto-maternal interface. Journal of Reproductive Immunology 47: 87–103

Salamonsen, L.A., Stuchbery, S.J., Ogrady, C.M., Godkin, J.D., Finlay, J.K. (1988) Interferon a mimics effects of ovine trophoblast protein-ion prostaglandin and protein secretion by ovine endometrial cells in vitro. Journal of Endocrinology 117: R1–R4

Salamonsen, L.A., Nagase, H., Woolley, D.E. (1995) MMPs and their tissue inhibitors at the ovine-trophoblast-uterine interface. Journal of Reproduction and Fertility Supplement 49: 29–37

Samuel, C.A., Perry, J.S. (1972) The ultrastructure of pig trophoblast transplanted to an ectopic site in the uterine wall. Journal of Anatomy 113: 139–149

Samuel, C.A., Allen, W.R., Steven, D.H. (1974) Studies on the equine placenta. 1. Development of the microcotyledons. Journal of Reproduction and Fertility 41: 441–445

Samuel, C.A., Allen, W.R., Steven, D.H. (1976) Studies on the equine placenta. II. Ultrastructure of the placental barrier. Journal of Reproduction and Fertility 48: 257–264

Sands, J., Dobbing, J. (1985) Continued growth and development of the third trimester human placenta. Placenta 6: 13–22

Sargent, I.L., Borzychowski, A.M., Redman, G.W.G. (2007) uNKs and reeclampsia. Journal of Reproductive Immunology (in press) doi:10.1016/jri.2007.03.009

Sasser, R.G., Crock, J., Runder, A., Montgomery, C.A. (1989) Characteristics of PSPB in cattle. Journal of Reproduction Fertility Supplement 37: 109–113

Schaerer, E., Neutra, M.R., Kraehenbuhl, J.-P. (1991) Topical review: molecular and cellular mechanisms involved in transepithelial transport. Journal of Molecular Biology 123: 93–103

Schiebler, T.H., Kaufmann, P. (1981) Reifeplazenta, in Die Plazenta des Menschen (eds. V. Becker, T.H. Schiebler, F. Kubli), Thieme, New York

Schindler, V.F., DeVries, U. (1987) Protein uptake and transport by trophotaeniae absorptive cells in two goodeid embryos. Journal of Experimental Zoology 241: 17–29

Schindler, J.F., Hamlett, W.C. (1993) Maternal: embryonic relations in viviparous teleosts. Journal of Experimental Zoology 266: 378–393

Schlabritz-Loutsevitch, N., Ballesteros, B., Dudley, C., Jenkins, S., Hubbard, G., Burton, G.J., Nathanielsz, P. (2007) Moderate maternal nutrient restriction, but not glucocorticoid administration, leads to placental morphological changes in the baboon (*Papio* sp.). Placenta 28: 783–792

Schlafke, S., Enders, A.C. (1975) Cellular basis of interaction between trophoblast and uterus at implantation. Biology of Reproduction 12: 41–65

Schlafke, S., Welsh, A.O., Enders, A.C. (1985) Penetration of the basal lamina of the uterine luminal epithelium during implantation in the rat. Anatomical Record 212: 47–56

Schlernitzauer, D.A., Gilbert, P.W. (1966) Placentation and associated aspects of gestation in the bonnethead shark *Sphyrna tiburo*. Journal of Morphology 120: 219–232

Schonecht, P.A., Currie, W.B., Bell, A.W. (1992) Kinetics of PL in mid and late gestation fetuses. Journal of Endocrinology 133: 95–100

Schuhmann, R.A. (1982) Histocheinicaland electron microscopical studies of maternofetal circulatory units of mature human placentas. Obstetrics and Gynecology Annual 11: 1–31

Schulman, J.D., Mann, L., Doores, L., Dunchi, S., Halverstam, J., Mastroantonia, J.G. (1975) Amino acid metabolism by the fetal brain during normal and hypoglycemic conditions. Biology of the Neonate 25: 15–75

Selwood, L., Johnson, M.H. (2006) Trophoblast and hypoblast in the monotreme, marsupial and eutherian mammal: evolution and origins. BioEssays 28: 128–145

Settle, P., Mynett, K., Speake, P., Champion, E., Doughty, I.M., Sibley, C.P., D'Souza, S.W., Glazier, J. (2004) Polarized lactate transporter activity and expression in the syncytiotrophoblast of the term human placenta. Placenta 25: 496–504

Shanklin, D.R., Sibai, B.M. (1989) Ultrastructural aspects of preeclampsia. 1. Placental bed and uterine boundary vessels. American Journal of Obstetrics and Gynecology 161: 735–741

Sharp, D.C., McDowell, K.J., Weithenauer, J., Thatcher, W.W. (1989) The continuum of events leading to maternal recognition of pregnancy in mares. Journal of Reproduction and Fertility Supplement 37: 101–107

Shemesh, M. (1990) Production and regulation of progesterone in bovine CL and placenta in mid and late gestation: a personal review. Reproduction, Fertility and Development 2: 129–135

Shennan, D.B., Boyd, C.A.R. (1987) Ion transport' by the placenta: a review of membrane transport systems. Biochemica et Biophysica Acta 906: 437–457

Sherwood, O.D. (1988) Relaxin, in Physiology of Reproduction, Chap. 16 (eds. E. Knobil, J. Neill), Raven Press, New York

Shi, W., Krella, A., Orth, A., Yu, Y., Fundele, R. (2005) Widespread disruption of genomic imprinting in adult interspecies mouse hybrids. Genesis 43: 100–108

Shigeru, S., Akitoshi, N., Subaru, M.H., Shiozaki, A. (2007) The balance between cytotoxic NK cells and regulatory NK cells in human pregnancy Journal of Reproductive Immunology (in press) doi:10.1016 j.jri.2007.04.007

Shimizu, T., Tsutsui, T., Orima, H. (1990) Incidence of transuterine migration of embryos in the dog. Japanese Journal of Veterinary Science 52: 1273–1275

Shin, B., Fujikura, K., Suzuki, T., Tanaka, S., Takata, K. (1997) Glucose transporter-3 in the rat placental barrier: a possible machinery for the transplacental transfer of glucose. Endocrinology 138: 3997–4004

Shine, R., Bull, J.J. (1979) The evolution of live bearing in lizards and snakes. American Naturalist 113: 905–923

Shine, R., Guillette, L.J. (1988) The evolution of viviparity in reptiles; a physiological model and its ecological consequences. Journal of Theoretical Biology 132: 43–50

Sibley, C.P., Boyd, R.D.H. (1988) Control of transfer across the mature placenta. Oxford Reviews of Reproductive Biology 10: 382–435

Sibley, C.P., Coan, P.M., Ferguson-Smith, A.C., Dean, W., Hughes, J., Smith, P., Reik, W., Burton, G.J., Fowden, A.L., Constancia, M. (2004) Placental-specific IGF2 regulates the diffusional exchange characteristics of the mouse placenta. Proceedings of the National Academy of Sciences 101: 8204–8208

Silver, M., Comline, R.S. (1975) Transfer of gases and metabolites in the equine placenta: a comparison with other species. Journal of Reproduction and Fertility Supplement 23: 589–594

Silver, M., Steven, D. (1975) Placental exchange of blood gases, in Comparative Placentation (ed. D. Steven), Academic Press, London, pp. 161–188

Silver, M., Steven, D.H., Comline, R.S. (1973) Placental exchange and morphology in ruminants and the mare, in Foetal and Neonatal Physiology (eds. R.S. Comline, K.W. Cross, G.S. Dawes, P.W. Nathanielz), Cambridge University Press, Cambridge, pp. 245–271

Silver, M., Barnes, R.J., Comline, R.S., Burton, G.J. (1982) Placental blood flow: some fetal and maternal cardiovascular adjustments during gestation. Journal of Reproduction and Fertility Supplement 31: 139–160

Silver, M., Fowden, A.L., Taylor, P.M., Knox, J., Hill, C.M. (1994) Blood amino acids in the pregnant mare and fetus: the effects of maternal fasting and intrafetal insulin. Experimental Physiology 79: 423–433

Simister, N.E. (2003) Placental transport of IgG. Vaccine 29: 3365–3369

Simmons, D.G., Fortier, A.L., Cross, J.C. (2007) Diverse subtypes and developmental origins of trophoblast giant cells in the mouse placenta. Developmental Biology 304: 567–578

Simpson, R.A., Mayhew, T.M., Barnes, P.R. (1992) From 13 weeks to term the trophoblast of the human placenta grows by the continuous recruitment of new proliferative units: a study of nuclear number using the disector. Placenta 13: 501–512

Simpson, N.A.B., Nimrod, C., De Vermette, R., Fournier, J. (1997) Determination of intervillous flow in early pregnancy. Placenta 18: 287–293

Sinha, A.A., Erickson, A.W. (1974) Ultrastructure of the placenta of the Antarctic seals during the first third of pregnancy. American Journal of Anatomy 141: 263–280

Sinha, A.A., Seal, U.S., Erickson, A.W., Mossman, H.W. (1969) Morphogenesis of the fetal membranes of the white tailed deer. American Journal of Anatomy 126: 201–242

Sinowatz, F., Friess, A.E. (1983) Uterine glands of the pig during pregnancy. An ultrastructural and cytochemical study. Anatomy and Embryology 166: 121–134

Skidmore, J.A., Wooding, F.B.P, Allen, W.R. (1996) Implantation and early placentation in the one-humped camel (Camelus dromedarius). Placenta 17: 253–262

Skidmore, J.A., Billah, M., Binns, M., Short, R.V., Allen, W.R. (1999) Hybridising old and new world camelids: Camelus dromedarius × Llama guanicoe. Proceedings of the Royal Society B 266: 649–656

Skolek-Winnisch, R., Lipp, W., Sinowatz, F., Friess, A.E. (1985) Enzymhistochemische untersuchungen an der Schweine placenta. II. Histotopik von Enzymen in den areolaren placenta epithelien. Acta Histochemica 76: 131–143

Slukvin, I.I., Breburda, E.E., Golos, T.G. (2004) Dynamic changes in primate endometrial leukocyte populations: differential distribution of macrophages and natural killer cells at the rhesus monkey implantation site and in early pregnancy. Placenta 25: 297–307

Smith, C.A., Moore, H.D.M., Hearn, J.P. (1987) The ultrastructure of early implantation in the marmoset monkey (*Callithrix jacchus*). Anatomy and Embryology 175: 399–410

Smith, C.E., Cullen, W.C., Godkin, J.D. (1990) Ultrastructural morphometric analysis of the uterine epithelium during early pregnancy in the sheep. Journal of Reproduction and Fertility 89: 517–525

Snow, M.A.L., Ansell, J.D. (1974) The chromosomes of giant trophoblast cells of the mouse. Proceedings of the Royal Society, B 187: 93–98

Soares, M.J. (2004) The PRL and GH families: pregnancy specific hormones/cytokines at the maternal-fetal interface. Reproductive Biology and Endocrinology 2: 51

Soares, M.J., Colosi, P., Ogren, L., Talamantes, F. (1983) Identification and partial characterisation of a lactogen from the midpregnant mouse conceptus. Endocrinology 112: 1313–1317

Soares, M.J., Faria, T.N., Roby, K.F., Deb, S. (1991) Pregnancy and the Prolactin family of hormones. Endocrinological Reviews 12: 402–423

Soares, M.J., Chapman, B.M., Rasmussen, C.A., Dai, G., Kamei, T., Orwig, K.E. (1996) Differentiation of trophoblast endocrine cells. Placenta: 17: 277–289

Soares, M.J., Alam, S.M.K., Konno, T., Ho-Chen, J.K., Ain, R. (2006) The prolactin family and pregnancy-dependent adaptations. Animal Science Journal 77: 1–9

Sooranna, S.R., Contractor, S.F. (1991) Vectorial transcytosis of IgG by human term trophoblast cells in culture. Experimental Cell Research 192: 41–46

Southard, J.N., Talamantes, F. (1991) Placental PRL-like proteins in rodents: variations on a structural theine. Molecular and Cellular Endocrinology 79: C133–C140

Sparks, J.W., Lynch, A., Glinsmann, W.H. (1976) Regulation of rat liver glycogen synthesis and activities of glycogen cycle enzymes by glucose and galactose. Metabolism 25: 47–55

Speake, B.K., Herbert, J.F., Thompson, M.B. (2004) Evidence for the transfer of lipids during gestation in the viviparous lizard, *Pseudemoia entrecasteauxii*. Comparative Biochemistry and Physiology A 139: 213–220

Spencer, T.E., Burghardt, R.C., Johnson, G.A., Bazer, F.W. (2004a) Conceptus signals for establishment and maintenance of pregnancy. Animal Reproduction Science 82–83: 539–550

Spencer, T.E., Johnson, G.A., Bazer, F.W., Burghardt, R.C. (2004b) Implantation mechanisms: insights from the sheep. Reproduction 128: 657–668

Spencer, T.E., Johnson, G.A., Bazer, F.W., Burghardt, R.C., Palmarini, M. (2007) Pregnancy recognition and conceptus implantation in domestic ruminants: roles of progesterone, interferons and endogenous retroviruses. Reproduction Fertility and Development 19: 65–78

Starck, D. (1949) Einbeitrag zur Kenntnis der Placentation bei der Macrosceliden. Zeitschrift fur Anatomie Entwicklungs Geschichte114: 319–339

Steer, H. (1971) Implantation of the rabbit blastocyst: the adhesive phase of implantation. Journal of Anatomy 109: 215–227

Steingrímsson, E., Tessarollo, L., Reid, S.W., Jenkins, N.A., Copeland, N.G. (1998) The bHLH-Zip transcription factor Tfeb is essential for placental vascularization. Development 125: 4607–4616

Stephens, R.J. (1962) Histology and histochemistry of the placenta and fetal membranes of the bat *Tadarida brasiliensis cynocephala*. American Journal of Anatomy 111: 239–285

Stephens, R.J. (1969) The development and fine structure of the allantoic placental barrier in the bat *Tadarida brasiliensis cynocephala*. Journal of Ultrastructure Research 28: 371–398

Stephens, R.J., Cabral, L. (1971) Direct contribution of the cytotrophoblast to the syncytiotrophoblast in the diffuse labyrinthine endotheliochorial placenta of the bat. Anatomical Record 169: 243–252

Stephens, R.J., Cabral, L.J. (1972) The diffuse labyrinthine endotheliodichorial placenta of the free tailed bat: a light and electron microscopy study. Anatomical Record 172: 221–252

Steven, D.H. (1966) Further observations on placental circulation in the sheep. Journal of Physiology London 187: 18P–19P

Steven, D.H. (ed.) (1975a) Comparative Placentation. Academic Press, New York

Steven, D.H. (1975b) Separation of the placenta in the ewe: an ultrastructural study. Quarterly Journal of Experimental Physiology 60: 37–44

Steven, D.H. (1982) Placentation in the mare. Journal of Reproduction and Fertility Supplement 31: 41–55

Steven, D.H. (1983) Interspecies differences in the structure and function of trophoblast, in Biology of Trophoblast (eds. C. Loke, A. Whyte), Elsevier, Amsterdam, pp. 111–136

Steven, D.H., Samuel, C.A. (1975) Anatomy of the placental barrier in the mare. Journal of Reproduction and Fertility Supplement 23: 579–582

Steven, D.H., Bass, F., Jansen, C.J.M., Krane, E.J., Mallon, K., Samuel, C.A., Thomas, A.L., Nathanielz, P.W. (1978) Ultrastructural changes in the placenta of the ewe after fetal pituitary stalk section. Quarterly Journal of Experimental Physiology 634: 221–231

Steven, D.H., Jeffcoat, L.B., Mallon, K.A, Ricketts, S.W., Rossdale, P.O., Samuel, C.A. (1979) Ultrastructural studies of the equine uterus and placenta following parturition. Journal of Reproduction and Fertility 17: 579–586

Stewart, I. (1985) Granulated me trial gland cells in the lungs of mice in pregnancy and pseudo-pregnancy. Journal of Anatomy 140: 551–563

Stewart, J.R. (1993) Yolk sac placentation in reptiles: structural innovation in a fundamental vertebrate fetal nutritional system. Journal of Experimental Zoology 266: 431–449

Stewart, I., Peel, S. (1977) The structure and differentiation of granulated metrial gland cells of the pregnant mouse uterus. Cell and Tissue Research 184: 517–527

Stewart, J.R., Blackburn, D.G. (1988) Reptilian placentation: structural diversity and terminology. Copeia 4: 839–852

Stewart, J.R., Thompson, M.B. (2000) Evolution of placentation among squamate reptiles: recent research and future directions. Comparative Biochemistry and Physiology A 127: 411–431

Stewart, J.R., Brasch, K.R. (2003) Ultrastructure of the placentae of the natricine snake *Virginia striatula* (Reptilia, Squamata). Journal of Morphology 255: 177–201

Stewart, J.R., Thompson, M.B. (2004) Placental ontogeny of the Tasmanian scincid lizard, *Niveoscincus ocellatus* (Reptilia, Squamata). Journal of Morphology 259: 214–237

Stewart, F., Lennard, S.N., Allen, W.R. (1995) Mechanisms controlling the formation of the equine chorionic girdle. Biology of Reproduction Monograph Series 1: 151–159

Stoffel, M.H., Friess, E., Hartmann, S.H. (2000) Ultrastructural evidence for transplacental transport of immunoglobulin G in bitches. Journal of Reproduction and Fertility 118: 315–326

Strickland, S., Richards, W.G. (1992) Invasion of the trophoblasts. Cell 71: 355–357

Stroband, H.W.J., Taverne, N., Largenfeld, K., Barends, P.M.G. (1986) The ultrastructure of the uterine epithelium of the pig during the estrous cycle and early pregnancy. Cell and Tissue Research 246: 81–89.

Stulc, J. (1989) Extracellular transport pathways in the hemochorial placenta. Placenta 10: 113–119

Stulc, J. (1997) Placental transfer of inorganic ions and water. Pediatric Research 77: 805–840

Sturgess, I. (1948) The early embryology and placenta of *Procavia capensis* (Hyrax). Acta Zoologica 29: 393–479

Summers, P.M., Shephard, A.M., Hodges, J.K., Kydd, J., Boyle, M.S, Allen, W.R. (1987) Successful transfer of the embryos of Przewalski's horses (*Equus przewalskii*) and Grant's zebra (*E. birchelli*) to domestic mares (*E. caballus*). Journal of Reproduction and Fertility 80: 13–20

Swanson, W.F., Roth, T.L., Brown, J.L., Wildt, D.E. (1995) Relationship of circulating steroid hormones, luteal luteinizing hormone receptor and progesterone concentration, and embryonic mortality during early embryogenesis in the domestic cat. Biology of Reproduction 53: 1022–1029

Symonds, M.E., Stephenson, T., David, S., Gardner, D.S., Budge, H. (2007) Long-term effects of nutritional programming of the embryo and fetus: mechanisms and critical windows. Reproduction Fertility and Development 19: 53–63

Tachi, S., Tachi, C. (1979) Ultrastructural studies on maternal embryonic cell interactions during experimentally induced implantation of rat to the endometrium of the mouse. Developmental Biology 68: 203–223

Tanaka, S., Oda, M., Toyoshima, Y., Wakayama, T., Tanaka, M., Yoshida, N., Hattori, N., Ohgane, J., Yanagimachi, R., Shiota, K. (2001) Placentomegaly in cloned mouse concepti caused by expansion of the spongiotrophoblast layer. Biology of Reproduction 65: 1813–1821

Tarachand, U. (1986) Decidualisation: origin and role of the associated cells. Biologie Cellulaire 57: 9–16

Tavolga, W.N., Rugh, K. (1947) Development of the platyfish *Platypoecilus maculatus*. Zoologica 32: 1–16

Tawfik, O.W., Hunt, D.S., Wood, G.W. (1986) Implication of PGE_2 in soluble factor mediated immune suppression by murine decidual cells. American Journal of Reproduction and Immunological Microbiology 12: 111–117

Taylor, M.J., Webb, K, Mitchell, M.D., Robinson, J.S. (1982) Effect of progesterone withdrawal in sheep during late pregnancy. Journal of Endocrinology 92: 85–93

Tedde, G., Tedde-Piras, A. (1978) Mitotic index of the Langhans cells in the normal human placenta from the early stages of pregnancy to term. Acta Anatomica 100: 114–119

Thompson, M.B., Speake, B.K. (2006) A review of the evolution of viviparity in lizards: structure, function and physiology of the placenta. Journal of Comparative Physiology B 176: 179–189

Thorburn, G.D. (1991) The placenta, prostaglandins and parturition. Reproduction, Fertility and Development 3: 271–294

Thornburg, K.L., Bissonnette, J.M., Faber, J.J. (1976) Absence of fetal placental waterfall phenomenon in chronically prepared fetal lambs. American Journal of Physiology 230: 886–892

Tiedemann, K., Minuth, W.W. (1980) The pig yolk sac: fine structure of the posthematopoietic organ. Histochemistry 68: 133–146

Toth, M., Asboth, G., Hertelendy, F. (1983) Turnover of lipid bound arachidonic acid and biosynthesis of prostanoids in the endometrium and myometrium of the laying hen. Archives of Biochemistry and Biophysics 226: 27–36

Tsoulos, N.G., Colwell, J.R., Battaglia, F.C, Makowski, E.L., Meschia, G. (1971) Comparison of glucose, fructose and oxygen uptakes by fetuses of fed and starved ewes. American Journal of Physiology 221: 234–237

Tsutsumi, Y. (1962) The vascular pattern of the placenta in farm animals. Journal of the Faculty of Agriculture Hokkaido University Sapporo 52: 372–482

Turner, C.L. (1947) Viviparity in the teleost fishes. Science Monthly 65: 508–518

Tyndale-Biscoe, H., Renfree, M. (1987) Reproductive Physiology of Marsupials. Cambridge University Press, Cambridge

Uekita, T., Yamanouchi, K., Sato, H., Tojo, H., Seiki, M., Tachi, C. (2004) Expression and localisation of MT1-MMPand MMP-2 and TIMP-2 during synepitheliochorial placentation in goats. Placenta 25: 810–819

Ullman, M.B., Reimers, T.J. (1989) Progesterone production by binucleate trophoblastic cells of cows. Journal of Reproduction and Fertility Supplement 37: 173–179

Ulysses, S.S., Akhouri, A., Doe, S., Doe, R.P. (1972) Placental iron transfer: relationship to placental anatomy and phylogeny of the mammals. American Journal of Anatomy 134: 263–269

Ushizawa, U., Takahashi, T., Kaneyama, K., Hosoe, M., Hashizume, K. (2006) Cloning of the bovine antiapoptotic regulator, BCL2-related protein A1, and its expression in trophoblastic binucleate cells of bovine placenta. Biology of Reproduction 74: 344–351

Vajta, G., Gjerrie, M. (2006) Science and technology of farm animal cloning: state of the art. Animal Reproduction Science 92: 211–230

Vandaele, L., Verberckmoes, S., El Amiri, B., Sulon, J., Duchateau, L., Van Soom, A., Beckers, J.F., De Kruif, A. (2005) Use of a homologous radioimmunoassay (RIA) to evaluate the effect of maternal and foetal parameters on pregnancy-associated glycoprotein (PAG) concentrations in sheep. Theriogenology 63: 1914–1924

Vanderpuy, O.A., Smith, C.A. (1987) Proteins of the apical and basal plasma membranes of the human placental syncytium: immunochemical and electrophoretic studies. Placenta 8: 591–608

Van Dorsser, F.J.D., Swanson, W.F., Lasano, S., Steinetz, B.G. (2006) Development, validation, and application of a urinary relaxin radioimmunoassay for the diagnosis and monitoring of pregnancy in felids. Biology of Reproduction 74: 1090–1095

Van Wyk, L.C., Van Niekerk, C.H., Belonje, P.C. (1972) Evolution of the post partum uterus of the ewe. Journal of South African Veterinary Association 43: 19–26

Varmuza, S., Prideaux, V., Kothary, R., Rossant, J. (1988) Polytene chromosomes in mouse trophoblast giant cells. Development 102: 127–134

Venuto, R.C., Cox, J.W., Stein, I.H., Ferris, T.F. (1976) The effect of changes in perfusion pressure on uteroplacental blood flow in the pregnant rabbit. Journal of Clinical Investigation 57: 938–944

Vieira, S., Perez, G., Ramirez-Pinilla, M.P. (2007) Invsive cells in the placentome of Andean Populations of Mabuya: an endotheliochorial contribution to the Placenta? Anatomical Record 290: 1508–1518

Vercruysse, L., Caluwaerts, S., Luyten, C., Pijnenborg, R. (2006) Interstitial trophoblast invasion in the decidua and mesometrial triangle during the last third of pregnancy in the rat. Placenta 27: 22–33

Verstegen, J., Fellmann, D., Beckers, J.F. (1985) Immunodetection of the bovine chorionic somatomammotropin. Acta Endocrinologia 109: 403–410

Vogel, P. (2005) The current molecular phylogeny of Eutherian mammals challenges previous interpretations of placental evolution. Placenta 26: 591–596

Vonnahme, K.A., Ford, S.P. (2004) Differential expression of the vascular endothelial growth factor-receptor system in the gravid uterus of Yorkshire and Meishan pigs. Biology of Reproduction 71: 163–169

Vrana, P.B., Fossella, J.A., Matteson, P., del Rio, T., Oneill, M.J., Tilghman, M. (2000) Genetic and epigenetic incompatibility underlie hybrid dysgenesis. Nature Genetics 25: 120–124

Wadsworth, P.F., Lewis, D.J., Heywood, R. (1980) The ultrastructural features of progestagen induced decidual cells in the rhesus monkey (Macaca mulatta). Contraception 22: 189–198

Wagner, W.C., Hansel, W. (1969) Reproductive physiology of the post partum cow. 1. Clinical and histological findings. Journal of Reproduction and Fertility 18: 493–500

Wake, M.H. (1993) Evolution of oviductal gestation in Amphibians. Journal of Experimental Zoology 266: 394–413

Wales, R.G., Cuneo, C.L. (1989) Morphological and chemical analysis of the sheep conceptus from the 13th to the 19th day of pregnancy. Reproduction Fertility and Development 1: 31–39

Walls, E.W. (1938) Myrmecophaga jubata: an embryo with a placenta. Journal of Anatomy 73: 311–317

Walter, I., Schonkypi, S. (2006) Extracellular matrix components and matrix degrading enzymes in the feline placenta during gestation. Placenta 27: 291–306

Wang, X., Matsumoto, H., Zhao, X., Das, S.K., Paria, B.C. (2004) Embryonic signals direct the formation of tight junctional permeability barrier in the decidualizing stroma during embryo implantation. Journal of Cell Science 117: 53–62

Wango, E.O., Wooding, F.B.P., Heap, R.B. (1990a) The role of trophoblastic cells in implantation in the goat: a morphological study. Journal of Anatomy 171: 241–257

Wango, E.O., Wooding, F.B.P., Heap, R.B. (1990b) The role of trophoblastic cells in implantation in the goat: a quantitive study. Placenta 11: 381–394

Wango, E.O., Wooding, F.B.P., Heap, R.B. (1991) Progesterone and 513 pregnanediol production by isolated fetal binucleate cells from sheep and goats. Journal of Endocrinology 129: 283–289

Ward, J.W., Wooding, F.B.P., Fowden, A.L. (2002) Effects of cortisol on the binucleate cell population in the ovine placenta during late gestation. Placenta 23: 451–458

Ward, J.W., Forhead, A.J., Wooding, F.B.P., Fowden, A.L. (2006) Functional Significance and cortisol dependence of the gross morphology of ovine placentomes during late gestation. Biology of Reproduction 74: 137–145

Waters, M.J., Oddy, V.H., McCloghny, C.E., Gluckman, P.D., Duplock, R., Owens, P.C. Brinsmead, M.W. (1985) An examination of the proposed roles of placental lactogen in the ewe by means of antibody neutralisation. Journal of Endocrinology 106: 377–386.

Wathes, D.E., Wooding, F.B.P. (1980) An electron microscopic study of implantation in the cow. American Journal of Anatomy 159: 285–306

Watkins, A.J., Ursell, E., Panton, R., Papenbrock, T., Hollis, L., Cunningham, C., Wilkins, A., Perry, V.H., Sheth, B., Wing Yee Kwong, W.Y., Eckert, J.J., Wild, A.E., Hanson, M.A., Osmond, C., Tom, P. Fleming, T.P. (2008) Adaptive responses by mouse early embryos to maternal diet protect fetal growth but predispose to adult onset disease. Biology of Reproduction 78: 299–306

Weekes, H.L.W. (1935) A review of the placentation among reptiles with particular regard to the function and evolution of the placenta. Proceedings of the Herpetological Society, Part 2: 625–645

Wehrenberg, W.B., Chaicharoen, D.P., Dierschke, D.J., Rankin, J.H., Ginther, O.J. (1977) Vascular dynamics of the reproductive tract in the female rhesus monkey: relative contributions of ovarian and uterine arteries. Biology of Reproduction 17: 148–153

Weier, J.F., Weier, H-U.G., Jung, C.J., Gormely, M., et al. (2005) Human cytotrophoblasts acquire aneuploidies as they differentiate to an invasive phenotype. Developmental Biology 279: 420–432

Welsh, A.O., Enders, A.C. (1983) Occlusion and reformation of rat uterine epithelium during pregnancy. American Journal of Anatomy 167: 463–477

Welsh, A.O., Enders, A.C. (1985) A light and electron microscopic examination of the mature decidual cells of the rat with emphasis on the antimesometrial decidua and its degeneration. American Journal of Anatomy 172: 1–30

Welsh, A.O., Enders, A.C. (1987) Trophoblast decidual cell interactions and establishment of maternal blood circulation in the parietal yolk sac placenta of the rat. Anatomical Record 217: 203–219

Welsh, A.O., Enders, A.C. (1991a) Chorioallantoic placental formation in the rat. I. Luminal epithelial cell death and ECM modifications in the mesometrial region of the implantation chamber. American Journal of Anatomy 192: 215–231

Welsh, A.O., Enders, A.C. (1991b) Chorioallantoic placental formation in the rat. II. Angiogenesis and maternal blood circulation in the mesometrial region of the implantation chamber prior to placental formation. American Journal of Anatomy 192: 346–365

White, F.J., Burghardt, R.C., Hu, J., Joyce, M.M., Spencer, T.E., Johnson, G.A. (2006) Secreted phosphoprotein 1 (osteopontin) is expressed by stromal macrophages in cyclic and pregnant endometrium of mice, but is induced by estrogen in luminal epithelium during conceptus attachment for implantation. Reproduction 132: 919–929

Whitwell, K.E., Jeffcott, L.B. (1975) Morphological studies on the fetal membranes of the normal singleton foal at term. Research in Veterinary Science 19: 44–55

Whyte, A., Allen, W.R. (1985) Equine endometrium at preimplantation stages of pregnancy has specific glycosylated regions. Placenta 6: 537–542

Wichtrup, L.G., Greven, H. (1985) Uptake of ferritin by the trophotaeniae of the goodeid fish *Ameca splendens* Miller and Fitzsimons 1971 (Teleostei; Cyprinodontiformes). Cytobiology 42: 33–40

Wigglesworth, J.S. (1969) Vascular anatomy of the human placenta and its significance for placental pathology. Journal of Obstetrics and Gynaecology of the British Commonwealth 76: 979–989

Wigmore, P.M.E., Strickland, N.E. (1985) Placental growth in the pig. Anatomy and Embryology 173: 263–268

Wild, A.E. (1981) Endocytotic mechanisms in protein transfer across the placenta. Placenta Supplement 1: 165–186

Wildman, D.E., Chen, C., Erez, O., Grossman, G.I., Goodman, M., Romero, R. (2006) Evolution of the mammalian placenta revealed by phylogenetic analysis. Proceedings of the National Academy of Sciences 103: 3203–3208

Wiley, A.A., Baktol, F.F., Barron, D.H. (1987) Histogenesis of the ovine uterus. Journal of Animal Science 64: 1262–1269

Wilkening, R.B., Meschia, G. (1983) Fetal oxygen uptake, oxygenation and acidbase balance as a function of uterine blood flow. American Journal of Physiology 244: H749–H755

Wilkening, R.B., Meschia, G. (1992) Current topic: comparative physiology of placental oxygen transport. Placenta 13: 1–15

Willadsen, S.M. (1986) Nuclear transplantation in sheep embryos. Nature 320: 63–65

Wilmut, I., Sales, D.I., Ashworth, C.J. (1986) Maternal and embryonic factors associated with prenatal loss in mammals. Journal of Reproduction and Fertility 76: 851–864

Wilmut, I., Schnieke, A.E., McWhir, J., Kind, A.J., Campbell, K.H. (1987) Viable offspring derived from adult mammalian cells. Nature 385: 810–813

Wilsher, S., Allen, W.R. (2003) The effects of maternal age and parity on placental and fetal development in the mare. Equine Veterinary Journal 35: 476–483

Wilson, I.B. (1963) A tumour tissue analogue of the implanting mouse embryo. Proceedings of the Zoological Society, London. 141: 137–152

Wilson, M.E., Biensen, N.J., Youngs, C.R., Ford, S.P. (1998) Development of Meishan and Yorkshire littermate conceptuses in either a Meishan or Yorkshire uterine environment to day 90 of gestation and to term. Biology of Reproduction 58: 905–910

Wimsatt, W.A. (1951) Observations on the morphogenesis cytochemistry and significance of the binucleate giant cells of the placenta of ruminants. American Journal of Anatomy 89: 233–282

Wimsatt, W.A. (1974) Morphogenesis of the fetal membranes and placenta of the black bear *Ursus americanus* (Pallas). American Journal of Anatomy 140: 471–496

Wimsatt, W.A., Enders, A.C. (1980) Structure and morphology of uterus, placenta and parapla-cental organs of the neotropical disc winged bat *Thyroptera tricolor* spix (Microchiroptera: Thyropteridae). American Journal of Anatomy 159: 209–243

Wimsatt, W.A., Enders, A.C., Mossman, H.W. (1973) A re-examination of the chorioallantoic placental membrane of a shrew, *Blarina brevicauda*: resolution of a controversy. American Journal of Anatomy 138: 207–234

Wintenberger-Torres S., Flechon, J. (1974) Ultrastructural evolution of the trophoblast cells in the preimplantation sheep blastocyst from day 8 to day 18. Journal of Anatomy 118: 143–153

Winther, H., Dantzer, V. (2001) Co-localization of vascular endothelial growth factor and its two receptors Flt-1 and KDR in the mink. Placenta 22: 457–465

Winther, H., Ahmed, A., Dantzer, V.B. (1999a) Immunohistochemical location of VEGF and its two specific receptors Fit-1 and KDR in the porcine placenta and non pregnant uterus. Placenta 20: 35–43

Winther, H., Leiser, R., Pfarrer, C., Dantzer, V. (1999b) Localization of micro- and intermediate filaments in non-pregnant uterus and placenta of the mink suggests involvement of maternal endothelial cells and periendothelial cells in blood flow regulation. Anatomy and Embryology 200: 253–263

Wislocki, G.B. (1929) On the placentation of primates with a consideration of the phylogeny of the placenta. Carnegie Institute Contributions to Embryology 20: 51–80

Wislocki, G.B. (1930) A series of placental stages of a platyrrhine monkey *Ateles geoffroyi*, with some remarks on age, sex and breeding period in platyrrhines. Carnegie Contributions to Embryology 22: 173–200

Wislocki, G.B., Dempsey, E.W. (1946) Histochemical reactions in the placenta of the cat. American Journal of Anatomy 78: 1–46

Wislocki, G.B., Streeter, G.L. (1938) On the placentation of the Macaque (*Macaca mulatta*) from the time of implantation until the formation of the definitive placenta. Carnegie Institute Contributions to Embryology 27: 1–66

Wislocki, G.B., Westhuysen, O.P.V. (1940) The placentation of *Procavia capensis*, with a discus-sion of the placental affinities of the Hyracoides. Carnegie Institute Contributions to Embryology 27: 67–88

Wooding, F.B.P. (1980) Electron microscopic localisation of binucleate cells in the sheep placenta using phosphotungstic acid. Biology of Reproduction 22: 357–365

Wooding, F.B.P. (1981) Localisation of ovine placental lactogen in sheep placentomes by electron microscope immunocytochemistry. Journal of Reproduction and Fertility 62: 15–19

Wooding, F.B.P. (1982a) Structure and function of placental binucleate (giant) cells. Bibliotheca Anatomica 22: 134–139

Wooding, F.B.P. (1982b) The role of the binucleate cell in ruminant placental structure. Journal of Reproduction and Fertility Supplement 31: 31–39

Wooding, F.B.P. (1983) Frequency and localisation of binucleate cells in the placentomes of ruminants. Placenta 4: 527–540

Wooding, F.B.P. (1984) Role of binucleate cells in fetomatemal cell fusion at implantation in the sheep. American Journal of Anatomy 170: 233–250

Wooding, F.B.P. (1987) Ultrastructural evidence for placental lactogen transport and secretion in ruminants. Journal of Physiology 386: 26P

Wooding, F.R.P. (1992) The synepitheliochorial placenta of ruminants: binucleate cell fusions and hormone production. Placenta 13: 101–113

Wooding, F.B.P., Beckers, J.F. (1987) Trinucleate cells and the ultrastructural localisation of bovine placental lactogen. Cell and Tissue Research 247: 667–673

Wooding, F.B.P., Flint, A.P.F. (1994) Placentation, in Marshalls Physiology of Reproduction, 4th edn, Vol. III, Part I, London, Chapman and Hall, pp. 230–466

Wooding, F.B.P., Fowden, A.L. (2006) Nutrient transfer across the equine placenta: correlation of structure and function. Equine Veterinary Journal 38: 175–183

Wooding, F.B.P., Staples, L.D. (1981) Functions of the trophoblast papillae and binucleate cells in implantation in the sheep. Journal of Anatomy 133: 110–112

Wooding, F.B.P., Wathes, D.C. (1980) Binucleate cell migration in the bovine placentome. Journal of Reproduction and Fertility 59: 425–430

Wooding, F.B.P., Chambers, S.G., Perry, J.S., George, M., Heap, R.B. (1980) Migration of binucleate cells in the sheep placenta during normal pregnancy. Anatomy and Embryology 158: 361–370

Wooding, F.B.P., Flint, A.P.F., Heap, R.B., Hobbs, T. (1981) Autoradiographic evidence for migration and fusion of cells in the sheep placenta; resolution of a problem in placental classification. Cell Biology International Reports 5 821–827

Wooding, F.B.P., Staples, L.D., Peacock, M.A.P. (1982) Structure of trophoblast papillae on the sheep conceptus at implantation. Journal of Anatomy 134: 507–516

Wooding, F.B.P., Flint, A.P.F., Heap, R.B., Morgan, G., Buttle, H.L., Young, L.R. (1986) Control of binucleate cell migration in the placenta of ruminants. Journal of Reproduction and Fertility 76: 499–512

Wooding, F.B.P., Morgan, G., Forsyth, L.A., Butcher, G., Hutchings, A., Billingsley, S.A, Gluckman, P.D. (1992) Light and electron microscopic studies of cellular localisation of OPL with monoclonal and polyclonal antibodies. Journal of Histochemistry and Cytochemistry 40: 1001–1009

Wooding, F.B.P., Hobbs, T., Morgan, G., Heap, R.B., Flint, A.P.F. (1993) Cellular dynamics of growth in sheep and goat synepitheliochorial placentomes: an autoradiographic study. Journal of Reproduction and Fertility 98: 275–283

Wooding, F.B.P., Morgan, G., Jones, G.V., Care, A.D. (1996) Calcium transport and the localisation of calbindin- D9K in the ruminant placenta during the second half of pregnancy. Cell and Tissue Research 285: 477–489

Wooding, F.B.P., Morgan, G., Adam, C.L. (1997) Structure and function in the ruminant synepitheliochorial placenta: central role of the trophoblast binucleate cell in deer. Microscopy Research and Technique 38: 88–99

Wooding, F.B.P., Morgan, G., Fowden, A.L., Allen, W.R. (2000) Separate sites and mechanisms for placental transport of calcium, iron and glucose in the equine placenta. Placenta 21: 635–645

Wooding, F.B.P., Morgan, G., Fowden, A.L., Allen, W.R. (2001) A structural and immunological study of chorionic gonadotropin production by equine trophoblast and cup cells. Placenta 22: 749–767

Wooding, F.B., Ozturk, M., Skidmore, J.A., Allen, W.R. (2003) Developmental changes in localization of steroid synthesis enzymes in camelid placenta. Reproduction 126: 239–247

Wooding, F.B.P., Fowden, A.L., Bell, A.W., Ehrhardt, R., Limesand, S.W., Hay, W.W. (2005a) Localisation of glucose transport in the ruminant placenta: implications for the sequential use of transporter isoforms. Placenta 26: 626–640

Wooding, F.B.P., Roberts, R.M., Green, J.A. (2005b) LM and EM immunocytochemical studies of the distribution of PAGs throughout pregnancy in the cow: possible functional implications. Placenta 26: 807–827

Wooding, F.B.P., Stewart, F., Mathias, S., Allen, W.R. (2005c) Placentation in the African elephant, *Loxodonta africanus*: III. Ultrastructural and functional features of the placenta. Placenta 26: 449–470

Wooding, F.B.P., Kimura, J.K., Fukuta, K., Forhead, A.J. (2007a) A light and electron microscopical study of the Tragulid (mouse deer) placenta. Placenta 28: 1039–1048

Wooding, F.B.P., Dantzer, V.B., Klisch, K., Jones, C.J.P., Forhead, A.J. (2007b) Glucose transporter 1 localisation throughout pregnancy in the carnivore placenta: light and electron microscope studies. Placenta 28: 453–464

Woollacott, R.M. and Zimmer, R.L. (1975) A simplified placenta-like system for the transport of extraembryonic nutrients during embryogenesis of *Rugula neritina* (Bryozoa). Journal of Morphology 147: 355–378

Wourms, J.P. (1981) Viviparity: the maternal fetal relationship in fishes. American Zoologist 21: 473–515

Wourms, J.P. (1993) Maximisation of the evolutionary trends for placental viviparityin the spadenose shark, Scoliodon laticaudus. Environmental Biology of Fishes 38: 269–294

Wourms, J.P., Grove, B.D., Lombardi, J. (1988) The maternal-embryonic relationship in viviparous fishes . In Fish Physiology Vol 11 B: 1–134. Academic Press, London

Wu, M.C., Chen, Z.Y., Jarrell, V.L., Dzuik, P.J. (1989) Effect of initial length of uterus per embryo on fetal survival and development in the pig. Journal of Animal Science 67: 1764–1772

Wu, W.X., Brooks, J., Millar, M.R, Ledger, W.L., Saunders, P.T.K., Glasier, A.F., McNeilly, A.S. (1991) Localisation of the sites of synthesis and action of PRL by immunocytochemistry and in situ hybridisation within the human uteroplacental unit. Journal of Molecular Endocrinology 7: 241–247

Wulff, C., Wilson, H., Dickson, S.E., Wiegand, S.J., Fraser, H.M. (2002) Hemochorial placentation in the primate: expression of vascular endothelial growth factor, angiopoietins, and their receptors throughout pregnancy. Biology of Reproduction 66: 802–812

Wynn, R.M. (1967) Comparative electron microscopy of placental junctional zone. American Journal of Obstetrics and Gynecology 19: 644–661

Wynn, R.M. (1974) Ultrastructural development of the human decidua. American Journal of Obstetrics and Gynecology 118: 652–670

Wynn, R.M., Hoschner, J.A., Oduor-Okelo, D. (1990) The interhemal membrane of the spotted hyena: an immunohistological reappraisal. Placenta 11: 215–221

Xavier, F. (1976) An exceptional reproductive strategy in Anura: *Nectophrynoides occidentalis* Angel (Bufonidae), in Major Patterns in Vertebrate Evolution (eds. M.K. Hecht, B.M. Hecht, P.C. Goody), Plenum, New York

Yamaguchi, M., Ogren, L., Endo, H., Thordarson, G., Bigsby, R.M., Talamantes, F. (1992) Production of mPL I and II by the same giant cell. Endocrinology 131: 1595–1602

Yaron, Z. (1977) Embryo maternal interactions in the lizard *Xantusia vigilis*, in Reproduction and Evolution (eds. J.H. Calaby, C.H. Tyndale-Biscoe) Australian Academy of Science, Canberra

Yoder, J.A., Walsh, C.P., Bestor, T.H. (1997) Cytosine methylation and the ecology of intragenomic parasites. Trends in Genetics 13: 335–340

Young, M. (1981) Placental amino acid transfer and metabolism. Placenta Supplement 2: 177–184

Young, M. (1988) Handling of amino acids by the fetoplacental unit, in Placental Nutrition (ed. B.S. Lindblad), Academic Press, New York, pp. 2543

Young, M., Allen, W.R., Deutz, N.E.P. (2003) Free amino acid concentrations in the equine placenta: relationship to maternal and fetal concentrations. Research in Veterinary Science 74: 279–328

Zarnani, A., Moazzeni, S., Shokri, F., Salehnia, M., Jeddi-Tehrani, M. (2007) Kinetics of murine decidual dendritic cells. Reproduction 133: 275–283

Zavy, M.T., Bazer, F.W., Sharp, D.C., Wilcox, C.J. (1979) Uterine luminal proteins in the cycling mare. Biology of Reproduction 20: 689–698

Zechner, U., Reule, M., Orth, A., Bonhomme, F., Strack, B., Guénet, J-L., Hameister, H., Fundele, R. (1996) An X-chromosome linked locus contributes to abnormal placental development in mouse interspecific hybrids. Nature Genetics 12: 398–403

Zeh, W.Z., Zeh, J.A. (2000) Reproductive mode and speciation: the viviparity-driven conflict hypothesis. Bioessays 22: 938–945

Zeller, U., Freyer, C. (2001) Early ontogeny and placentation of the grey short tailed possum *Monodelphis domestica*. Journal of Evolution Research and Zoological Systematics 39: 137–158

Zhou, J., Bondy, C. (1992) IGF II and its binding proteins in placental development. Endocrinology 131: 1230–1240

Zhou, J., Bondy, C. (1993) Placental transporter gene expression and metabolism in the rat. Journal of Clinical Investigation 91: 845–852

Zoli, A.P., Demez, P., Beckers, J.F., Reznik, M., Beckers, A. (1992) Light and electronmicroscopic immunolocalisation of bPAG in the bovine placentome. Biology of Reproduction 46: 623–629

Index

Printing: Krips bv, Meppel, The Netherlands
Binding: Stürtz, Würzburg, Germany